Springer Tracts in Advanced Robotics

Volume 38

Editors: Bruno Siciliano · Oussama Khatib · Frans Groen

Margaret E. Jefferies and Wai-Kiang Yeap (Eds.)

Robotics and Cognitive Approaches to Spatial Mapping

Springer

Professor Bruno Siciliano, Dipartimento di Informatica e Sistemistica, Università di Napoli Federico II, Via Claudio 21, 80125 Napoli, Italy, E-mail: siciliano@unina.it

Professor Oussama Khatib, Robotics Laboratory, Department of Computer Science, Stanford University, Stanford, CA 94305-9010, USA, E-mail: khatib@cs.stanford.edu

Professor Frans Groen, Department of Computer Science, Universiteit van Amsterdam, Kruislaan 403, 1098 SJ Amsterdam, The Netherlands, E-mail: groen@science.uva.nl

Editors

Dr. Margaret E. Jefferies
University of Waikato
Dept. of Computer Science
Private Bag 3105
Hamilton 2001
New Zealand

Dr. Wai-Kiang Yeap
Auckland University of Technology
School of Computer and Information Sciences
Private Bag 92006
Auckland 1020
New Zealand

ISBN 978-3-642-09462-0 e-ISBN 978-3-540-75388-9
ISSN print edition: 1610-7438
ISSN electronic edition: 1610-742X

Springer is a part of Springer Science+Business Media
springer.com

STAR (Springer Tracts in Advanced Robotics) has been promoted under the auspices of EURON (European Robotics Research Network)

Preface

At the dawn of the new millennium, robotics is undergoing a major transformation in scope and dimension. From a largely dominant industrial focus, robotics is rapidly expanding into the challenges of unstructured environments. Interacting with, assisting, serving, and exploring with humans, the emerging robots will increasingly touch people and their lives.

The goal of the new series of *Springer Tracts in Advanced Robotics (STAR)* is to bring, in a timely fashion, the latest advances and developments in robotics on the basis of their significance and quality. It is our hope that the wider dissemination of research developments will stimulate more exchanges and collaborations among the research community and contribute to further advancement of this rapidly growing field.

The collection edited by Margaret Jefferies and Albert Yeap is the fourth one in the series on mapping, and keenly focuses on the common core problems between cognitive and robot spatial mapping. Such cross-fertilisation was made possible thanks to a thematic workshop held in early 2003 at Auckland University of Technology, where scientists from the two communities, psychologists and roboticists, met and discussed freely in a meeting-of-the-mind environment.

The ambitious goal of the volume following the workshop is to show how cognitive researchers should give more thoughts to the perceptual and localization problems while robotics researchers should consider implementing autonomous systems not having the sole task of building a map of the environment. A number of significant applications are described where this sort of gap is effectively bridged.

The material is nicely organised in three parts; namely, robot mapping, cognitive mapping, and robot cognitive mapping, each accompanied by an introduction by a distinguished researcher in the field. Gathering some of the authorities working on spatial mapping, this volume is dedicated to the memory of Margaret Jefferies, who sadly passed away while completing the work. Here I warmly recall her profound passion and devotion during the revision and edition of the collection, despite her health problems. A very fine addition to our STAR series!

Naples, Italy,
July 2007

Bruno Siciliano
STAR Editor

List of Contributors

Jesse Baker
University of Waikato
Department of Computer Science
New Zealand

Byron Boots
Duke University
Center for Cognitive Neuroscience
USA

Raja Chatila
Laboratoire d'Analyse et
d'Architecture des Systèmes
Centre National de la Recherche
Scientifique
France

Ken Cheng
Macquarie University
Centre for the Integrative
Study of Animal Behaviour
Australia

Eric Chown
Bowdoin College
Department of Computer Science
USA

Hugh Durrant-Whyte
The University of Sydney
ARC Centre of Excellence
in Autonomous Systems
Australia

Matthias O. Franz
MPI für biologische Kybernetik
Germany

Christian Freksa
Universität Bremen
SFB-TR 8
Germany

Charles R. Gallistel
Rutgers University
Department of Psychology and
Rutgers Center for Cognitive Science
USA

Sabine Gillner
Universität Tübingen
Cognitive Neuroscience
Department of Zoology
Germany

Richard Grover
The University of Sydney
ARC Centre of Excellence
in Autonomous Systems
Australia

Verena V. Hafner
Humboldt-Universität zu Berlin
Department of Computer Science
Germany

Stephen C. Hirtle
University of Pittsburgh
School of Information Sciences
USA

Vincent Hok
Université de Provence
Laboratory of Neurobiology
and Cognition
France

Wolfgang Hübner
Universität Tübingen
Kognitive Neurowissenschaften
Germany

Margaret E. Jefferies
University of Waikato
Department of Computer Science
New Zealand

Benjamin Kuipers
University of Texas at Austin
Department of Computer Sciences
USA

Longin Jan Latecki
Temple University
Department of Computer and
Information Sciences
USA

Hanspeter A. Mallot
Universität Tübingen
Cognitive Neuroscience
Department of Zoology
Germany

Ulrich Nehmzow
University of Essex
Department of Computer Science
UK

Carole Parron
Université de la Méditerranée
Mediterranean Institute for
Cognitive Neuroscience
France

Bruno Poucet
Université de Provence
Laboratory of Neurobiology
and Cognition
France

Sophie Renaudineau
Université de Provence
Laboratory of Neurobiology
and Cognition
France

Etienne Save
Université de Provence
Laboratory of Neurobiology
and Cognition
France

Steven Scheding
The University of Sydney
ARC Centre of Excellence
in Autonomous Systems
Australia

Jochen Schmidt
Auckland University of Technology
Centre for Artificial Intelligence
Research
New Zealand

Wolfgang Stürzl
Universität Tübingen
Kognitive Neurowissenschaften
Germany

Sebastian Thrun
Stanford University
Computer Science Department
USA

Nicola Tomatis
Ecole Polytechnique Fédérale de
Lausanne
Autonomous Systems Lab
Switzerland

BlueBotics SA, Lausanne
Switzerland

Wengrong Weng
University of Waikato
Department of Computer Science
New Zealand

Diedrich Wolter
Universität Bremen
Cognitive Systems
SFB-TR 8
Germany

Chee Kit Wong
Auckland University of Technology
Centre for Artificial Intelligence
Research
New Zealand

Wai K. Yeap
Auckland University of Technology
Centre for Artificial Intelligence
Research
New Zealand

University of Malaya
Department of Artificial Intelligence
Malaysia

Contents

Robot and Cognitive Approaches to Spatial Mapping

Margaret E. Jefferies and Wai K. Yeap

Introduction

Imagine designing an autonomous mobile robot to mingle with us in our environment. Although the robot, like us, could search a map to locate places and consult a global positioning system to find out where things are, it should also, like us, be able to compute, or update, its representation of its surroundings as observed through its sensors.

Robot spatial mapping, in this book, is about the problem of a robot computing such a representation from data gathered by its sensors. This problem has been studied since the creation of the first autonomous mobile robot, SHAKEY [8], in the late nineteen-sixties. For some recent examples of work in this area, see [5].

People (and animals) compute a representation of their environment, too. Their representation is commonly referred to as a cognitive map [9]. Cognitive spatial mapping, in this book, is about the problem of computing a cognitive map. It has been studied extensively by many researchers of disparate backgrounds. For some major reviews and collections of works in this area, see [2, 3, 4].

It is not surprising that the robot and cognitive mapping problems share some common core problems. One would reasonably expect some cross-fertilisation of research between the two areas to have occurred, and this has happened but only recently. One reason for this is that much of the early research in both fields has focussed on opposite ends of the mapping problem. On the one hand, roboticists were working hard on what can be called "the sensor problems". Robots were not equipped with powerful sensors in the early days and still aren't; although better routines were developed to deal with sensor errors and the use of lasers have become more common. Robots still lack powerful vision systems. On the other hand, cognitive researchers focused mainly on "the knowledge problems". They investigated what people remembered most when visiting new places and how their conceptually rich knowledge of their environment is organised into hierarchies. They discussed the use of landmarks and high-level cognitive capabilities, such as the use of short-cuts, and the ability to orient oneself in complex spaces, such as places far apart in a city.

M.E. Jefferies et al. (Eds.): Robot. & Cogn. Approach. to Spat. Map., STAR 38, pp. 1–5, 2008.
springerlink.com

"Never the twain shall meet" is not the case. There are signs that both fields have matured and that efforts to cross-fertilise are happening. For instance, roboticists speak of landmarks and place recognition, and cognitive scientists have shown some interests in how the shape of local environments affects long-term memory of space. This cross-fertilization, however, is neither complete nor common yet. Some usage of the term landmark by roboticists does not have a corresponding usage in the cognitive science community. Roboticists may be asking the same kind of questions (how can someone recognize their immediate location) but their solutions are often highly dependent on the poor sensors with which their robot is equipped.

Nevertheless, autonomous mobile robots have now been created that are mapping an ever increasing environment. Stimulants like the DARPA Grand Challenge should encourage the development of robots that can move outside the office environment and into less structured and more chaotic ones. Researchers (such as [1, 7, 11]) interested in developing computational theories of cognitive maps are beginning to test their ideas using real robots instead of relying on simulation studies. Doing so should force them to think about how to fill the gaps between a cognitive theory and its implementation on a practical platform.

In 2000, we published a paper, titled "On early cognitive mapping", in the Spatial Cognition and Computation Journal [12]. In it, we reviewed two distinct paradigms for cognitive mapping, discussing works from both the psychological and the robotics literature. In writing the paper, we felt the need to more actively promote the cross-fertilisation between the two fields. In early 2002, we decided to organise a special workshop on "Robotics and Cognitive Approaches to Spatial Mapping". Our idea was to invite researchers from both fields to attend a single meeting. The workshop was subsequently organized and held at AUT Technology Park from 27th February to 1st March in 2003. Those who accepted our invitation and were able to attend were: Ken Cheng, Matthias Franz, Christian Freksa, Stephen Hirtle, Hanspeter Mallot, Ulrich Nehmzow, Sebastian Thrun, Nicola Tomatis, and Steve Scheding.

It was an interesting meeting. We brought together, on the one hand, researchers who have dedicated their life to studying human and animal wayfinding behavior, and on the other hand, roboticists fascinated with creating robots which try to solve a similar problem. It was clear from the outset that few in the group were aware of the "story" from the other side. and thus sitting through the talks gave both sides a fresh perspective on a familiar problem. Now cognitive researchers might give more thoughts to the perceptual and localization problems while roboticists might consider implementing an autonomous system that does not have the sole task of building a map of its environment.

This book is one result of the workshop. Note that participants' invitations were based upon their past work and their continuing interests in spatial mapping. No papers were submitted or reviewed beforehand. This was not necessary as our participants have, in the past, contributed significant ideas about spatial mapping. We wanted a meeting-of-the-mind to cross-fertilize the fields with new ideas, not a presentation of new findings from their respective research areas.

Each participant did submit an abstract prior to the meeting, and produced a working paper for the workshop. The working papers, in most cases, have been revised significantly and presented as chapters in this book.

In completing the book, we were delighted that Eric Chown, Ben Kuipers and Verena Hafner accepted our invitation to contribute a chapter each to the book. Catherine Blanc, who originally accepted our invitation to attend the workshop, was, sadly, unable to join us. On Catherine's recommendation, Etienne Save was invited to write a chapter for the book, instead. We were honored to have three distinguished researchers, Raja Chatila, Charles Gallistel and Ben Kuipers to write an introduction each for the three parts of the book.

Organisation of the Book

The book is organized into three parts, namely: Part I on robot mapping, Part II on cognitive mapping, and Part III on cognitive robot mapping.

Part I consists of five chapters which together address a cross section of problems with, (such as uncertainty, localization, unstructured environments and control architectures), and approaches to (such as topological and geometric), robot mapping. Thrun provides a comprehensive introduction to one of the key problems roboticists face in developing autonomous mobile robots, namely the famously known SLAM problem. Tomatis discusses the use of a hybrid topological/metric representation for robot mapping whereby each node in it could be a local metric map of a local space (say, a room) visited by the robot. As evident in the other two parts, cognitive mapping researchers are increasingly in favour of using such a representation for cognitive maps.

Scheding et al. and Nehmzow both discuss a variety of issues related to robot spatial mapping. As noted above, robots have begun to tackle the very difficult problem of mapping unstructured outdoor environments. Scheding et al. discuss various mapping issues related to the design of such robots and show how they overcome some of the issues raised. Nehmzow focuses on the use of neural networks in robot spatial mapping whereby neat solutions could emerge from allowing the robot to interact with its environment. Wolter et al. provide a detailed analysis of the self-localization problem using a formal qualitative spatial reasoning approach to handling incomplete, imprecise, and partially conflicting (fuzzy) information about spatial situations. This is an important problem in spatial mapping and the analysis covers a wide range of situations in which self-localization is needed.

Part II consists of five chapters about cognitive mapping from researchers of disparate background. These researchers review their earlier work and provide interesting comments about how their findings could benefit researchers interested in robot mapping. Cheng and Save et al. both discuss cognitive mapping based upon their past experiments with rats. Cheng re-examines his earlier work on the role of geometric information in spatial cognition. He discusses his own work with rats and that of others with children which demonstrate a preferred reliance on the overall shape of their environment, sometimes to the exclusion of non-geometric cues. Save et al. review the work on how spatial information is encoded in the brain. They discuss the importance of cue and goal encoding in the brains of rodents.

Gillner and Mallot, and Hirtle discuss ideas about cognitive mapping from observing human spatial behaviors in both natural and artificial settings. Gillner and Mallot discuss a hierarchy of spatial tasks and review their earlier work on a series of behavioral experiments on human cognition using virtual environments which they referred to as "Hexatown". Hirtle discusses the notion of landmarks in both robot and cognitive mapping and reviewed a tripartite theory of landmarks that can be applied to navigation by humans in real and electronic spaces.

Chown and Boots end this part with an interesting discussion on the cross-fertilization of ideas between robot and cognitive spatial mapping. They argue that, for now, the best strategy might be for the roboticists to extract abstract strategies from cognitive studies and "apply them to robotics in a manner that best suits the underlying hardware and challenge at hand". They extended the PLAN model of cognitive mapping [1] and introduced C-Plan. This chapter leads the readers nicely to the next part.

Part III consists of five chapters discussing implementations of cognitive mapping theories on a mobile robot. In 1977, Kuipers developed the first computational model of cognitive maps [6]. We are honored to have him contribute a chapter which reflects on how his ideas have developed since his early days at MIT. His work has been, and continues to be, very influential in the study of spatial mapping. In our own chapters, we continue our effort to implement our ideas of cognitive mapping onto mobile robots. Jefferies et al. describe hybrid topological/metric mapping on a robot approached from a cognitive mapping perspective. The important difference from other hybrid approaches in this book is that the topological and absolute metric maps have equal status; each is required to improve the quality of the information in the other. A key idea underlying Yeap's theory of cognitive mapping [10] is that a cognitive agent that has some form of range sensing of its environment will need to compute Absolute Space Representations (ASR) as a basis for developing a richer cognitive map. Yeap et al. show how ASRs are computed by a mobile robot equipped with sonar sensors. Like Chown and Boots, Yeap et al. show how a cognitive mapping theory when applied to a mobile robot must adapt to the hardware of the robot instead of that of humans. Unlike other approaches in this part, Franz describes an implementation that does not begin with a theory of cognitive mapping. Rather, it begins with a hierarchy of needs in way-finding and the implementation starts by developing solutions at the lower level which could then be extended or re-used to solve a problem at a higher level in the hierarchy. Such an approach is known as the biomimetic approach. Hafner discusses the neural basis for cognitive mapping and shows how such findings could be used to develop a neural model of cognitive mapping. Two particularly interesting problems were encountered in the implementation, namely how metric information is encoded in a neural model and how best to evaluate the resulting map which is computed at the neural level. Hafner shows how her model is tested via simulation studies and via implementations on an actual robot.

Acknowledgement

We are grateful to the Institute for IT Research @ AUT for funding the workshop. We thank the editor of STAR for accepting this book as part of their series on robotics. We thank Saide Lo and Jochen Schmidt for their great help in preparing the manuscript and to Saide too for helping to organise the workshop. We thank all the authors for their contributions. We thank everyone for their hard work and patience in developing this book.

While working on this book, Margaret Jefferies was plagued with health problems which were later diagnosed as cancer. She sadly succumbed to it in early January 2006. Despite her poor health, she worked hard on the book, insisting we both thoroughly reviewed all chapters. Her devotion, dedication, gentleness and friendliness will be dearly missed by all of us who worked with her on this book. We dedicate this book to her memory. Goodbye Margaret.

References

1. Chown, E., Kaplan, S., Kortenkamp, D.: Prototypes, location, and associative networks (PLAN): Towards a unified theory of Cognitive Mapping. Cognitive Science 19, 1–52 (1995)
2. Downs, R.M., Stea, D.: Image and Environment: Cognitive Mapping and Spatial Behavior. Aldine Publishing Coy (1973)
3. Gallistel, C.R.: The organization of learning. MIT Press, Cambridge (1990)
4. Golledge, R.G.: Wayfinding Behavior: Cognitive Mapping and Other Spatial Processes. John Hopkins University Press, Baltimore (1999)
5. Kortenkamp, D., Bonasso, R., Murphy, R. (eds.): Artificial Intelligence and Mobile Robots: Case Studies of Successful Robot Systems. AAAI/MIT Press (1998)
6. Kuipers, B.J.: Representing knowledge of large-scale space. PhD thesis, Massachusetts Institute of Technology (1977)
7. Kuipers, B.J.: The Spatial Semantic Hierarchy. Artificial Intelligence 19, 191–233 (2000)
8. Nilsson, N.J.: A Mobile Automaton: An Application of Artificial Intelligence Techniques. In: Proc. of the First International Joint Conference on Artificial Intelligence, Washington, USA, pp. 509–519 (1969)
9. Tolman, E.C.: Cognitive Maps in Rats and Men. Psychological Review 55, 189–208 (1948)
10. Yeap, W.K.: Towards a Computational Theory of Cognitive Maps. Artificial Intelligence 34(3), 297–360 (1988)
11. Yeap, W.K., Jefferies, M.E.: Computing a Representation of the Local Environment. Artificial Intelligence 107(2), 265–301 (1999)
12. Yeap, W.K., Jefferies, M.E.: On Early Cognitive Mapping. Spatial Cognition and Computation 2(2), 85–116 (2000)

Part I

Robot Mapping

Robot Mapping: An Introduction

Raja Chatila

How to reach a designated location? This question defines a basic problem in robotics. The location may be arbitrarily remote and out of sight in general. To specify it to the robot one can use *coordinates* in a given reference system, or some distinguishable perceptual *feature* which, when recognized by the robot, defines a termination condition for the navigation task. In general, erratic motion in the environment to reach the goal location, with simple local obstacle detection and avoidance will lead to inefficient trajectories, if not to failure of reaching the target. A real robot in general situations has to determine a path to reach its goal location that should be efficient with respect to distance, time, energy consumption or other specific criteria related to the context in which the motion is achieved (e.g. avoid some dangerous areas, or stay in sight of some features). Thus, an efficient motion requires knowledge about the environment layout, which is generally not available to the robot *a priori*, and must therefore be acquired through perception. Hence this **environment mapping** problem appears as one key issue of the navigation problem and a prerequisite for motion planning and execution.

The first question that pops up then is how to represent the environment? And first of all, is any representation necessary at all? And if yes, is there a unique representation or many? And what should be represented? There has been a great deal of discussions and controversies on these issues in the past twenty years.

If we consider natural systems, data from neuroscience indicate that there are neuronal organizations and interactions that amount to representations of the environment's layout. In robotics, the total absence of representations would result in inefficient motions, and there is today a good understanding of the different kinds of representations and what knowledge they should convey to the robot.

SLAM

Representations being defined (see below), the mapping problem can be stated as follows. The robot discovers its environment gradually while moving in it. Objects and regions perceived from a given location must therefore be related to previously perceived ones and integrated with them in a consistent manner. The result of this integration is a map representing the layout of the environment, or of parts of it.

However, for this integration to be consistent, it is necessary to know – or be able to estimate correctly – the transform between the newly observed part and

M.E. Jefferies et al. (Eds.): Robot. & Cogn. Approach. to Spat. Map., STAR 38, pp. 9–12, 2008.
springerlink.com

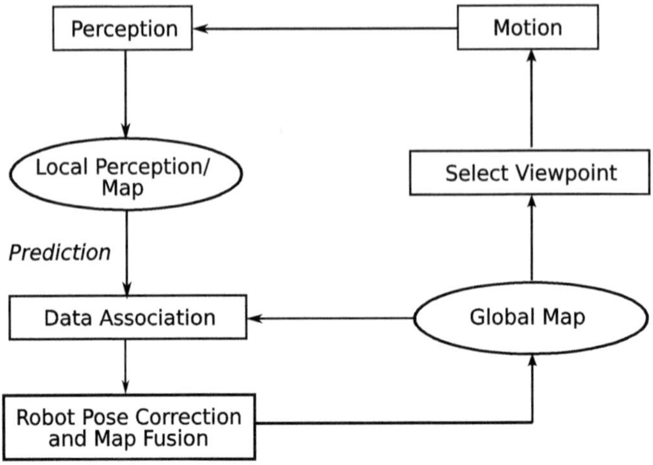

Fig. 1. The general SLAM process. After an observation, the local perception or map has to be associated with the global one to enable an update of both robot pose and environment model. Selecting the next best view for completing the model is a research issue by itself.

the already built model. Relating the robot's new position to the previous ones provides this transform. In other words the **localization** of the robot appears necessary to correctly relate the newly perceived areas with the already known ones. In the absence of external means of localization – such as GPS, which would put a strong constraint on the feasibility of mapping – there are two ways to localize the robot. One is incremental motion integration (e.g., through odometry or inertial units), which is known to eventually diverge because of error accumulation and integration. The other means of localization is to recognize known features in the environment and use them as landmarks. However this latter solution requires knowing the environment map (the landmark positions). Hence incremental *environment mapping* and *robot localization* appear to be two intimately related processes. The problem to be solved is then **simultaneous localization and mapping** (SLAM). The general SLAM process is depicted in Fig. 1. Solving this problem requires identifying same environment elements perceived from different positions. It appeared very clearly since the mid eighties that the issue of uncertainties was central in the mapping process. Sensors are indeed always imperfect; their data are incomplete, noisy and inaccurate. Environment representations should therefore explicitly take into account uncertainties to solve the data association problem, i.e., the ability to recognize the same feature from different perceptions.

Representations

For its navigation the robot needs to know the position and metrics of physical objects to plan its motions (a geometrical model) in a given space region, and the relationships between regions of space to decide for the general roadmap it will follow. Hence both a geometrical and a topological model are useful, and complementary. In addition, semantics, which defines the nature of objects or of space regions, would be an important knowledge, although work on mapping has rather mostly focused on trying to capture geometry and topology.

There are three main different widely used representations related to geometry: appearance, grids and features. "Appearance" designates the practically raw data provided by sensors, such as 3D points from a laser rangefinder or stereo. Such data require minimal processing but are almost impossible to use individually. They are noisy and difficult to be recognized from one perception to the next. Therefore they are rather often used globally, i.e., as sets of points with statistical properties. Grid representations capture the presence or probability of presence of objects in space areas organized as preset grid cells. This kind of representation is easy to construct but requires that grid cells be recognized as such for updating. Features are structures of the environment that require some processing to build. The advantage is that they have some stability that makes them easier to recognize from a perception to another. Some of the simplest structures are pixels with invariance properties such as SIFT, or Harris interest points. More complex ones are geometrical features (e.g., linear approximations), which have the drawback of requiring a segmentation that might not be adapted to the actual structure of the environment.

Whatever the representation, correct data association is a pre-requisite to enable correct updating of the map. This consists in identifying that an environment element perceived from different positions is actually the same, and this is complicated by uncertainties. The data association problem has drawn a lot of attention, and is often solved by a statistical test. One observation can be however made: the less information a feature conveys, the easier it is to be confused with other features. Invariance over observations and uniqueness of features (or of their layout) make data association easier. In other words, there is a compromise to find between the processing necessary to build informative features and the risk of bad data association, which could jeopardize the whole mapping process and end up with inconsistent maps.

Now if we consider that data association is solved correctly, the fusion step is the central part of the whole process. Addressing this problem requires a solid mathematical formalism to represent and handle uncertainties: the probabilistic framework. The basic problem amounts then to a Bayesian update. The Kalman filter, as a Bayesian filter, in its extended form because of the non-linearity of the transforms, was historically the process at the core of the map fusion. In this case the robot coordinates and the environment representations define together the *state* that is updated by *observations*. It is also established that there are correlations between observations and between observations and state, and that they imperatively should be considered in the filter formalization and application. However, the Gaussian noise hypothesis and the difficulty of data association in many applications where the representations are rather poor drafted attention to other kinds of more general filters, namely particle filtering, which also enable to maintain several hypotheses on the state simultaneously.

Open Issues

The main mathematical framework for SLAM in Robotics is well established today. Some of the important open issues, which are the focus of ongoing research, are: complexity (i.e., how to deal with a large – and growing – number of environment features); data association (which stems partly from the poor representations as mentioned before); and dealing with complex and dynamic environments (i.e., outdoors, non structured, 3D).

Simultaneous Localization and Mapping is actually not a new problem. When we look at old maps of coastlines and continents (e.g., from the Sixteenth Century), we can understand why they are inaccurate and twisted: the explorers and cartographers had to solve exactly a SLAM problem with inaccurate sensors. Robots and humans face the same problem.

1

Simultaneous Localization and Mapping

Sebastian Thrun

Summary. This article provides a comprehensive introduction into the *simultaneous localization and mapping problem*, better known in its abbreviated form as *SLAM*. SLAM addresses the problem of a robot navigating an unknown environment. While navigating the environment, the robot seeks to acquire a map thereof, and at the same time it wishes to localize itself using its map. The use of SLAM problems can be motivated in two different ways. One might be interested in detailed environment models, or one might seek to maintain an accurate sense of a mobile robot's location. SLAM servers both of these purposes.

We review three major paradigms of algorithms from which a huge number of recently published methods are derived. First comes the traditional approach, which relies in the extended Kalman filter (EKF) for representing the robot's best estimate. The second paradigm draws its intuition from the fact that the SLAM problem can be viewed as a sparse graph of constraints, and it applies nonlinear optimization for recovering the map and the robot's locations. Finally, we survey the particle filter paradigm, which applies non-parametric density estimation and efficient factorization methods to the SLAM problem. This article discusses extensions of these basic methods. It elucidates variants of the SLAM problem and poses a taxonomy for the field. Relevant research is referenced extensively, and open research problems are discussed.

1.1 Introduction

This chapter provides a comprehensive introduction into one of the key enabling technologies of mobile robot navigation: *simultaneous localization and mapping*, or in short *SLAM*. SLAM addresses the problem of acquiring a spatial map of a mobile robot environment while simultaneously localizing the robot relative to this model. The SLAM problem is generally regarded as one of the most important problems in the pursuit of building truly autonomous mobile robots. Despite significant progress in this area, it still poses great challenges. At present, we have robust methods for mapping environments that are static, structured, and of limited size. Mapping unstructured, dynamic, or large-scale environments remains largely an open research problem.

The historical roots of SLAM can be traced back to Gauss [31], who is largely credited for inventing the least squares method, for calculating planetary orbits. In the Twentieth Century, a number of fields outside robotics have studied the making of environment models from a moving sensor platform, most notably in *photogrammetry* [44] and *computer vision* [88, 79]. SLAM builds on this work, often extending the basic paradigms into more scalable algorithms.

M.E. Jefferies et al. (Eds.): Robot. & Cogn. Approach. to Spat. Map., STAR 38, pp. 13–41, 2008.
springerlink.com

This article begins with a definition of the SLAM problem, which shall include a brief taxonomy of different versions of the problem. The centerpiece of this article is a layman introduction into the three major paradigms in this field, and the various extensions that exist. As the reader will quickly recognize, there is no single best solution to the SLAM method. The method chosen by the practitioner will depend on a number of factors, such as the desired map resolution, the update time, and the nature of the features in the map, and so on. Nevertheless, the three methods discussed in this article cover the major paradigms in this field. For an in-depth study of SLAM algorithms, we refer the reader to a recent textbook on probabilistic robotics, which dedicates a number of chapters to the topic of SLAM [82].

1.2 SLAM: Problem Definition

1.2.1 Mathematical Basis

The SLAM problem is defined as follows: A mobile robot roams an unknown environment, starting at a location with known coordinates. Its motion is uncertain, making it gradually more difficult to determine its global coordinates. As it roams, the robot can sense its environment. The SLAM problem is the problem of building a map the environment while simultaneously determining the robot's position relative to this map.

Formally, SLAM is best described in probabilistic terminology. Let us denote time by t, and the robot location by x_t. For mobile robots on a flat ground, x_t is usually a three-dimensional vector, comprising its 2-dimensional coordinate in the plane plus a single rotational value for its orientation. The sequence of locations, or *path*, is then given as

$$X_T = \{x_0, x_1, x_2, \ldots x_T\}. \tag{1.1}$$

Here T is some terminal time (T might be ∞). The initial location x_0 is known; other positions cannot be sensed.

Odometry provides relative information between two consecutive locations. Let u_t denote the odometry that characterized the motion between time $t - t$ and time t; such data might be obtained from the robot's wheel encoders or from the controls given to those motors. Then the sequence

$$U_T = \{u_1, u_2, u_3 \ldots x_T\} \tag{1.2}$$

characterizes the relative motion of the robot. For noise-free motion, U_T would be sufficient to recover the past X_T from the initial location x_0. However, odometry measurements are noisy, and path integration techniques inevitably diverge from the truth.

Finally, the robot senses objects in the environment. Let m denote the "true" map of the environment. The environment may be comprised of landmarks, objects, surfaces, etc., and m describes their locations. The environment map m is typically assumed to be time-invariant (i.e., static).

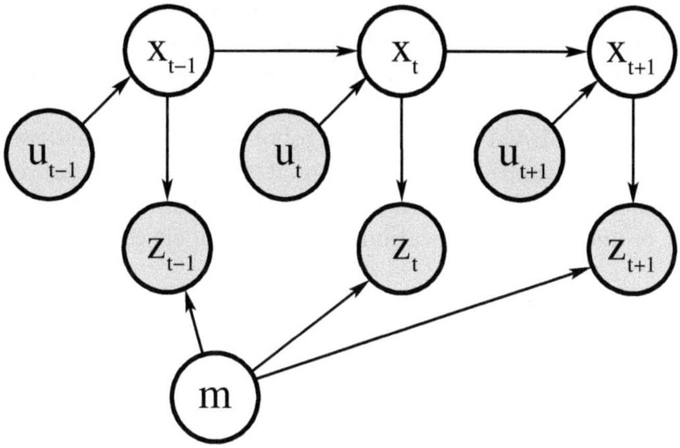

Fig. 1.1. Graphical model of the SLAM problem. Arcs indicate causal relationships, and shaded nodes are directly observable to the robot. In SLAM, the robot seeks to recover the unobservable variables.

The robot measurements establish information between features in m and the robot location x_t. If we, without loss of generality, assume that the robot takes exactly one measurement at each point in time, the sequence of measurements is given as

$$Z_T = \{z_1, z_2, z_3, \ldots, z_T\}. \tag{1.3}$$

Figure 1.1 illustrates the various variables involved in the SLAM problem. It shows the sequence of locations and sensor measurements, and the causal relationships between these variables. Such a diagram is known as a *graphical model*. It is useful in understanding the dependencies in the SLAM problem.

The SLAM problem is now the problem of recovering a model of the world m and the sequence of robot locations X_T from the odometry and measurement data. The literature distinguishes two main forms of the SLAM problem, which are both of equal practical importance. One is known as the *full SLAM problem*: it involves estimating the posterior over the entire robot path together with the map:

$$p(X_T, m \mid Z_T, U_T). \tag{1.4}$$

Written in this way, the full SLAM problem is the problem of calculating the joint posterior probability over X_T and m from the available data. Notice that the variables right of the conditioning bar are all directly observable to the robot, whereas those on the left are the ones that we wanted. As we shall see, algorithms for the offline SLAM problem are often batch, that is, they process all data at the same time.

The second, equally important SLAM problem is the *online SLAM problem*. This problem is defined defined via

$$p(x_t, m \mid Z_T, U_T). \tag{1.5}$$

Online SLAM seeks to recover the present robot location, instead of the entire path. Algorithms that address the line problem are usually incremental and can process one data item at a time. In the engineering literature, such algorithms are called *filters*.

To "solve" either SLAM problem, the robot needs to be endowed with two more models: a mathematical model the relates odometry measurements u_t to robot locations x_{t-1} and x_t; and a model that relates measurements z_t to the environment m and the robot location x_t. These models corresponds to the arcs in Fig. 1.1.

In SLAM, it is common to think of those mathematical models as probability distributions: $p(x_t \mid x_{t-1}, u_t)$ characterizes the probability distribution of the location x_t assuming that a robot started at a known location x_{t-1} and measured the odometry data u_t. And likewise, $p(z_t \mid x_t, m)$ is the probability for measuring z_t if this measurement is taking at a known location x_t in a known environment m. Of course, in the SLAM problem we do *not* know the robot location, neither do we know the environment. As we shall see, Bayes rule takes care of that, by transforming these mathematical into a form where we can 'recover' probability distributions over those latent variables from the measured data.

1.2.2 Example: SLAM in Landmark Worlds

One common setting of SLAM involves an assumption that the environment is populated by point-landmarks. When building 2-D maps, point-landmarks may correspond to door posts and corners of rooms, which, when projected into a 2-D map are characterized by a point coordinate. In a 2-D world, each point-landmark is characterized by two coordinate values. Hence the world is a vector of size $2N$, where N is the number of point-landmarks in the world.

In a commonly studied setting, the robot can sense three things: the relative range to nearby landmarks, their relative bearing, and the identity of these landmarks. The range and bearing may be noisy, but in the most simple case the sensed landmarks identity will not be noisy.

To model such a setup, one begins with defining the *exact*, noise-free measurement function. The measurement function h describes the workings of the sensors: it accepts as input a description of the environment m and a robot location x_t, and it computes the measurement:

$$h(x_t, m). \tag{1.6}$$

Computing h is straightforward in our simplified landmark setting; it is a simple exercise in trigonometry.

The probabilistic measurement model is derived from this measurement function by adding a noise term. It is a probability distribution that peaks at the noise-free value $h(x_t, m)$ but allows for measurement noise:

$$p(z_t \mid x_t, m) \sim \mathcal{N}(h(x_t, m), Q_t). \tag{1.7}$$

Here $\mathcal{N}$ denotes the 2-dimensional normal distribution, which is centered at $h(x_t, m)$. The 2-by-2 matrix Q_t is the noise covariance, indexed by time.

The motion model is derived from a kinematic model of robot motion. Given the location vector x_{t-1} and the motion u_t, textbook kinematics tells us how to calculate x_t. Let this function be denoted by g:

$$g(x_{t-1}, u_t). \tag{1.8}$$

The motion model is then defined by a normal distribution centered at $g(x_{t-1}, u_t)$ but subject to Gaussian noise:

$$p(x_t \mid x_{t-1}, u_t) = \mathcal{N}(g(x_{t-1}, u_t), R_t). \tag{1.9}$$

Here QR_t is a covariance. It is of size 3-by-3, since the location is a three-dimensional vector.

With these definitions, we have all we need to develop a SLAM algorithm. While in the literature, point-landmark problems with range- bearing sensing are by far the most studied, SLAM algorithms are not confined to landmark worlds. But no matter what the map representation and the sensor modality, any SLAM algorithm needs a similarly crisp definition of the features in m, the measurement model $p(z_t \mid x_t, m)$ and the motion model $p(x_t \mid x_{t-1}, u_t)$.

1.2.3 Taxonomy of the SLAM Problem

SLAM problems are distinguished along a number of different dimensions. Most important researcher papers identify the type of problems by making the underlying assumptions explicit. We already encountered one such distinction: full versus online. Other common distinctions are as follows:

- **Volumetric versus feature-based.** In volumetric SLAM, the map is sampled at a resolution high enough to allow for photo-realistic reconstruction of the environment. The map m in volumetric SLAM is usually quite high-dimensional, with the result that the computation can be quite involved. Feature-based SLAM extracts sparse features from the sensor stream. The map is then only comprised of features. Our point-landmark example is an instance of feature-based SLAM. Feature-based SLAM techniques tend to be more efficient, but their results may be inferior to volumetric SLAM due to the fact that the extraction of features discards information in the sensor measurements.
- **Topological versus metric.** Some mapping techniques recover only a qualitative description of the environment, which characterizes the relation of basic locations. Such methods are known as topological. A topological map might be defined over a set of distinct places and a set of qualitative relations between these places (e.g., place A is adjacent to place B). Metric SLAM methods provide metric information between the relation of such places. in recent years, topological methods have fallen out of fashion, despite ample evidence that humans often use topological information for navigation.
- **Known versus unknown correspondence.** The correspondence problem is the problem of relating the identity of sensed things to other sensed things.

In the landmark example above, we assumed that the identity of landmarks is known. Some SLAM algorithms make such an assumption, other do not. The ones that do not provide special mechanisms for estimating the correspondence of measured features to previously observed landmarks in the map. The problem of estimating the correspondence is known as *data association problem.* It is one of the most difficult problems in SLAM.

- **Static versus dynamic.** Static SLAM algorithms assume that the environment does not change over time. Dynamic methods allow for changes in the environment. The vast literature on SLAM assumes static environments. Dynamic effects are often treated just as measurement outliers. Methods that reason about motion in the environment are more involved, but they tend to be more robust in most applications.
- **Small versus large uncertainty.** SLAM problems are distinguished by the degree of location uncertainty that they can handle. The most simple SLAM algorithms allow only for small errors in the location estimate. They are good for situations in which a robot goes down a path that does not intersect itself, and then returns along the same path. In many environments it is possible to reach the same location from multiple directions. Here the robot may accrue a large amount of uncertainty. This problem is known as the *loop closing problem.* When closing a loop, the uncertainty may be large. The ability to close loops is a key characteristic of modern-day SLAM algorithms. The uncertainty can be reduced if the robot can sense information about its position in some absolute coordinate frame, e.g., through the use of a satellite-based global positioning receiver (GPS).
- **Active versus passive.** In passive SLAM algorithms, some other entity controls the robot, and the SLAM algorithm is purely observing. The vast majority of algorithms are of this type; they give the robot designer the freedom to implement arbitrary motion controllers, and pursue arbitrary motion objectives. In active SLAM, the robot actively explores its environment in the pursuit of an accurate map. Active SLAM methods tend to yield more accurate maps in less time, but they constrain the robot motion. There exist hybrid techniques in which the SLAM algorithm controls only the pointing direction of the robot's sensors, but not the motion direction.
- **Single-robot versus multi-robot.** Most SLAM problems are defined for a single robot platform, although recently the problem of multi-robot exploration has gained in popularity. Multi-robot SLAM problems come in many flavors. In some, robots get to observe each other, in others, robots are told their relative initial locations. Multi-robot SLAM problems are also distinguished by the type communication allowed between the different robots. In some, the robots can communicate with no latency and infinite bandwidth. More realistic are setups in which only nearby robots can communicate, and the communication is subject to latency and bandwidth limitations.

As this taxonomy suggests, there exists a flurry of SLAM algorithms. Most modern-day conferences dedicate multiple session to SLAM. This article focuses on the very basic SLAM setup. In particular it assumes a static environment with a single robot. Extensions are discussed towards the end of this article, in which the relevant literature is discussed.

1.3 SLAM: Problem Definition

This section reviews three basic SLAM paradigms, from which most others are derived. The first, known as EKF SLAM, is historically the earliest but has recently become slightly unpopular due to its limiting computational properties. The second, which is based on graphical representations, successfully applies sparse nonlinear optimization methods to the SLAM problem, and has become the main paradigm for solving the full SLAM problem. The third and final method uses non-parametric statistical filtering techniques known as particle filters. It is a popular method for online SLAM, and provides a fresh new solution to the data association problem in SLAM.

1.3.1 Extended Kalman Filters

Historically EKF SLAM is the earliest, and perhaps the most influential SLAM algorithm. EKF SLAM is based on the extended Kalman filter, or EKF [41, 42, 52]. It solves the online SLAM problem by applying the well-known EKF to the estimation of the robot location and the map. The EKF method for SLAM was introduced through a series of seminal papers [7, 78, 77]; early implementation results were reported in [47, 58, 59].

The EKF algorithm represents the robot estimate by a multivariate Gaussian:

$$p(x_t, m \mid Z_T, U_T) \sim \mathcal{N}(\mu_t, \Sigma_t) \tag{1.10}$$

The high-dimensional vector μ_t contains the robot's best estimate of its own location and the location of the features in the environment. In our point-landmark example, the dimension of μ_t would be $3 + 2N$, since we need three variables to represent the robot location and $2N$ variables for the N landmarks in the map.

The matrix Σ_t is the covariance of the robot's assessment of its expected error in the guess μ_t. As a quadratic matrix, Σ_t is of size $(3+2N) \times (3+2N)$. In SLAM, this matrix is usually distinctly non-sparse. The off-diagonal elements capture the correlations in the estimates of different variables. Non-zero correlations come along because the robot's location is uncertain, and as result the locations of the landmarks in the maps are uncertain. The importance of maintaining those off-diagonal elements is one of the key properties of EKF SLAM [9].

The EKF SLAM algorithm is easily derived for our point-landmark example. Suppose, for a moment, the motion function g and the measurement function h were *linear* in their arguments. Then the vanilla Kalman filter, as described in any textbook on Kalman filtering, would be applicable. EKF SLAM linearizes the functions g and h using Taylor series expansion—again, this is standard textbook material. Thus, in its most basic form (and in the absence of any data association problems), EKF SLAM is nothing but the application of the vanilla EKF to the online SLAM problem.

Figure 1.2 illustrates the EKF SLAM algorithm for an artificial example. The robot navigates from a start pose that serves as the origin of its coordinate system. As it moves, its own pose uncertainty increases, as indicated by uncertainty ellipses of growing diameter. It also senses nearby landmarks and maps them with an uncertainty that combines the fixed measurement uncertainty with the increasing pose uncertainty. As a result, the uncertainty in the landmark locations grows over time. The interesting transition happens in Fig. 1.2d: Here the robot observes the landmark it saw in the very beginning of mapping, and whose location is relatively well known. Through this observation, the robot's pose error is reduced, as indicated in Fig. 1.2d—notice the very small error ellipse for the final robot pose. This observation also reduces the uncertainty for other landmarks in the map. This phenomenon arises from a correlation that is expressed in the covariance matrix of the Gaussian posterior. Since most of the uncertainty in earlier landmark estimates is caused by the robot pose, and since this very uncertainty persists over time, the location estimates of those landmarks are correlated. When gaining information on the robot's pose, this information spreads to previously observed landmarks. This effect is probably the most important characteristic of the SLAM posterior [9]. Information that helps localize the robot is propagated through map, and as a result improves the localization of other landmarks in the map.

EKF SLAM also addresses the *data association problem.* If the identity of observed features is unknown, the basic EKF idea becomes inapplicable. The solution here is to reason about the most likely data association when a landmark is observed. This is usually done based on proximity: *which of the landmarks in the map corresponds most likely to landmark just observed?* The proximity calculation considers the measurement noise and the actual uncertainty in the poster estimate, and the metric used in this calculation is known as a Mahalanobis distance, which is a weighted quadratic distance. To minimize the chances of false data associations, many implementations use visible features to distinguish individual landmarks and associate groups of landmarks observed simultaneously [61, 62]. Modern-day implementations also maintains *provisional landmark list* and only add landmarks to the internal map when they have been observed sufficiently frequently [1, 13, 14, 91]. With an appropriate landmark definition and careful implementation of the data association step, EKF SLAM becomes a powerful method for feature-based SLAM.

EKF SLAM has been applied successfully to a large range of navigation problems, involving airborne, underwater, indoor, and various other vehicles. Figure 1.3a shows an example result obtained with a state-of-the-art implementation of EKF SLAM. Shown there is an underwater map obtained with the underwater robot Oberon, developed at the University of Sydney, Australia, and shown in Fig. 1.3b. This vehicle is equipped with a pencil sonar, a sonar that can scan at very high resolutions and detect obstacles up to 50 meters away. To facilitate the mapping problem, researchers have deposited long, small vertical objects in the water, which can be extracted from the sonar scans with relative ease. In this specific experiment, there is a row of such objects, spaced approximately

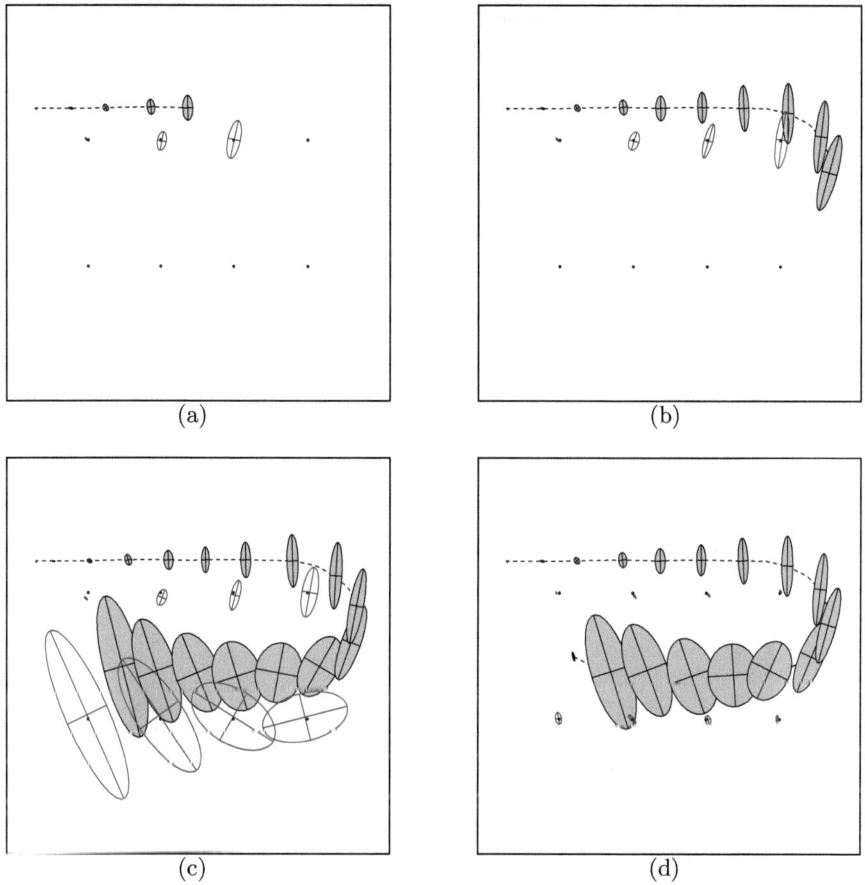

Fig. 1.2. EKF a applied to the online SLAM problem. The robot's path is a dotted line, and its estimates of its own position are shaded ellipses. Eight distinguishable landmarks of unknown location are shown as small dots, and their location estimates are shown as white ellipses. In (a)–(c) the robot's positional uncertainty is increasing, as is its uncertainty about the landmarks it encounters. In (d) the robot senses the first landmark again, and the uncertainty of *all* landmarks decreases, as does the uncertainty of its current pose. Image courtesy of Michael Montemerlo, Stanford University.

10 meters apart. In addition, a more distant cliff offers additional point features that can be detected using the pencil sonar.

The map shown in Fig. 1.3a depicts the robot path, marked by the triangles connected by a line. Around each triangle one can see an ellipse, which corresponds to the covariance matrix of the Kalman filter estimate projected into the robot location. The ellipse shows the variance; the larger it is, the less certain the robot is about its current pose. Various small dots in Fig. 1.3 show landmark sightings, obtained by searching the sonar scan for small and highly reflective objects. The majority of these sightings is rejected using statistical outlier rejection

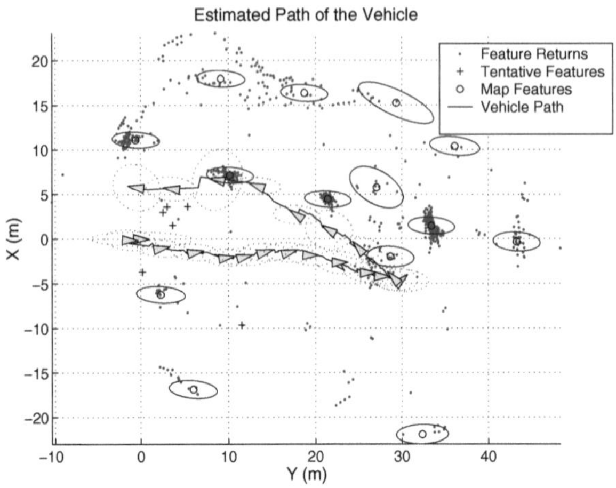

(a)

(b)

Fig. 1.3. (a) Example of Kalman filter estimation of the map and the vehicle pose. (b) Underwater vehicle Oberon, developed at the University of Sydney. Images courtesy of Stefan Williams and Hugh Durrant-Whyte, Australian Centre for Field Robotics.

techniques [13]. However, some are believed to correspond to a landmark and are added to the map. At the end of the run, the robot has classified 14 such objects as landmarks, each of which is plotted with the projected uncertainty ellipse in Fig. 1.3. These landmarks include the artificial landmarks put out by the researchers, but they also include various other terrain features in the vicinity of the robot. The residual pose uncertainty is small.

A key limitation of the EKF solution to the SLAM problem lies in the quadratic nature of the covariance matrix. A number of researchers have proposed extensions to the EKF SLAM algorithms that gain remarkable scalability through decomposing the map into submaps, for which covariances are maintained separately. Relevant literature can be found in [1, 27, 33, 48, 93, 80]. Other researchers have developed hybrid SLAM techniques, which combine EKF-style SLAM techniques with volumetric map representation; see [2, 3, 35, 68]. Finally, researchers have developed data association techniques for SLAM [6, 76, 83] based on advanced statistical techniques such as Dempster's EM algorithm [11].

1.3.2 Graph-Based Optimization Techniques

A second family of algorithms solves the SLAM problem through nonlinear sparse optimization. They draw their intuition from a graphical representation of the SLAM problem. Graph-based techniques were first mentioned in [7, 18], but a seminal paper [51] provided a first working solution. The representation in this section is closely related to a series of recent papers [10, 16, 17, 25, 26, 28, 29, 32, 45, 54]. We note that most contemporary techniques are offline and address the full SLAM problem, although some online versions exist that will be discussed below.

The basic intuition of graph-based SLAM is a follows. Landmarks and robot locations can be thought of as nodes in a graph. Every consecutive pair of locations x_{t-1}, x_t is tied together by an arc that represents the information conveyed by the odometry reading u_t. Other arcs exist between locations x_t and landmarks m_i, assuming that at time t the robot sensed landmark i. Arcs in this graph are soft constraints. Relaxing these constraints yields the robot's best estimate for the map and the full path.

The construction of the graph is illustrated in Fig. 1.4. Suppose at time $t = 1$, the robot senses landmark m_1. This adds an arc in the (yet highly incomplete) graph between x_1 and m_1. When caching the edges in a matrix format (which happens to correspond to a quadratic equation defining the resulting constraints), a value is added to the elements between x_1 and m_1, as shown on the right hand side of Fig. 1.4a.

Now suppose the robot moves. The odometry reading u_2 leads to an arc between nodes x_1 and x_2, as shown in Fig. 1.4b. Consecutive application of these two basic steps leads to a graph of increasing size, as illustrated in Fig. 1.4c. Nevertheless this graph is *sparse,* in that each node is only connected to a small number of other nodes. The number of constraints in the graph is (at worst) *linear* in the time elapsed and in the number of nodes in the graph.

If we think of the graph as a spring-mass model [32], computing the SLAM solution is equivalent to computing the state of minimal energy this model. To see, we note that the graph corresponds to the log-posterior of the full SLAM problem (cf. (1.4)):

$$\log p(X_T, m \mid Z_T, U_T). \tag{1.11}$$

Without derivation, we state that this logarithm is of the form

$$\begin{aligned} &\log p(X_T, m \mid Z_T, U_T) \\ &= \text{const} + \sum_t \log p(x_t \mid x_{t-1}, u_t) + \sum_t \log p(z_t \mid x_t, m) \end{aligned} \tag{1.12}$$

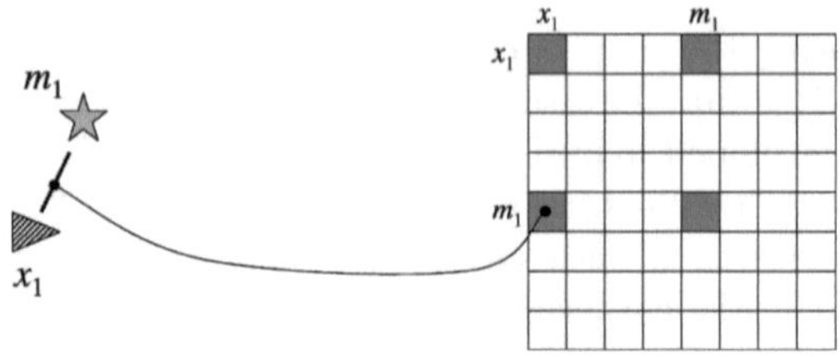

(a) Observation ls landmark m_1

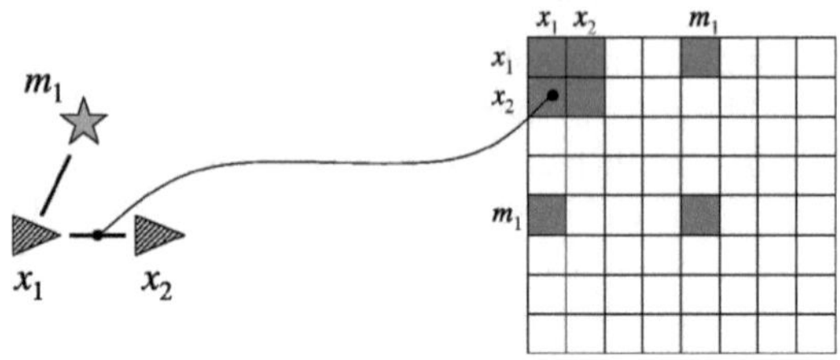

(b) Robot motion from x_1 to x_2

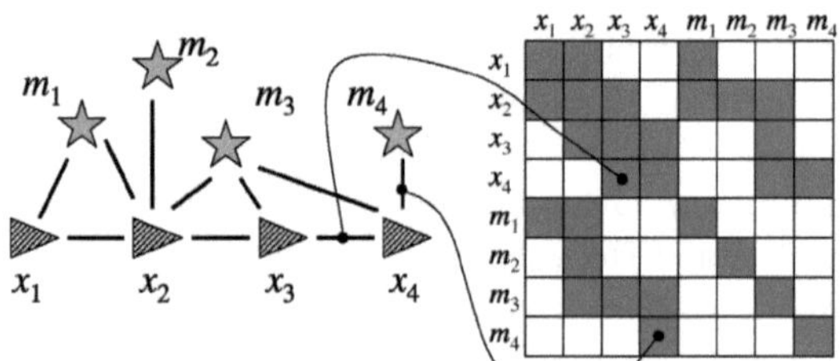

(c) Several steps later

Fig. 1.4. Illustration of the graph construction. The left diagram shows the graph, the right the constrains in matrix form.

Each constraint of the form $\log p(x_t \mid x_{t-1}, u_t)$ is the result of exactly one robot motion event, and it corresponds to an arc in the graph. Likewise, each constraint of the form $\log p(z_t \mid x_t, m)$ is the result of one sensor measurement, to which we can also find a corresponding arc in the graph. The SLAM problem is then simply to find the mode of this equation.

$$X_t^*, m^* = \operatorname*{argmax}_{X_T, m} \log p(X_T, m \mid Z_T, U_T) \tag{1.13}$$

Without derivation, we note that under the Gaussian noise assumptions, which was made in the point-landmark example, this expression resolves to the following quadratic form:

$$\begin{aligned} &\log p(X_T, m \mid Z_T, U_T) \\ &= \text{const} + \sum_t [x_t - g(x_{t-1}, u_t)]^T \, R_t^{-1} \, [x_t - g(x_{t-1}, u_t)] \\ &\quad + \sum_t [z_t - h(x_t, m)]^T \, Q_t^{-1} \, [z_t - h(x_t, m)] \end{aligned} \tag{1.14}$$

This is a sparse function. A number of efficient optimization techniques can be applied. Common choices include gradient descent, conjugate gradient, and others. Most SLAM implementations rely on some sort of iterative linearizing the functions g and h, in which case the objective in (1.14) becomes quadratic in all of its variables.

The graphical paradigm is easily extended to handle data association problems. This is because (1.14) is easily extended to integrate additional knowledge on data association. Suppose some oracle informed us that landmarks m_i and m_j in the map corresponded to one and the same physical landmark in the world. Then we can either remove m_j from the graph and attach all adjacent arcs to m_i, or we can add a *soft correspondence constraint* [50] of the form

$$[m_j - m_i]^T \, \Gamma \, [m_j - m_i] \tag{1.15}$$

Here Γ is 2-by-2 diagonal matrix whose coefficients determine the penalty for *not* assigning identical locations to two landmarks (hence we want Γ to be large). Since graphical methods are usually used for the full SLAM problem, the optimization can be interleaved with the search for the optimal data association. State-of-the-art implementations rely on RANSAC [24] or branch-and-bound methods [37, 46].

Graphical SLAM methods have the advantage that they scale to much higher-dimensional maps than EKF SLAM. The key limiting factor in EKF SLAM is the covariance matrix, which takes space (and update time) quadratic in the size of the map. No such constraint exists in graphical methods. The update time of the graph is constant, and the amount of memory required is linear (under some mild assumptions). Performing the optimization can be expensive, however. Technically, finding the optimal data association is suspected to be an NP-hard problem, although in practice the number of plausible assignments is usually small. The continuous optimization of the log likelihood function in (1.14) depends among other things on the number and size of loops in the map.

Figure 1.5 shows the result of a state-of-the-art SLAM algorithm based on analyzing the constraint graph and a nested search of the best data association. The data for this map was acquired by CMU's Groundhog robot [87], built to explore and map abandoned underground mines. Groundhog is equipped with a laser range finder which measures the range to obstacles along a horizontal slice of the world. The specific map shown here covers an area of 250 by 150 meters. The form of the map is known as *occupancy grid map*, which is due to Elfes and Moravec [19, 57]. Occupancy grid maps use Bayesian reasoning to estimate the posterior probability that a cell is free, thereby accommodating noise in range finders.

As a baseline for comparison, Fig. 1.5a shows a map constructed in a much simpler way: Here scans are localized relative to slightly older scans and, once localized, are added to the map under the assumption that the estimated location is correct. Such a technique is called *scan matching* [51]. Scan matching is a SLAM method, but it can only accommodate very small amounts of location uncertainties. The failure to close loops is obvious from Fig. 1.5a. In fact, pairwise

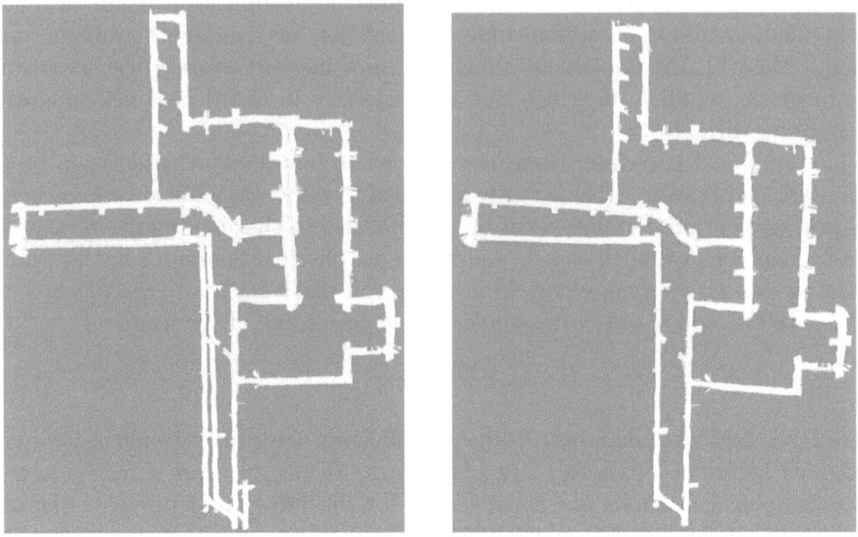

Fig. 1.5. An occupancy grid map of an abandoned mine. The left map estimates the data association incrementally, and only in reference to the most recent sensor measurement. The right map is the result of a global data association search and a graphical optimization. Images courtesy of Dirk Hähnel, University of Freiburg.

scan matching can be thought of as a version of a graphical SLAM algorithm, but correspondence is only established (and constraints inserted in the graph) between immediately consecutive scans.

To map this data into a graph of manageable size, the algorithm decomposes the map into small local submaps, one for each five meters of robot travel. Within these five meters, the maps are sufficiently accurate, as general drift is small and hence scan matching performs essentially flawlessly. Each submap coordinates become a pose node in the GraphSLAM. Adjacent submaps are linked through the relative motion constraints between them. The resulting structure is shown in Fig. 1.5b.

On this graph, we can now perform a branch-and-bound recursive search for correspondences. For finding good submaps tat might correspond, this algorithm uses a correlation analysis for two overlaying maps. Once two suitable maps are found, a soft constraint of the type stated in (1.15) is added to the graph, followed by an optimization step of the resulting set of constraints. Figure 1.6 illustrates the process of data association: each circle corresponds to a new constraint that would be found in the search. The figure illustrates the iterative nature of the search: certain correspondences are only discovered when others have been propagated, and others are dissolved in the process of the search. The final model is stable, in that additional search for new data association induces no further changes. The resulting grid map is shown in Fig. 1.5b.

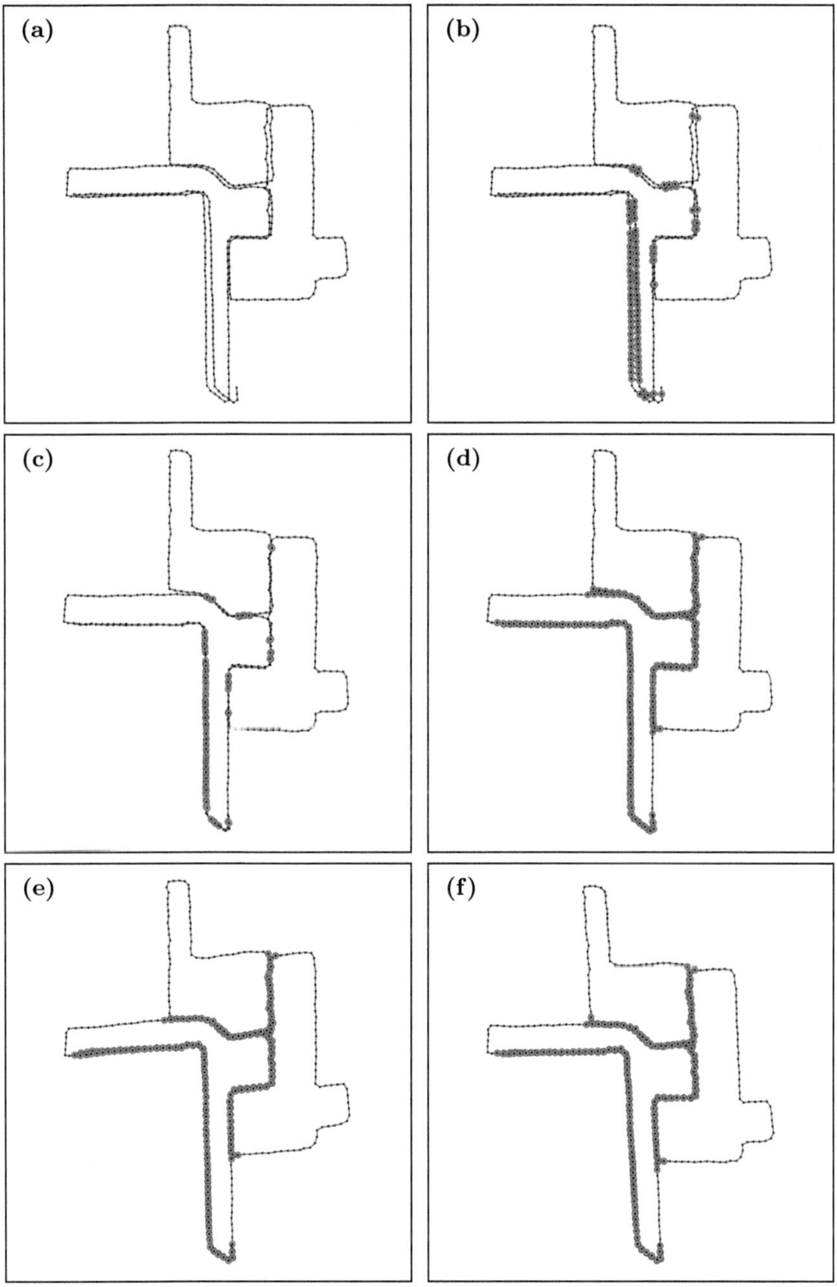

Fig. 1.6. Data association search. See text.

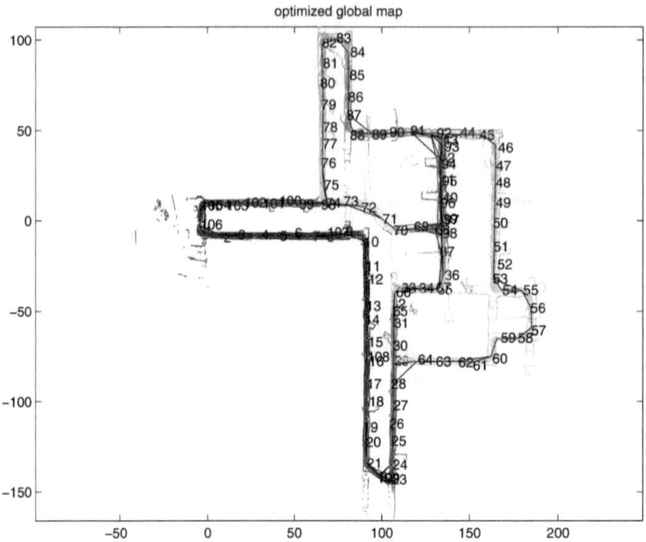

Fig. 1.7. Mine map generated by the *Atlas* SLAM algorithm by [5]. Image courtesy of Michael Bosse, Paul Newman, John Leonard, and Seth Teller, MIT.

Other graph-based techniques for SLAM have produced similar results. Figure 1.7 shows a map of the same data set generated by [5], using an algorithm called *Atlas*. This algorithm decomposes maps into submaps whose relation is maintained through information-theoretic relative links.

We note that the graph-based paradigm is very closely link to information theory, in that the soft constraints constitute the information the robot has on the world (in an information-theoretic sense [8]). Most methods in the field are inherently offline, that is, they optimize for the entire robot path. If the robot path is long, the optimization may become cumbersome. This is one of the disadvantages of the graph-based paradigm. There exists a number of cross-overs that manipulate the graph online so as to factor our past robot location variables. The resulting algorithms are filters; see [5, 66, 69, 84], and they tend to be intimately related to information filter methods [9, 63, 65, 67, 84, 86]. Many of the original attempts to decompose EKF SLAM representations into smaller submaps to scale up are based on motivations that are not dissimilar to the graphical approach; see [33, 48, 92].

As this article is being written, graphical and optimization-based SLAM algorithm are subject of intense research. Recent results show that the paradigm scales to maps with 10^8 features [5, 10, 16, 17, 25, 26, 28, 29, 45, 54, 87]. Arguably, the graph-based paradigm has generated some the largest SLAM maps ever built, but usually in an offline fashion.

1.3.3 Particle Methods

The third principal SLAM paradigm is based on particle filters. Particle filters can be traced back to [53], but they have become popular only in recent years. Particle filter represent a posterior through a set of *particles*. For the novice in SLAM, each particle is best thought as a concrete guess as to what the true value of the state may be. By collecting many such guesses into a set of guesses, or set of particles, the particle filters captures a representative sample from the posterior distribution. The particle filter has been shown under mild conditions to approach the true posterior as the particle set size goes to infinity. It is also a non-parametric representation that represents multimodal distributions with ease. In recent years, the advent of extremely efficient microprocessors has made particle filters a popular algorithm [75, 15, 43, 49, 71].

The key problem with the particle filter in the context of SLAM is that the space of maps and robot paths is huge! Suppose we have a map with 1000 features. How many particles would it take to populate that space? In fact, particle filters scale exponentially with the dimension of the underlying state space. Three or four dimensions are thus acceptable, but 100 dimensions are generally not.

The trick to make particle filters amenable to the SLAM problem goes back to [4, 72]. The trick has been introduced into the SLAM literature in [60], followed by [55], who coined the name FastSLAM. Let us first explain the basic FastSLAM algorithm on the simplified point-landmark example, and then discuss the justification for this approach.

At any point in time, FastSLAM maintains K particles of the type:

$$X_t^{[k]}, \qquad \mu_{t,1}^{[k]}, \ldots, \mu_{t,N}^{[k]}, \qquad \Sigma_{t,1}^{[k]}, \ldots, \Sigma_{t,N}^{[k]} \tag{1.16}$$

Here $[k]$ is the index of the sample. This expression states that a particle contains

- a sample path $X_t^{[k]}$, and
- a set of N 2-dimensional Gaussians with means $\mu_{t,n}^{[k]}$ and variances $\Sigma_{t,n}^{[k]}$), one for each landmark in the environment.

Here n is the index of the landmark (with $1 \leq n \leq N$). From that it follows that K particles possess K path samples. It also possesses KN Gaussians, each of which models exactly one landmark for one of the particles.

Initializing FastSLAM is simple: just set each particle's robot location to its known starting coordinates, and zero the map. The particle update then proceeds as follows.

- When an odometry reading is received, new location variables are generated stochastically, one for each of the particles. The distribution for generating those location particles is based on the motion model:

$$x_t^{[k]} \sim p(x_t \mid x_{t-1}^{[k]}, u^t). \tag{1.17}$$

  Here $x_{t-1}^{[k]}$ is the previous location, which is part of the particle. This probabilistic sampling step is easily implemented for any robot whose kinematics can be computed.

- When a measurement z_t is received, two things happen: first, FastSLAM computes for each particle the probability of the new measurement z_t. Let the index of the sensed landmark be n. Then the desired probability is defined as follows:

$$w_t^{[k]} := \mathcal{N}(z_t; \mid x_t^{[k]}, \mu_{t,n}^{[k]}, \Sigma_{t,n}^{[k]}). \tag{1.18}$$

  The factor $w_t^{[k]}$ is called the *importance weight*, since it measures how "important" the particle is in the light of the new sensor measurement. As before, $\mathcal{N}$ denotes the normal distribution, but this time it is calculated for a specific value, z_t. The importance weights of all particles are then normalized so that they sum to 1.

  Next, FastSLAM draws with replacement from the set of existing particles a set of new particles. The probability of drawing a particle is its normalized importance weight. This step is called *resampling.* The intuition behind resampling is simple: particles for which the measurement is more plausible have a higher chance of surviving the resampling process.

  Finally, FastSLAM updates for the new particle set the mean $\mu_{t,n}^{[k]}$ and covariance $\Sigma_{t,n}^{[k]}$, based on the measurement z_t. This update follows the standard EKF update rules.

This all sounds complex, but FastSLAM is quite easy to implement. Sampling from the motion model is usually straightforward, since it involves a simple kinematic calculation. Computing the importance of a measurement is, too, straightforward, especially for Gaussian measurement noise. And updating a low-dimensional particle filter is also straightforward. This makes FastSLAM one of the easiest-to-implement algorithm presently available.

FastSLAM has been shown to approximate the full SLAM posterior. The derivation of FastSLAM exploits three techniques: Rao-Blackwellization, conditional independence, and resampling. Rao-Blackwellization is the following concept. Suppose we would like to compute a probability distribution $p(a, b)$, where a and b are arbitrary random variables. The vanilla particle filter would draw particles from the joint distributions, that is, each particle would have a value for a and one for b. However, if the conditional $p(b \mid a)$ can be described in closed form, it is equally legitimate to just draw particles from $p(a)$, and attach to each particle a closed-form description of $p(b \mid a)$. This trick is known as Rao-Blackwellization, and it yields better results than sampling from the joint. FastSLAM applies this technique, in that it samples from the path posterior $p(X_t^{[k]} \mid U_t, Z_t)$ and represents the map $p(m \mid X_t^{[k]}, U_t, Z_t)$ in Gaussian form.

FastSLAM also breaks down the posterior over maps (conditioned on paths) into sequences of low-dimensional Gaussians. The justification for this decomposition is subtle. It arises from a specific conditional independence assumption that is native to SLAM. Fig. 1.8 illustrates the concept graphically. In SLAM, knowledge of the robot path renders all landmark estimates independent. This is easily shown for the graphical network in Fig. 1.8: we find that if we remove the path variables from Fig. 1.8 then the landmark variables are all disconnected [70].

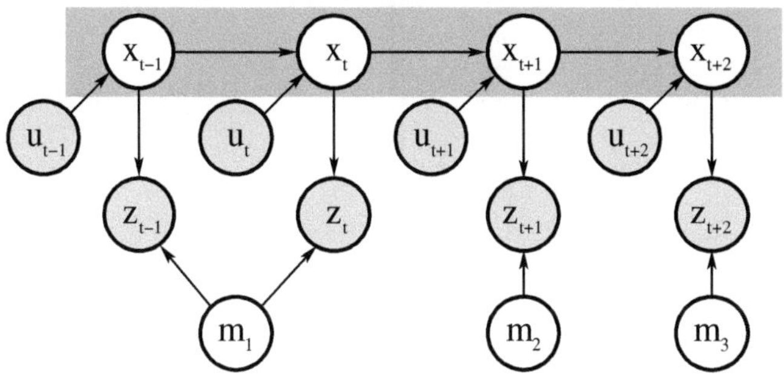

Fig. 1.8. The SLAM problem depicted as Bayes network graph. The robot moves from location x_{t-1} to location x_{t+2}, driven by a sequence of controls. At each location x_t it observes a nearby feature in the map $m = \{m_1, m_2, m_3\}$. This graphical network illustrates that the location variables "separate" the individual features in the map from each other. If the locations are known, there remains no other path involving variables whose value is not known, between any two features in the map. This lack of a path renders the posterior of any two features in the map conditionally independent (given the locations).

Thus, in SLAM *any dependence between multiple landmark estimates is mediated through the robot path.* This subtle but important observation implies that even if we used a large, monolithic Gaussian for the entire map (one per particle, of course), the off-diagonal element between different landmarks would simply remain zero. It is therefore legitimate to implement the map more efficiently, using N small Gaussians, one for each landmark. This explains the efficient map representation in FastSLAM.

We also note that FastSLAM uses a particle filter. Derivations of the particle filter can be found in the literature referenced above. Here we note that both steps—the motion and the measurement steps—retain the property that (asymptotically) samples are drawn from the full SLAM posterior. This is quite easy to see for the motion update step. For the measurement step, the property is retained through resampling, which adjusts the particle population in response to the new information added by the measurement.

Figure 1.9 shows results for a point-feature problem; here the point features are the centers of tree trunks as observed by an outdoor robot. The dataset used here is known as the Victoria Park dataset [34]. Figure 1.9a shows the path of the vehicle obtained by integrating the vehicle controls, without perception controls are a poor predictor of location for this vehicle; after 30 minutes of driving, the estimated position of the vehicle is well over 100 meters away from its GPS position.

The FastSLAM algorithm has a number of remarkable properties, which may not be intuitive to the untrained eye. First, it solves both the full and the online SLAM problems. Each particle has a sample of an entire path (and in fact, conditioning on the entire path is required for its mathematical derivation), but the actual update equation only uses the most recent pose. This makes FastSLAM a

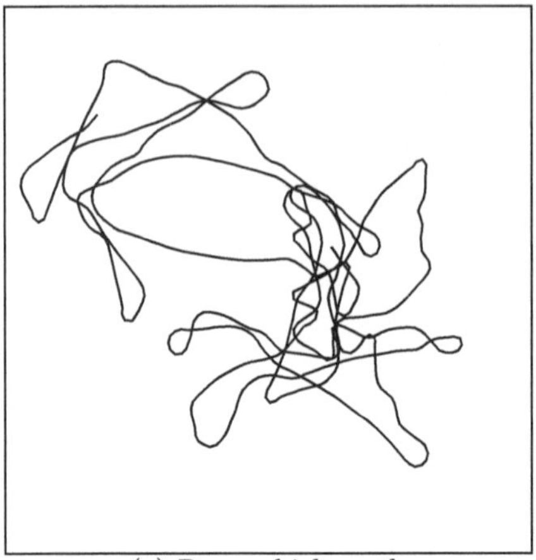

(a) Raw vehicle path

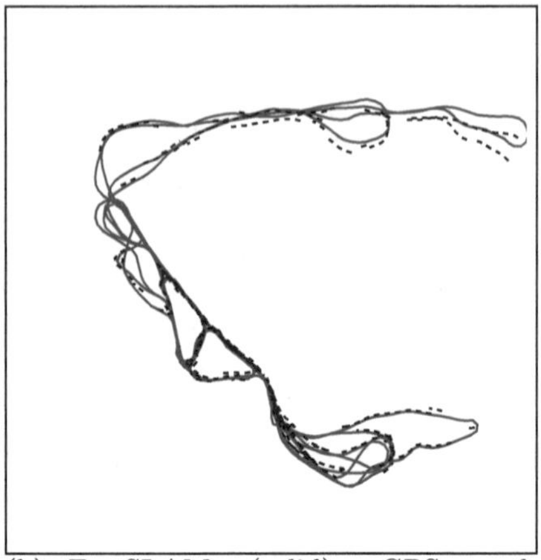

(b) FastSLAM (solid), GPS path (dashed)

(c) Path and map with aerial image

Fig. 1.9. (a) Vehicle path predicted by the odometry; (b) True path (dashed line) and FastSLAM 1.0 path (solid line); (c) Victoria Park results overlaid on aerial imagery with the GPS path in blue (dashed), average FastSLAM 1.0 path in yellow (solid), and estimated features as yellow circles. Data and aerial image courtesy of José Guivant and Eduardo Nebot, Australian Centre for Field Robotics.

filter, similar to the EKF. Second, FastSLAM makes it easy to pursue multiple data association hypotheses. It is straightforward to make data association decisions on a per-particle basis, instead of having to adopt the same hypothesis for the entire filter. While we will not give any mathematical justification, we note that the resulting FastSLAM algorithm samples the correct posteriors even for SLAM problems with unknown data association—something that neither of the previous two algorithms can claim. And third, FastSLAM can be implemented very efficiently. using advanced tree methods to represent the map estimates,

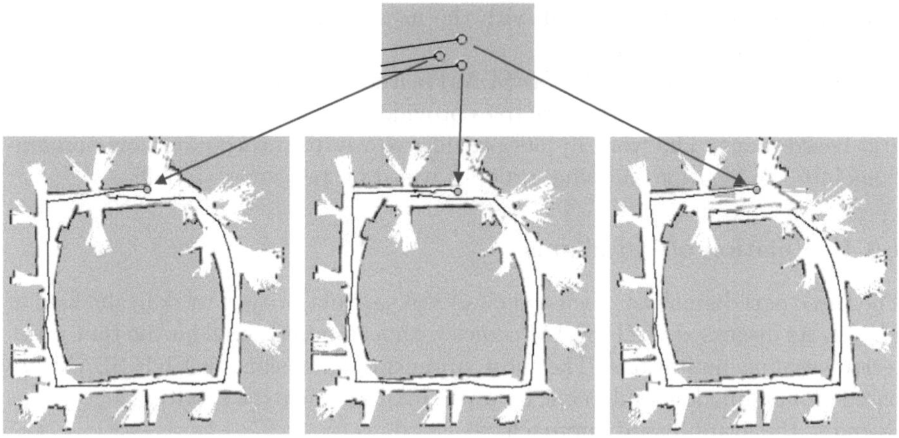

Fig. 1.10. Application of the grid-based variant of the FastSLAM algorithm. Each particle carries its own map and the importance weights of the particles are computed based on the likelihood of the measurements given the particle's own map.

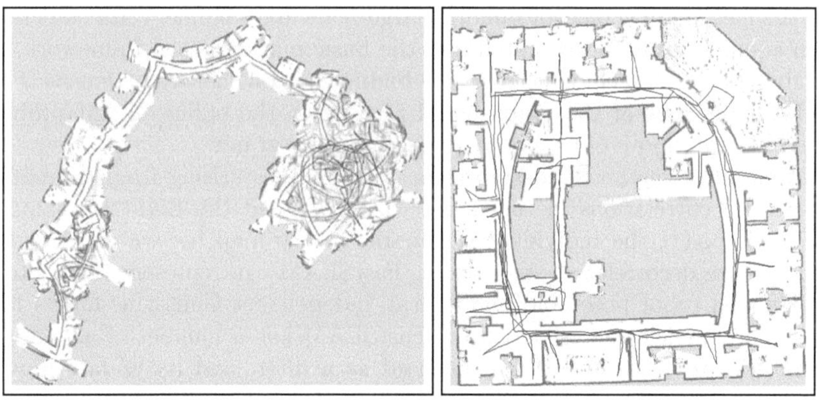

Fig. 1.11. Occupancy grid map generated from laser range data and based on pure odometry. All images courtesy of Dirk Hähnel, University of Freiburg.

the update can be performed in time logarithmic in the size of the map N, and linear in the number of particles M. These properties, along with the relative ease of implementation, has made FastSLAM a popular choice.

FastSLAM has been extended in great many ways. One important variant is a grid-based version of FastSLAM, in which the Gaussians are replaced by an occupancy grid map [38]. This variant is illustrated in Figs. 1.10 and 1.11. Figure 1.10 shows the situation just before closing a large loop. The three different particles each stand for different paths, and they also posses their own local maps. When the loop is closed importance resampling selects those particles whose maps are most consistent with the measurement. A resulting large-scale map is shown in Fig. 1.11.

Significant extensions of the FastSLAM method can be found in [20, 21], whose methods *DP-SLAM* and *ancestry trees* provide efficient tree update methods for grid-based maps. The work in [56] provides a way to incorporate new observations into the location sampling process, based on prior work in [89].

1.3.4 Relation of Paradigms

The three paradigms just discussed cover the vast majority of work in the field of SLAM. As discussed, EKF SLAM comes with a computational hurdle that poses serious scaling limitations. The most promising extensions of EKF SLAM are based on building local submaps; however, in many ways the resulting algorithms resemble the graph-based approach.

Graph-based methods address the full SLAM problem, hence are by nature not online. They draw their intuition that SLAM can be modeled by a sparse graph of soft constraints, where each constraint either corresponds to a motion or a measurement event. Due to the availability of highly efficient optimization methods for sparse nonlinear optimization problems, graph-based SLAM has become the method of choice for building large-scale maps offline. Data association search is quite easily incorporated into the basic mathematical framework, and a number of search techniques exist for finding suitable correspondences. There are also extensions of the graph-based SLAM for the online SLAM problem. Those tend to remove old robot locations from the graph.

Particle filter methods sidestep some of the issues arising from the natural inter-feature correlations in the map—which plagued the EKF. By sampling from robot poses, the individual landmarks in the map become independent, and hence are decorrelated. As a result, FastSLAM can represent the posterior by a sampled robot pose, and many local, independent Gaussians for its landmarks. The particle representation of FastSLAM has a number of advantages. Computationally, FastSLAM can be used as a filter, and its update requires linear-logarithmic time where EKF needed quadratic time. Further, FastSLAM can sample over data association, which makes it a prime method for SLAM with unknown data association. On the negative side, the number of necessary particles can grow very large, especially for robots seeking to map multiple nested loops. We discussed extensions of FastSLAM that use occupancy grid maps instead of Gaussian landmarks, and showed state-of-the-art examples in large map building.

1.4 Conclusion and Future Challenges

This article provided a comprehensive introduction into the SLAM problem and its primary solutions. The SLAM problem was defined as the problem faced by a mobile platform roaming an unknown environment, and seeking to localize and map its environment at the same time. The article discussed three main paradigms in SLAM, which are based on the extended Kalman filter, graph-based sparse optimization techniques, and particle filters. It pointed out some of the advantages and disadvantages of those methods. For a more in-depth discussion, the interested reader shall be referred to a recent textbook covering SLAM [82].

Interestingly, the field of SLAM is still relatively young, and it has made enormous progress within just the past decade. In fact, nearly every method described here has been developed within the past few years. Despite all this progress, there remains a great number of open research issues that warrant future research.

In particular, SLAM techniques mostly deal with static environments, yet nearly every actual robot environment is dynamic. Early applications of SLAM methods to dynamic environments can be found in [39, 90, 94]. More work is needed to understand the interaction of moving and non-moving objects in SLAM.

Most SLAM work addresses single-robot mapping, yet sometimes one is given a team of robots. Early and highly restrictive work on multi-robot SLAM can be found in [36, 64]. More recent methods include those in [23, 73, 85]. Multi-robot SLAM has benefited greatly from a recent DARPA project focused on this topic; nevertheless, the existing methods have not yet matured to a level where they can be used by non-experts in the field.

One of the primary challenges in SLAM is to pursue significant implementations. While the theory of SLAM is now quite well-developed, SLAM has not yet been used extensively in industrial or commercial applications. There exist promising proto-types, including methods for building large-scale 3-D volumetric maps [30, 12, 40, 81], detailed underwater reconstruction [22, 74, 93], and mapping of abandoned underground mines [87]. Nevertheless, the authors feel that more work is needed to mature the technology into industrial-strength applications.

References

1. Bailey, T.: Mobile Robot Localisation and Mapping in Extensive Outdoor Environments. PhD thesis, University of Sydney, Sydney, NSW, Australia (2002)
2. Betgé-Brezetz, S., Chatila, R., Devy, M.: Object-based modelling and localization in natural environments. In: ICRA. Proceedings of the IEEE International Conference on Robotics and Automation, Osaka, Japan, IEEE Computer Society Press, Los Alamitos (1995)
3. Betgé-Brezetz, S., Hébert, P., Chatila, R., Devy, M.: Uncertain map making in natural environments. In: ICRA. Proceedings of the IEEE International Conference on Robotics and Automation, Minneapolis, IEEE Computer Society Press, Los Alamitos (1996)
4. Blackwell, D.: Conditional expectation and unbiased sequential estimation. Annals of Mathematical Statistics 18, 105–110 (1947)
5. Bosse, M., Newman, P., Leonard, J., Soika, M., Feiten, W., Teller, S.: Simultaneous localization and map building in large-scale cyclic environments using the atlas framework. International Journal of Robotics Research 23(12), 1113–1139 (2004)

6. Burgard, W., Fox, D., Jans, H., Matenar, C., Thrun, S.: Sonar-based mapping of large-scale mobile robot environments using EM. In: Proceedings of the International Conference on Machine Learning, Bled, Slovenia (1999)
7. Cheeseman, P., Smith, P.: On the representation and estimation of spatial uncertainty. International Journal of Robotics 5, 56–68 (1986)
8. Cover, T.M., Thomas, J.A.: Elements of Information Theory. Wiley, Chichester (1991)
9. Csorba, M.: Simultaneous Localisation and Map Building. PhD thesis, University of Oxford (1997)
10. Dellaert, F.: Square root SAM. In: Thrun, S., Sukhatme, G., Schaal, S., Brock, O. (eds.) Proceedings of the Robotics Science and Systems Conference, Cambridge, MA (2005)
11. Dempster, A.P., Laird, A.N., Rubin, D.B.: Maximum likelihood from incomplete data via the EM algorithm. Journal of the Royal Statistical Society, Series B 39(1), 1–38 (1977)
12. Devy, M., Parra, C.: 3-d scene modelling and curve-based localization in natural environments. In: ICRA. Proceedings of the IEEE International Conference on Robotics and Automation, Leuven, Belgium, IEEE Computer Society Press, Los Alamitos (1998)
13. Dissanayake, G., Newman, P., Clark, S., Durrant-Whyte, H.F., Csorba, M.: A solution to the simultaneous localisation and map building (SLAM) problem. IEEE Transactions of Robotics and Automation 17(3), 229–241 (2001)
14. Dissanayake, G., Williams, S.B., Durrant-Whyte, H., Bailey, T.: Map management for efficient simultaneous localization and mapping (SLAM). Autonomous Robots 12, 267–286 (2002)
15. Doucet, A.: On sequential simulation-based methods for Bayesian filtering. Technical Report CUED/F-INFENG/TR 310, Cambridge University, Department of Engineering, Cambridge, UK (1998)
16. Duckett, T., Marsland, S., Shapiro, J.: Learning globally consistent maps by relaxation. In: Proceedings of the IEEE International Conference on Robotics and Automation, San Francisco, pp. 3841–3846. IEEE Computer Society Press, Los Alamitos (2000)
17. Duckett, T., Marsland, S., Shapiro, J.: Fast, on-line learning of globally consistent maps. Autonomous Robots 12(3), 287–300 (2002)
18. Durrant-Whyte, H.F.: Uncertain geometry in robotics. IEEE Transactions on Robotics and Automation 4(1), 23–31 (1988)
19. Elfes, A.: Sonar-based real-world mapping and navigation. IEEE Journal of Robotics and Automation RA-3(3), 249–265 (1987)
20. Eliazar, A., Parr, R.: DP-SLAM: Fast, robust simultaneous localization and mapping without predetermined landmarks. In: IJCAI. Proceedings of the Sixteenth International Joint Conference on Artificial Intelligence, Acapulco, Mexico (2003)
21. Eliazar, A., Parr, R.: DP-SLAM 2.0. In: ICRA. Proceedings of the IEEE International Conference on Robotics and Automation, New Orleans, USA, IEEE Computer Society Press, Los Alamitos (2004)
22. Eustice, R., Singh, H., Leonard, J., Walter, M., Ballard, R.: Visually navigating the RMS Titanic with SLAM information filters. In: Thrun, S., Sukhatme, G., Schaal, S., Brock, O. (eds.) Proceedings of the Robotics Science and Systems Conference, Cambridge, MA (2005)

23. Fenwick, J., Newman, P., Leonard, J.: Collaborative concurrent mapping and localization. In: ICRA. Proceedings of the IEEE International Conference on Robotics and Automation, Washington, DC, IEEE Computer Society Press, Los Alamitos (2002)
24. Fischler, M.A., Bolles, R.C.: Random sample consensus: A paradigm for model fitting with applications to image analysis and automated cartography. Communications of the ACM 24, 381–395 (1981)
25. Folkesson, J., Christensen, H.I.: Graphical SLAM: A self-correcting map. In: ICRA. Proceedings of the IEEE International Conference on Robotics and Automation, New Orleans, USA, IEEE Computer Society Press, Los Alamitos (2004)
26. Folkesson, J., Christensen, H.I.: Robust SLAM. In: Proceedings of the International Symposium on Autonomous Vehicles, Lisboa, PT (2004)
27. Folkesson, J., Christensen, H.I.: Outdoor exploration and slam using a compressed filter. In: ICRA. Proceedings of the IEEE International Conference on Robotics and Automation, pp. 419–427. IEEE Computer Society Press, Los Alamitos (2003)
28. Frese, U., Hirzinger, G.: Simultaneous localization and mapping—a discussion. In: Proceedings of the IJCAI Workshop on Reasoning with Uncertainty in Robotics, Seattle, WA, pp. 17–26 (2001)
29. Frese, U., Larsson, P., Duckett, T.: A multigrid algorithm for simultaneous localization and mapping. IEEE Transactions on Robotics 21(2), 196–207 (2005)
30. Frueh, C., Zakhor, A.: Constructing 3d city models by merging ground-based and airborne views. In: CVPR. Proceedings of the IEEE Computer Society Conference on Computer Vision and Pattern Recognition, Madison, Wisconsin, IEEE Computer Society Press, Los Alamitos (2003)
31. Gauss, K.F.: Theoria Motus Corporum Coelestium (Theory of the Motion of the Heavenly Bodies Moving about the Sun in Conic Sections). Little, Brown, and Co., Republished, and by Dover in 1963, 1809 (1857)
32. Golfarelli, M., Maio, D., Rizzi, S.: Elastic correction of dead-reckoning errors in map building. In: IROS. Proceedings of the IEEE/RSJ International Conference on Intelligent Robots and Systems, Victoria, Canada, pp. 905–911 (1998)
33. Guivant, J., Nebot, E.: Optimization of the simultaneous localization and map building algorithm for real time implementation. IEEE Transactions of Robotics and Automation 17(3), 242–257 (2001) (in press)
34. Guivant, J., Nebot, E., Baiker, S.: Autonomous navigation and map building using laser range sensors in outdoor applications. Journal of Robotics Systems 17(10), 565–583 (2000)
35. Guivant, J.E., Nebot, E.M., Nieto, J., Masson, F.: Navigation and mapping in large unstructured environments. International Journal of Robotics Research 23(4) (2004)
36. Gutmann, J.-S., Konolige, K.: Incremental mapping of large cyclic environments. In: CIRA. Proceedings of the IEEE International Symposium on Computational Intelligence in Robotics and Automation, IEEE Computer Society Press, Los Alamitos (2000)
37. Hähnel, D., Burgard, W., Wegbreit, B., Thrun, S.: Towards lazy data association in SLAM. In: ISRR 2003. Proceedings of the 11th International Symposium of Robotics Research, Sienna, Italy, Heidelberg (2003)
38. Hähnel, D., Fox, D., Burgard, W., Thrun, S.: A highly efficient FastSLAM algorithm for generating cyclic maps of large-scale environments from raw laser range measurements. In: IROS. Proceedings of the Conference on Intelligent Robots and Systems (2003)

39. Hähnel, D., Schulz, D., Burgard, W.: Mobile robot mapping in populated environments. Autonomous Robots 17(7), 579–598 (2003)
40. Iocchi, L., Konolige, K., Bajracharya, M.: Visually realistic mapping of a planar environment with stereo. In: Proceesings of the 2000 International Symposium on Experimental Robotics, Waikiki, Hawaii (2000)
41. Jazwinsky, A.M.: Stochastic Processes and Filtering Theory. Academic, New York (1970)
42. Kalman, R.E.: A new approach to linear filtering and prediction problems. Trans. ASME, Journal of Basic Engineering 82, 35–45 (1960)
43. Kitagawa, G.: Monte Carlo filter and smoother for non-Gaussian nonlinear state space models. Journal of Computational and Graphical Statistics 5(1), 1–25 (1996)
44. Konecny, G.: Geoinformation: Remote Sensing, Photogrammetry and Geographical Information Systems. Taylor & Francis, London (2002)
45. Konolige, K.: Large-scale map-making. In: Proceedings of the AAAI National Conference on Artificial Intelligence, San Jose, CA, pp. 457–463. AAAI, USA (2004)
46. Kuipers, B., Modayil, J., Beeson, P., MacMahon, M., Savelli, F.: Local metrical and global topological maps in the hybrid spatial semantic hierarchy. In: ICRA. Proceedings of the IEEE International Conference on Robotics and Automation, New Orleans, IEEE Computer Society Press, Los Alamitos (2004)
47. Leonard, J.J., Durrant-Whyte, H.F.: Mobile robot localization by tracking geometric beacons. IEEE Transactions on Robotics and Automation 7(3), 376–382 (1991)
48. Leonard, J.J., Feder, H.J.S.: A computationally efficient method for large-scale concurrent mapping and localization. In: Hollerbach, J., Koditschek, D. (eds.) Proceedings of the Ninth International Symposium on Robotics Research, Salt Lake City, Utah (1999)
49. Liu, J., Chen, R.: Sequential monte carlo methods for dynamic systems. Journal of the American Statistical Association 93, 1032–1044 (1998)
50. Liu, Y., Thrun, S.: Results for outdoor-SLAM using sparse extended information filters. In: ICRA. Proceedings of the IEEE International Conference on Robotics and Automation, IEEE Computer Society Press, Los Alamitos (2003)
51. Lu, F., Milios, E.: Globally consistent range scan alignment for environment mapping. Autonomous Robots 4, 333–349 (1997)
52. Maybeck, P.S.: The Kalman filter: An introduction to concepts. In: Cox, I.J., Wilfong, G.T. (eds.) Autonomous Robot Vehicles, Springer, Heidelberg (1990)
53. Metropolis, N., Ulam, S.: The Monte Carlo method. Journal of the American Statistical Association 44(247), 335–341 (1949)
54. Montemerlo, M., Thrun, S.: Large-scale robotic 3-d mapping of urban structures. In: ISER. Proceedings of the International Symposium on Experimental Robotics, Singapore (2004) (Springer Tracts in Advanced Robotics (STAR))
55. Montemerlo, M., Thrun, S., Koller, D., Wegbreit, B.: FastSLAM: A factored solution to the simultaneous localization and mapping problem. In: Proceedings of the AAAI National Conference on Artificial Intelligence, Edmonton, Canada, AAAI, USA (2002)
56. Montemerlo, M., Thrun, S., Koller, D., Wegbreit, B.: FastSLAM 2.0: An improved particle filtering algorithm for simultaneous localization and mapping that provably converges. In: IJCAI. Proceedings of the Sixteenth International Joint Conference on Artificial Intelligence, Acapulco, Mexico (2003)
57. Moravec, H.P.: Sensor fusion in certainty grids for mobile robots. AI Magazine 9(2), 61–74 (1988)

58. Moutarlier, P., Chatila, R.: An experimental system for incremental environment modeling by an autonomous mobile robot. In: 1st International Symposium on Experimental Robotics, Montreal (June 1989)
59. Moutarlier, P., Chatila, R.: Stochastic multisensory data fusion for mobile robot location and environment modeling. In: 5th Int. Symposium on Robotics Research, Tokyo (1989)
60. Murphy, K., Russell, S.: Rao-Blackwellized particle filtering for dynamic Bayesian networks. In: Doucet, A., de Freitas, N., Gordon, N. (eds.) Sequential Monte Carlo Methods in Practice, pp. 499–516. Springer, Heidelberg (2001)
61. Neira, J., Tardós, J.D.: Data association in stochastic mapping using the joint compatibility test. IEEE Transactions on Robotics and Automation 17(6), 890–897 (2001)
62. Neira, J., Tardós, J.D., Castellanos, J.A.: Linear time vehicle relocation in SLAM. In: ICRA. Proceedings of the IEEE International Conference on Robotics and Automation, IEEE Computer Society Press, Los Alamitos (2003)
63. Nettleton, E., Thrun, S., Durrant-Whyte, H.: Decentralised slam with low-bandwidth communication for teams of airborne vehicles. In: Proceedings of the International Conference on Field and Service Robotics, Lake Yamanaka, Japan (2003)
64. Nettleton, E.W., Gibbens, P.W., Durrant-Whyte, H.F.: Closed form solutions to the multiple platform simultaneous localisation and map building (slam) problem. In: Dasarathy, B.V. (ed.) Sensor Fusion: Architectures, Algorithms, and Applications IV, Bellingham, vol. 4051, pp. 428–437 (2000)
65. Newman, P.: On the Structure and Solution of the Simultaneous Localisation and Map Building Problem. PhD thesis, Australian Centre for Field Robotics, University of Sydney, Sydney, Australia (2000)
66. Newman, P., Rikoski, J.L.R.: Towards constant-time slam on an autonomous underwater vehicle using synthetic aperture sonar. In: Proceedings of the International Symposium of Robotics Research, Sienna, Italy (2003)
67. Newman, P.M., Durrant-Whyte, H.F.: A new solution to the simultaneous and map building (SLAM) problem. In: Proceedings of SPIE (2001)
68. Nieto, J., Guivant, J.E., Nebot, E.M.: The hybrid metric maps (HYMMs): A novel map representation for dense SLAM. In: ICRA. Proceedings of the IEEE International Conference on Robotics and Automation, New Orleans, USA, IEEE Computer Society Press, Los Alamitos (2004)
69. Paskin, M.A.: Thin junction tree filters for simultaneous localization and mapping. In: IJCAI. Proceedings of the Sixteenth International Joint Conference on Artificial Intelligence, Acapulco, Mexico (2003)
70. Pearl, J.: Probabilistic reasoning in intelligent systems: networks of plausible inference. Morgan Kaufmann Publishers, San Mateo, CA (1988)
71. Pitt, M., Shephard, N.: Filtering via simulation: auxiliary particle filter. Journal of the American Statistical Association 94, 590–599 (1999)
72. Rao, C.R.: Information and accuracy obtainable in estimation of statistical parameters. Bulletin of the Calcutta Mathematical Society 37, 81–91 (1945)
73. Rekleitis, I.M., Dudek, G., Milios, E.E.: Multi-robot collaboration for robust exploration. Annals of Mathematics and Artificial Intelligence 31(1-4), 7–40 (2001)

74. Rikoski, R., Leonard, J., Newman, P., Schmidt, H.: Trajectory sonar perception in the ligurian sea. In: ISER. Proceedings of the International Symposium on Experimental Robotics, Singapore (2004) (Springer Tracts in Advanced Robotics (STAR))
75. Rubin, D.B.: Using the SIR algorithm to simulate posterior distributions. In: Bernardo, M.H., an DeGroot, K.M., Lindley, D.V., Smith, A.F.M. (eds.) Bayesian Statistics 3, Oxford University Press, Oxford, UK (1988)
76. Shatkay, H., Kaelbling, L.: Learning topological maps with weak local odometric information. In: Proceedings of IJCAI 1997, IJCAI, Inc. (1997)
77. Smith, R., Self, M., Cheeseman, P.: Estimating uncertain spatial relationships in robotics. In: Cox, I.J., Wilfong, G.T. (eds.) Autonomous Robot Vehicles, pp. 167–193. Springer, Heidelberg (1990)
78. Smith, R.C., Cheeseman, P.: On the representation and estimation of spatial uncertainty. International Journal of Robotics Research 5(4), 56–68 (1986)
79. Soatto, S., Brockett, R.: Optimal structure from motion: Local ambiguities and global estimates. In: CVPR. Proceedings of the Conference on Computer Vision and Pattern Recognition, pp. 282–288 (1998)
80. Tardós, J.D., Neira, J., Newman, P.M., Leonard, J.J.: Robust mapping and localization in indoor environments using sonar data. International Journal of Robotics Research 21(4), 311–330 (2002)
81. Teller, S., Antone, M., Bodnar, Z., Bosse, M., Coorg, S., Jethwa, M., Master, N.: Calibrated, registered images of an extended urban area. In: CVPR. Proceedings of the Conference on Computer Vision and Pattern Recognition (2001)
82. Thrun, S., Burgard, W., Fox, D.: Probabilistic Robotics. MIT Press, Cambridge, MA (2005)
83. Thrun, S., Fox, D., Burgard, W.: A probabilistic approach to concurrent mapping and localization for mobile robots. Machine Learning 31, 29–53 (1998) (also appeared in Autonomous Robots 5, 253–271 (joint issue))
84. Thrun, S., Koller, D., Ghahramani, Z., Durrant-Whyte, H., Ng, A.Y.: Simultaneous mapping and localization with sparse extended information filters. In: Boissonnat, J.-D., Burdick, J., Goldberg, K., Hutchinson, S. (eds.) Proceedings of the Fifth International Workshop on Algorithmic Foundations of Robotics, Nice, France (2002)
85. Thrun, S., Liu, Y.: Multi-robot SLAM with sparse extended information filers. In: ISRR 2003. Proceedings of the 11th International Symposium of Robotics Research, Sienna, Italy, Springer, Heidelberg (2003)
86. Thrun, S., Liu, Y., Koller, D., Ng, A.Y., Ghahramani, Z., Durrant-Whyte, H.: Simultaneous localization and mapping with sparse extended information filters. International Journal of Robotics Research 23(7–8) (2004)
87. Thrun, S., Thayer, S., Whittaker, W., Baker, C., Burgard, W., Ferguson, D., Hähnel, D., Montemerlo, M., Morris, A., Omohundro, Z., Reverte, C., Whittaker, W.: Autonomous exploration and mapping of abandoned mines. IEEE Robotics and Automation 11(4) (2005)
88. Tomasi, C., Kanade, T.: Shape and motion from image streams under orthography: A factorization method. International Journal of Computer Vision 9(2), 137–154 (1992)
89. van der Merwe, R., de Freitas, N., Doucet, A., Wan, E.: The unscented particle filter. Advances in Neural Information Processing Systems 13 (2001)

90. Wang, C.-C., Thorpe, C., Thrun, S.: Online simultaneous localization and mapping with detection and tracking of moving objects: Theory and results from a ground vehicle in crowded urban areas. In: ICRA. Proceedings of the IEEE International Conference on Robotics and Automation, IEEE Computer Society Press, Los Alamitos (2003)
91. Williams, S., Dissanayake, G., Durrant-Whyte, H.F.: Constrained initialization of the simultaneous localization and mapping algorithm. In: Proceedings of the Symposium on Field and Service Robotics, Springer, Heidelberg (2001)
92. Williams, S.B.: Efficient Solutions to Autonomous Mapping and Navigation Problems. PhD thesis, ACFR, University of Sydney, Sydney, Australia (2001)
93. Williams, S.B., Dissanayake, G., Durrant-Whyte, H.: An efficient approach to the simultaneous localisation and mapping problem. In: ICRA. Proceedings of the IEEE International Conference on Robotics and Automation, Washington, DC, pp. 406–411. IEEE Computer Society Press, Los Alamitos (2002)
94. Wolf, D.F., Sukhatme, G.S.: Mobile robot simultaneous localization and mapping in dynamic environments. Autonomous Robots (March 2004)

2

Hybrid, Metric-Topological Representation for Localization and Mapping

Nicola Tomatis

Summary. This chapter describes an approach for indoor spatial representation, which is used to model the environment for the navigation of a fully autonomous robot. The metric and topological paradigms are integrated in a hybrid system for both localization and map building: A global topological map connects local metric maps avoiding the requirement of global metric consistency. This allows for a compact environment model, which permits both precision and robustness and allows the handling of loops in the environment during automatic mapping by means of the information of the multimodal topological localization.

The presented implementation uses a 360° laser scanner to extract corners and openings for the topological approach and lines for the metric method. This hybrid approach has been tested in a 50 x $25m^2$ portion of the institute building with the fully autonomous robot Donald Duck. Experiments are of four types: Maps created by a complete exploration of the environment are compared to estimate their quality; Test missions are randomly generated in order to evaluate the efficiency of the approach for both the localization and relocation; The fourth type of experiments shows the practicability of the approach for closing the loop.

2.1 Introduction

Research in mobile robot navigation has to focus on various issues. Environmental modeling, perception, localization and mapping are all needed in order to build a coherent working framework. Even though several successful approaches have been recently presented, solutions for consistent mapping allowing precise and robust localization in unmodified, dynamic, real-world environments are very rare. The problem is highly complex due to the fact that it requires the robot to remain localized with respect to the portion of the environment which has already been mapped in order to build a coherent map. The research has diverged to different category of approaches:

- Metric: Robot position defined by position and orientation $[x \quad y \quad \theta]^T$.
- Topological: Position defined by states or places.
- Hybrid: Combination of both the above mentioned.

Approaches using purely metric maps are vulnerable to inaccuracies in both map-making and odometry abilities of the robot. Even by taking into account

M.E. Jefferies et al. (Eds.): Robot. & Cogn. Approach. to Spat. Map., STAR 38, pp. 43–63, 2008.
springerlink.com

all relationships between features and the robot itself, in large environments the drift in the odometry makes the global consistency of the map difficult to maintain. To overcome this problem, one can rely on relative measurements. However, the observations must be associated with the map and this process still relies on the odometry.

Landmark-based approaches, which rely on the topology of the environment, can better handle this problem, because they only have to maintain topological global consistency, not metric. The advantage relies in the fact that topological relationships do not suffer of incremental drift as for the metric ones since they are qualitative, not quantitative. However, these approaches are either less precise than fully metric approaches, due to the discretization of the localization space, or computationally intractable for autonomous robots, when high resolution is used (centimeter range [11]).

More recently, approaches combining the topological and the metric paradigms have shown that positive characteristics of both can be integrated to compensate for the weakness of each single approach.

The approach presented here proposes a natural integration of both the metric and topological paradigms to combine the best characteristics of both. For this, the environmental model embodies both a metric and a topological representation. The metric model consists of infinite lines that belong to the same place. These places are related to each other by means of a topological map which is composed of nodes representing topological locations and edges between nodes. Connections between a node and a place are a special case: Traveling along these edges causes a switch from the topological to the metric paradigm. The approach allows for both the localization and the creation of maps. The automatic mapping permits also the handling of loops in the environment. This combination of the metric and the topological representation allows, without any loss of precision, efficient "planning in the large", advantageous symbolic representation for use of the robot (i.e. man-machine interaction) and robustness due to its multi-hypothesis tracking.

2.2 Spatial Representation

A robot, like a human or an animal, does not need to know its precise position with respect to the environment when traveling. Of course, it has to avoid obstacles during motion, and, therefore, to measure their relative distance, but the only moment when it really needs to know its precise position with respect to the environment, is when it has to interact with the environment (e.g. docking for recharging, manipulation of an object, human-robot interaction, etc.).

Given this and knowing that it is comparatively easier to maintain topological global consistence instead of metric, it seems suitable to have a hybrid approach using a global topological map and many local metric maps (Fig. 2.1) for the navigation of a mobile robot.

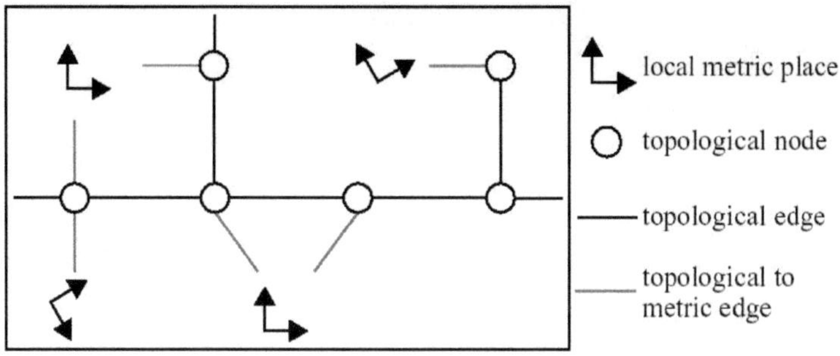

Fig. 2.1. The space is represented by places given by their metric maps and nodes representing topological locations. The graph represents the topological map, which is used for traveling. When interaction with the environment is needed, the local metric map is used.

2.3 Environment Model

In this section the environmental modeling is presented for the current implementation.

The environment is described by a global topological map, which permits moving in the whole environment, and local metric maps which can be used by the robot as soon as it needs further localization precision (Fig. 2.1). In order to change from topological to metric, the metric position of the robot in the local metric map has to be calculated (i.e. initialization of metric parameters). This requires the measurement of one or more metric features, which are known in the current local metric map. Thus, the only requirement specific to this model is to have a *detectable metric feature* when traveling from a topological node to a metric place. Given this, local metric maps can be placed anywhere in the environment.

Switching to topological does not require any specific characteristic: The robot navigates back metrically to the position, where it initialized the local metric map and then it resumes its topological navigation.

2.3.1 Global Topological Map

The perception required by a topological approach has to permit the distinction between places. A model which permits the optimization of the distinctiveness of the current location, not the precision, is required. In the model describe here, two different features are chosen for their distinctiveness. These are:

- Corners, which are characterized by their orientation.
- Openings, which are also used for model transition.

The topological map can be viewed as a graph. Topological locations are represented by nodes containing the information about the way to reach the connected topological location/metric place. Furthermore, the landmarks lying

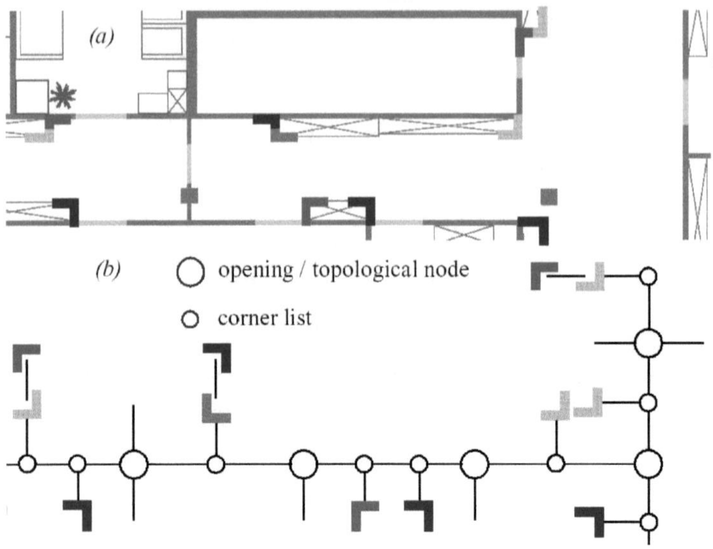

Fig. 2.2. This figure shows the topological model only, not the local metric maps as in Fig. 1. (a) A portion of a hallway with the extracted corner and opening features. (b) The topological map is represented by a graph. It contains nodes connected to each other with the list of corner features lying between them. Openings (topological nodes) can either be a transition to a room or be a connection to another hallway.

between two locations are represented as a list between the two nodes. In Fig. 2.2 the graph representing the topological model is shown for a portion of the environment

The corner extractor returns a set of $[x \quad y \quad \theta]$ parameters in robot coordinates, representing the position and orientation of the corners with respect to the robot. Furthermore, an extraction confidence parameter p_c is calculated for each corner by taking into account its size. Openings are either large discontinuities perpendicular to the direction of motion in hall-ways or transitions from rooms to hallways. They can either be a transition between a hallway and a room or between two perpendicular hallways. Due to the use of a 360° laser scanner, an observation contains many landmarks, which are transformed in a graph compatible to the environment model, as shown in Fig. 2.3.

2.3.2 Local Metric Maps

The features used for metric environmental representation are infinite lines. They are less informative than line segments, but have a better probabilistic model with an analytical solution and permit a very compact representation of structured geometric environments requiring only about 10 bytes per m^2 for a typical office environment. In Fig. 2.4 a typical office is shown with the lines used for its local metric map. The line model is $\rho\cos(\varphi - \alpha) - r = 0$, where (φ, α) is

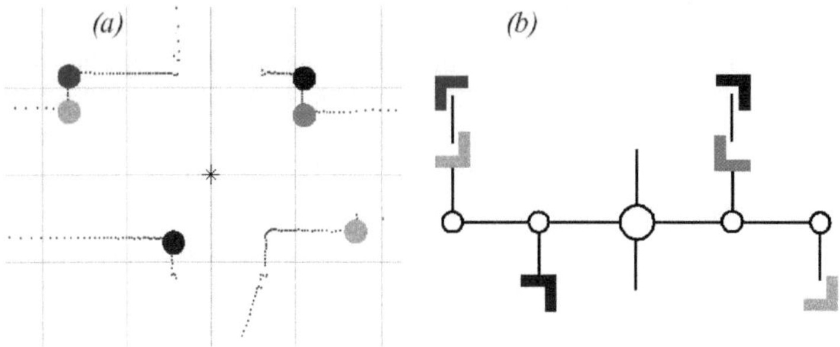

Fig. 2.3. (a) Laser data and the extracted features. (b) The resulting observation graph.

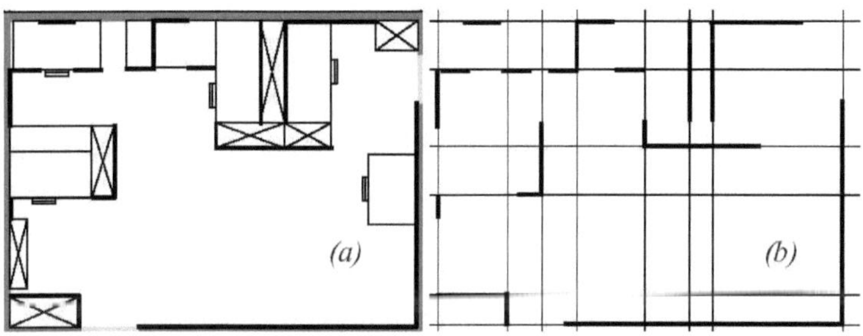

Fig. 2.4. An office of the institute (a) and the lines representing it in the local metric map (b). The black segments show the correspondence between the two figures.

the raw measurement and (α, r) the model parameters. The angle between the line and its perpendicular is α, r is its length. The used extraction algorithm has been described in [2]. Its result is a set of (α, r) parameters with their 2 x 2 covariance matrix, which is calculated by propagating the uncertainty from the laser measurements.

2.4 Localization and Map Building

The environment models allow the use of two different navigation methods with complementary characteristics. The metric approach is an *Extended Kalman Filter* (EKF). This method has already proven its strength for localization [3] . Map building can then be done with the *Stochastic Map approach* [21] . Topological navigation uses a *Partially Observable Markov Decision Process* (POMDP) [5] for state estimation. The metric localization permits thus a very precise positioning at the goal point [3, 25] whereas topological localization [5, 25] guarantees robustness against getting lost due to the multimodal representation of the robot's location.

2.4.1 Map Building Strategy

As explained earlier, the environment model is composed of a global topological map and a set of local metric maps. Local metric maps can be anywhere in the environment. Even if the approach is applicable to any structured environment, a suitable environment-dependent strategy has to be adopted. Here, it is assumed that the robot will have to be very precise in rooms, where most of its tasks have to be executed (e.g. docking for power recharging; manipulation tasks with objects on a table; human-robot interaction). While navigating in the large (i.e. hallways), precision with respect to the features is less important, but robustness and global consistency take an important role. Because of this, the two different levels of abstraction (metric and topological) are used in combination of the different type of environmental structures:

- While navigating in hallways, the robot firstly creates and then updates the global topological map.
- When it enters a room, it creates a new local metric map.

These two environmental structures are recognized with the laser sensor by means of a heuristic established by experience: Thin and long open spaces are assumed to be hallways, while other open spaces will be defined as rooms.

2.4.2 Exploration Strategy

The proposed exploration strategy is simple: The robot first explores all the hallways in a depth-first way. It then explores each room it encountered by backtracking. Note that, in general, for each hallway the room exploration reduces to a linear list traversal. Rooms with multiple openings cause two special cases, which are treated in the next paragraphs.

SRooms with an opening to another room

The robot continues building the current metric map. This leads to the next case if the other room has an opening to a hallway.

Rooms with multiple openings to a hallway

Due to the metric navigation mode during room exploration, the robot knows the direction of the opening and can therefore deduce if it opens to the same hallway, a known one or a new one. In the case of known hallways, the robot simply goes back to the hallway it was coming from and continues its exploration. If the robot reenters the same room from another opening without recognizing it, the result will be that two metric maps for the same metric place are created, one for each opening. In the case of a new hallway, the exploration continues in a hallway depth-first way.

2.4.3 Topological Localization and Map Building

The current experimental test bed is a part of the institute building. This environment is rectilinear and mainly composed of offices, meeting rooms and hallways. Therefore, only four directions of travel are employed: N, E, S, W. However, this is not an inherent loss of generality because it is not a general requirement of the POMDP algorithm.

Position Estimator

Given a finite set of environment states S, a finite set of actions A and a state transition model T, the model can be defined by introducing partial observability. This includes a finite set O of possible observations and an observation function OS, mapping S into a discrete probability distribution over O. $T(s, a, s')$ represents the probability that the environment makes a transition from state s to state s' when action a is taken. $OS(o, s, a)$is the probability of making an observation o in state s after having taken action a. The probability of being in state s' (*belief state of* s') after having made observation o while performing action a is then given by the equation:

$$SE_{s'}(k+1) = \frac{OS(o, s', a) \sum_{s \in S} T(s, a, s') SE_s(k)}{P(o|a, SE(k))} \tag{2.1}$$

where $SE_s(k)$ is the belief state of s for the last step, $SE(k)$ is the belief state vector of the last step and $P(o|a, SE(k))$ is a normalizing factor. The observation function OS is made robust by the fact that an observation is composed of many landmarks (Fig. 2.3), which give rise to its distinctiveness. The observation probability is calculated by graph matching between the graph representing the map and the observation graph (Fig. 2.3). When no openings are visible, $T(s, a, s) = 0.99$ while $T(s, a, s') = 0.01$ for $s \neq s'$, this means that it is highly probable that the robot is still in the same state. When the robot encounters an opening, the most probable state s' is searched by comparing the traveled distance d, measured starting from s, to the information saved in state node s during map building. In this case, $T(s, a, s') = 0.99$ while for $T(s, a, s'') = 0.01$.

Heading Estimator

Because the position estimator does not take into account the heading of the robot, this is done separately as in [12]. However, in this case, the orientation is estimated by a weighted mean of each observed l ine that is either horizontal or vertical with respect to the environment. The success of this method is guaranteed by the fact that, in general, lines given by the environmental structures are either parallel or perpendicular to the direction of travel. Infinite lines are matched by means of the validation test

$$\left(z^{[i]} - \widehat{z}^{[i]}\right) S_{ij}^{-1} \left(z^{[i]} - \widehat{z}^{[i]}\right)^T \leq \chi^2_{\alpha,n} \tag{2.2}$$

where prediction $\widehat{z}^{[j]}$ is directly the odometry state vector variable θ and $\chi^2_{\alpha,n}$ is a number taken from a χ^2 distribution with $n = 1$ degrees of freedom. This can be viewed as an EKF for heading only, where no map is required because, for prediction, θ is directly used instead of map features.

Control Strategy

Since it is computationally intractable to compute the optimal POMDP control strategy for a large environment [5], simple suboptimal heuristics are introduced. F or the system presented here the *most likely state* policy has been adopted: The world state with the highest probability is found and the action that would be optimal for that state is executed. However it can happen that the robot is not sure about its current state. This is calculated by means of the function $U(SE(k))$ (called the unconfident function), which is the entropy of the probability distribution over the states of the map. The POMDP is confident when

$$U(SE(k)) = -\sum_{s} SE_s(k) \log SE_s(k) < U_{\max} \tag{2.3}$$

where $U_{\max}$ is determined by experience. When the robot is unconfident, it follows the hallway in the direction where it expects to find more information. What has to be avoided at any cost is to switch from the multimodal topological navigation to the unimodal metric navigation when the robot is unconfident about its location, otherwise it could enter a false local metric place and therefore be lost. This is a critical situation because the robot would have to both detect its lost situation (unfeasible without arbitrary assumptions) and recover from it (relocation).

Map Building

Instead of using a complex scheme for model learning like in [11] and [23], where an extension of the Baum-Welch algorithm is adopted, here the characteristics of the observation graph (Fig. 2.3) are used. When the robot feels confident about its position, it can decide if an extracted landmark is new by comparing the observation graph to the node in the map corresponding to the *most likely* state. This can happen either in an unexplored portion of the environment or in a known portion, where new landmarks appear due to the environment dynamic. As already explained, the landmarks have an extraction confidence P_l. This characteristic is firstly used to decide if the new landmark can be integrated in the map. When an opening landmark is extracted (i.e. door or cross with another hallway), it is integrated in the map as a new state node (Fig. 2.2) with a rough measure of the distance to the last state node. Furthermore, for each integrated landmark, the confidence P_l is used to model the probability of seeing that landmark the next time P_{lmap}. When it is reobserved, the probability in the map is averaged with the confidence of the extracted one. If the robot does not see an expected landmark the probability $1 - P_{\text{lmap}}$ is used instead.

$$p_{\text{lmap}}(t_i) = \sum_{i=1}^{n} \frac{p_l(t_i)}{n} \tag{2.4}$$

where

$$p_l(t_i) = \begin{cases} p_l(t_i) & \text{observed} \\ 1 - p_{\text{lmap}}(t_{i-1}) & \neg\text{observed} \end{cases} \tag{2.5}$$

When the confidence P_{lmap} decreases and is below a minimum, the corresponding landmark is deleted from the map. This allows for dynamics in the environment, where landmarks that disappear in the real world will be deleted from the map too.

Closing the Loop

The problem of *closing the loop* can be defined as how to detect and correctly map the environment when coming back to an already visited place due to a loop. In [23] this is achieved by adding a topological mapper, which ensures global consistency. The information from this topological mapper is used to correct their global metric map which eventually converges to a global consistent map. The current approach differs in two main aspects with respect to the known methods:

- Instead of closing the loops only by means of the perception, loops are detected and closed by means of the *localization information*. This means that the robot decides by looking where it thinks it is thanks to its multi-hypothesis system.
- Even by allowing precise metric information, loops have to be closed only in the topological map because the metric model is represented by many disconnected local metric maps.

Loops can also exist in a local metric map. However, by assuming that the topology allows for small maps, the drift in odometry does not cause any problem to the local consistency, as it has been shown in [7]. The current method works as follows: The robot does not try to recognize if a single observation has already been seen somewhere else. However, as soon as the robot creates the map for a part of the environment which has already been visited, the probability distribution starts diverging into two peaks: One for the current map position; another for the previously created location representing the same physical place. The algorithm starts tracking the two highest probabilities as soon as the POMDP becomes unconfident because this is the first clue indicating a divergence of the probability distribution. A loop can then easily be detected when the distribution has converged into two peaks which move in the same way. The position where the loop has to be closed can be detected by turning off the automatic mapper and backtracking with localization until the distribution reconverges to a single peak. This will be the point where the robot started mapping the loop. An example is given in Fig. 2.5.

2.4.4 Metric Localization and Mapping

This section briefly describes the main characteristics of the *Stochastic Map* approach [21], which permits the simultaneous building of the map and localization using an *Extended Kalman Filter* [9, 18] for localization simultaneously.

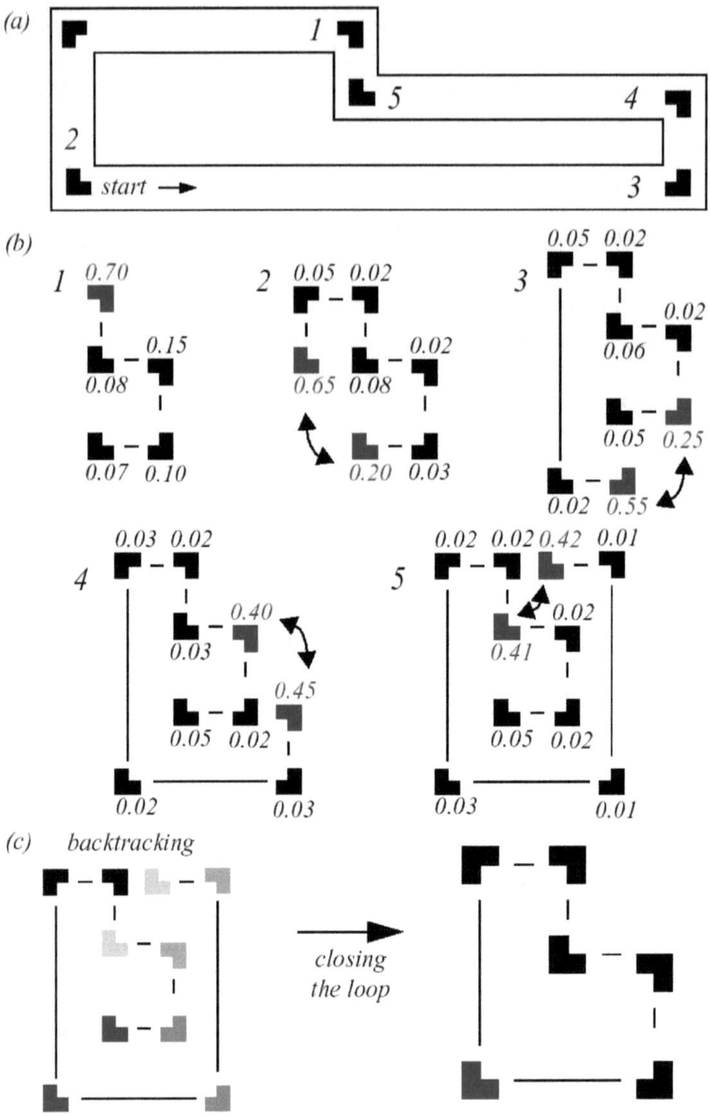

Fig. 2.5. a) A loop in the environment. (b) Mapping with the POMDP. 1) The map when the robot is at position 1 in the environment. 2) The robot is re-exploring the start point. The observation function $OS(o, s, a)$ gives high values for both the new node in the map and for the start node, but the probability distribution has not yet diverged because the transition function $T(s, a, s')$ gives a low probability of coming at the map start. 3) However, by moving in the same way on the map the distribution diverges and the POMDP becomes unconfident. 4) The distribution has diverged and the two peaks move in the same way to 5. (c) The mapping is stopped. The loop is closed by backtracking.

With this approach both the robot position $x_r = (x, y, \theta)^T$ and the features $x_i = (\alpha, r)^T$ are represented in the system state vector and its covari-ance:

$$x = \begin{bmatrix} x_r \\ x_i \\ \dots \\ x_n \end{bmatrix} \quad C(x) = \begin{bmatrix} C_{rr} & C_{r1} & \dots & C_{rn} \\ C_{1r} & C_{11} & \dots & C_{1n} \\ \dots & \dots & \dots & \dots \\ C_{nr} & C_{n1} & \dots & C_{nn} \end{bmatrix} \tag{2.6}$$

This represents the uncertain spatial relationship between objects in the map, which is changed by three actions:

- Robot displacement
- Observation of a new object
- Reobservation of an object already existing in the map

Robot Displacement

When the robot moves with an uncertain displacement u given by its two first moments (u, C_u), which are measured by the odometry, the robot state is updated to $g(x_r, u)$. The updated position and uncertainty of the robot pose are obtained by error propagation on g:

$$x_r(k+1) = g(x_r(k), u) - x_r(k) \oplus u \tag{2.7}$$

$$C_{rr}(k+1) = G \begin{bmatrix} C_{rr}(k) & C_{ru}(k) \\ C_{ur}(k) & C_{uu}(k) \end{bmatrix} G^T \tag{2.8}$$

where $\oplus$ is the compounding operator and G is the Jacobean of g with respect to x_r and u.

New Object

When a new object is found, a new entry must be created in the system state vector. A new row and column are also added to the system covariance matrix to describe the uncertainty in the object's location and the inter-dependencies with the other objects. The new object $(\widehat{x}_{\text{new}}, C_{\text{new}})$ can be integrated in the map by computing the following equations of uncertainty propagation:

$$x_{N+1}(k) = g(x_r(k), x_{\text{new}}) = x_r(k) \oplus x_{\text{new}} \tag{2.9}$$

$$C_{N+1,N+1}(k) = G_{x_r} C_{rr}(k) G_{x_r}^T + G_{x_{\text{new}}} C_{\text{new}} G_{x_{\text{new}}}^T \tag{2.10}$$

$$C_{N+1,i}(k) = G_{x_r} C_{ri}(k) \tag{2.11}$$

Re-observation

Let x_{new} be the new observation in the robot frame. The measurement equation is defined as:

$$z = h(x_r, x_{\text{new}}, x_i) = g(x_r, x_{\text{new}}) - x_i \tag{2.12}$$

x_i is temporarily included in the state to apply the EKF. However, if prediction x_i satisfies the validation test:

$$(x_{\text{new}} - x_i) S^{-1}_{\text{new},i} (x_{\text{new}} - x_i)^T \leq \chi^2_{\alpha,n} \tag{2.13}$$

where $S_{\text{new}i} = C_{\text{new,new}} + C_{ii} - C_{\text{new}i} - C_{i\text{new}}$, $\chi^2_{\alpha,n}$ is a number taken from a χ^2 distribution with $n = 2$ degrees of freedom and α the level on which the hypothesis of pairing correctness is rejected, then x_{new} is a re-observation of x_i.

Extended Kalman Filter

When a spatial relationship is re-observed, the updated estimate is a weighted average of the two estimates calculated by means of an EKF. It allows the update of a subset of the state vector while maintaining consistency by means of the covariance matrices. A measurement equation $z = h(x_1, \ldots, x_m)$ is considered as a function of m relationships included in x. All of the m estimates x_i of the state vector x are updated by a value which is proportional to the difference $\delta = z - \widehat{z}$ between the ideal measurement z and the actual measurement $\widehat{z}$:

$$x_i(k+1) = x_i(k) + \Gamma_{iz} \Gamma_{zz}^{-1} \delta \tag{2.14}$$

$$\Gamma_{iz} = E\left[x_i \delta^T\right] = \sum_{j=1}^{M} C_{ij} H_{x_j}^T \tag{2.15}$$

$$\Gamma_{zz} = E\left[\delta \delta^T\right] = \sum_{j=1}^{M} \sum_{k=1}^{M} H_{x_j} C_{jk} H_{x_k}^T \tag{2.16}$$

where H_{xj} is the Jacobean matrix of h with respect to x_j. The variance and covariance C_{ij} are also updated:

$$C_{ij}(k+1) = C_{ij}(k) - \Gamma_{iz} \Gamma_{zz}^{-1} \Gamma_{jz}^T \tag{2.17}$$

2.5 Experiments

The approach has been tested in the 50 x $25m^2$ portion of the institute build-ing shown in Fig. 2.6 with four different types of experiments for a total of more than 1.5km. For the experiments, Donald Duck has been used (Fig. 2.7). It is a fully autonomous mobile vehicle running XO/2, a deadline driven hard real-time operating system [4]. Donald navigates locally by means of a motion control algorithm, which plays the role of both position control and obstacle avoidance: It reaches the given $(x, y, \theta)^T$ or $(x, y)^T$ goal by planning a collision free path

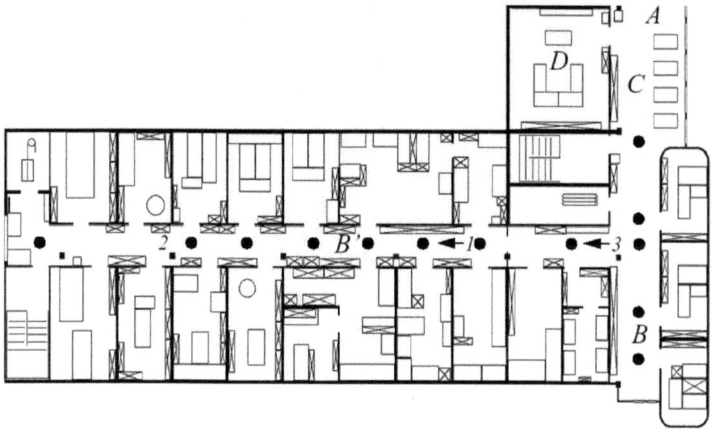

Fig. 2.6. The test environment. It is complex, dynamic and artificially closed in A so that the exploration procedure is finite. Black dots are the places where the automatic mapper is expected to extract state nodes (the other doors are closed). In B and B' the robot had problems distinguishing between the two neighbor locations. C and D are detected as rooms and represented by a single local metric map. A large loop does not exist in this environment. Therefore, for the experiments, a loop is "artificially created" by starting the exploration in 1, stopping it in 2, taking the robot manually to 3 and resuming.

(with respect to the current local data), and reacting to the dynamic environment either by merely re-planning the path or by changing heading direction and re-planning when an object appears in front of the robot. Its complete control architecture is briefly presented in Fig. 2.8.

2.5.1 Map Building

In this section, the automatic mapping capabilities of the presented approach are evaluated. Note that the environment is arbitrarily closed (Fig. 2.6), so that the exploration procedure is finite. Furthermore, local metric maps are taken from the a priori map used in [3], because the *stochastic map* is not yet implemented on the robot and therefore, runs only offline. For this evaluation, five maps generated by complete explorations of the environment shown in Fig. 2.6 are compared to evaluate their quality with respect to consistency and completeness. In order to evaluate the topological mapper, maps are compared before the backtracking step. By knowing which door is open during the exploration, it can be extrapolated how many state nodes should be extracted (see the black dots in Fig. 2.6). Their position (odometry) and type (opening or hallway) are stored during exploration to check whether the resulting model is consistent with the real environment. For the other features (corners), each resulting map is compared to the others to calculate the average difference between all pairings of the maps. The results are presented in Table 2.1. One of the problems

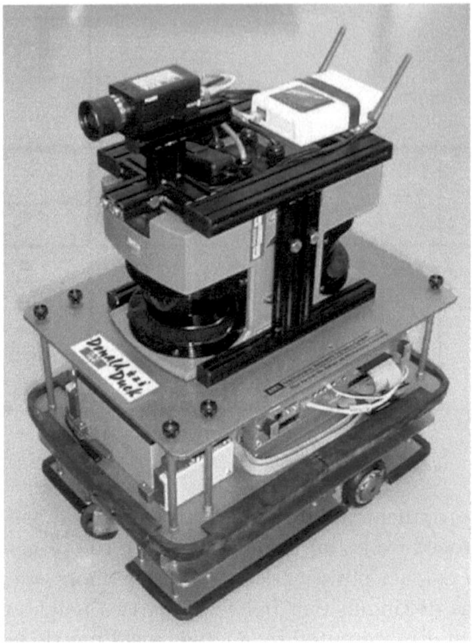

Fig. 2.7. The fully autonomous robot Donald Duck. Its controller consists of a VME standard backplane with a Motorola PowerPC 604 microprocessor clocked at 300MHz running XO/2. Among its peripheral devices, the most important are the wheel encoders, a 360 ° laser range finder and a grey-level CCD camera.

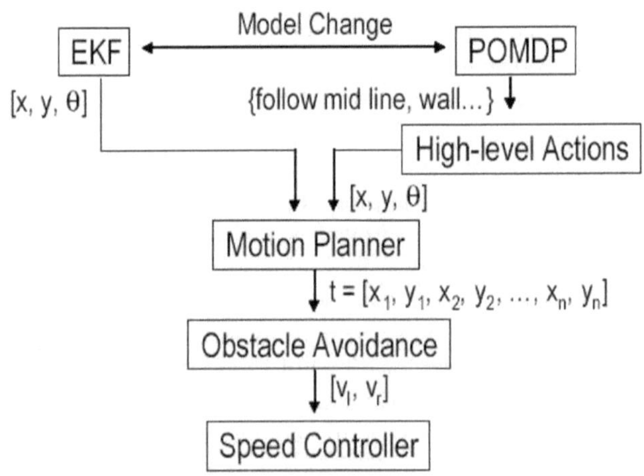

Fig. 2.8. The control architecture of the robot. The goal of this architecture is to share as may functionality as possible. This results in an emulation of high-level actions by means of a motion planner relying on a safe obstacle avoidance.

Table 2.1. Comparison of five maps generated by complete explorations of the environment shown in Fig. 2.6

Number of explorations	5
Total traveled distance	343m
Number of states in the environment	13
Mean detected states	12.8/98%
Mean detection errors for hallway/opening	1.2/9.2%
Mean detected features per exploration	78
Mean different features	18/23%

encountered during the exploration was the difficulty of distinguishing between openings and hallways. This leads to a mean of 1.2 false detections for each experiment. Nevertheless by visiting all the openings when traversing the environment by backtracking to add the local metric maps, these errors will be detected and corrected. In one experiment a state (opening) was not extracted at all.

For the corner features it is more difficult to define which features really exist in the environment. What can be measured is the difference between two maps. The mean number of extracted corners in a map is 78. An average of 18 of these are either corners or noise that are not always extracted. This means that almost 77% of the features are constant in the five maps showing that the perception delivers valuable information to the mapper.

2.5.2 Localization

The quality of a map can also easily be estimated by testing it for localization. For this, two types of localization experiments are performed: One for localization *(position tracking)* and the other for *relocation.* To test the topological localization, 25 randomly generated test missions for a total of about 900m and 28 000 estimates are performed. The robot knows in which state it is at the start point. A mission is successful when the robot reaches its goal location, is in front of the opening and is confident about its position. There it switches to the metric approach by measuring the door frame and using this information to initialize the EKF. To have more information about the experiments, the information associated with each state transition is stored in a log file which makes it possible to determine if each transition detected by the localization actually took place. The results are presented in Table 2.2. Even if all the missions are successful, using the log file allows finding 21 false state transitions. These caused 404 false estimates in B and B' (Fig. 2.6), where the peak probability moved forward and backward between two neighbor states. These false estimates represent only 1.4% of the total, meaning that the system recovers quite fast from these errors. The robot had also confident false estimates (0.5%) that could cause a mission failure if the goal state is estimated when the robot is in front of another opening.

Table 2.2. Localization experiments. All the test missions have been successfully performed. However the robot also made false state transitions that caused some false estimates (1.4%). This happened only by B and B' in Fig. 2.6. The reason leading to a success rate of 100% is that the system always recovered from its error without estimating the goal location in front of a false opening.

Number of missions	25
Successful missions	25/100%
Total traveled distance	899m
Mean traveled distance	36m
Mean travel speed	0.31m/s
Total real state transitions	181
False state transitions	21/12%
Total estimates	27 870
Unconfident states	3 413/12%
False estimates	404/1.4%
Confident false estimates	149/0.5%

Table 2.3. Recovering from a lost situation (i.e. overall constant belief state). The robot requires from 1 to 4 states to recover, depending on the distinctiveness of the part of the environment where it is moving.

Number of experiments	10
Total traveled distance	250m
Mean distance for recovering	13.7m
Min / max distance for recovering	1.21/20.31m
Mean number of states for recovering	2.11
Min / max of states for recovering	1/4

The second type of test is focused on recovering from a lost situation *(relocation)*. Ten experiments are started from a randomly defined position in the environment with a uniform belief state distribution (i.e. *lost situation*). The goal is to measure which distance or number of state transitions are required in order to converge to a correct confident state estimate. To avoid false interpretations, the robot is required to travel 3 state nodes further without estimate errors to fulfill the test. In Table 2.3 the ten tests are briefly outlined.

As expected the robot can always recover. Its policy is simple: Go forward until recovery or end of hallway; If end of hallway, turn. The system requires a minimum of 1 and a maximum of 4 states to recover. The interesting point is that this difference in the results is position dependent and repeatable. For example, the crossing between the two hallways permits recovery with a single state transition because it is global distinctive for the environment in Fig. 2.6. On the other hand, the right part of the horizontal hallway seems to be more distinctive than the left one where the robot required the maximum number of states to recover. Metric localization is used but not explicitly tested here, because the EKF used has already been extensively tested in [3] with a total of 6.4km. The mean 2σ-error bounds are approximately 1 centimeter in x and y and 1 degree for θ. Furthermore, the metric localization approach has also been tested with this hybrid method for localization on the same robot in [25], where ground truth measurements at the goal position resulted in an average error of less than 1cm.

2.5.3 Closing the Loop

In the test environment there are no large loops. In order to test the proposed approach, a loop is artificially created by displacing the robot during the exploration as shown in Fig. 2.6. As already explained, it can be assumed that when two peaks appear and move in the same way for three subsequent state transitions a loop has been discovered. In all the other experiments this has effectively never appeared, showing that this a good test for loops. This experiment has been performed three times. Each time the probability distribution has effectively diverged into two peaks allowing the detection of the loop. In order to close the loop the robot has turned off the mapping algorithm and has gone back until the distribution has converged to a single confident peak. This took place where the map has been started (1 in Fig. 2.6) proving that the loop could be correctly closed.

2.6 Related Work

Successful navigation of embedded systems for real applications relies on the precision that the vehicle can achieve, the capacity of not getting lost and the practicability of their algorithms on the limited resources of the autonomous system. Furthermore, the fact that a priori maps are rarely available and, even when given, not in the format required by the robot, and that they are mainly unsatisfactory due to imprecision, incorrectness and incompleteness, makes automatic mapping a real need for application-like scenarios.

Simultaneous localization and map building research can be divided into two main categories: Metric and topological. Metric approaches are defined here as methods, which allow the robot to estimate its $[x \;\; y \;\; \theta]^T$ position, while topological approaches are those where the position is given by a location without precise metric information.

After the first precise mathematical definition of the stochastic map [21], early experiments [9, 18], have shown the quality of fully metric simultaneous localization and map building. The resulting environment model permits highly precise localization, which is only bounded by the quality of the sensor data [3]. However, these approaches suffer some limitations. Firstly they rely strongly on odometry. For automatic mapping this makes the global consistency of the map difficult to maintain in large environments, where the drift in the odometry becomes too important. Furthermore they represent the robot pose with a single Gaussian distribution. This means that an unmodeled event (i.e. collision) could cause a divergence between the ground-truth and the estimated pose from which the system is unable to recover (lost situation). In [7] it has been shown that by taking into account all the correlations (off-diagonal cross-covariance in (2.6)), the global consistency is better maintained. However, this is not sufficient, as confirmed by a recent work [6], where a solution is proposed by extending the absolute localization to include localization relative to local frames. An alternative approach is to work with relative information only, as proposed in [19]. However, the problem of relying on odometry remains to be faced for the association problem.

On the other hand, topological approaches [16] can handle multi-hypothesis tracking and have a topological global consistency, which is easier to maintain. The robustness of such approaches has firstly been proven by the application of the state set progression [20], which has then been generalized to the POMDP approach [5, 12]. For automatic mapping in [15] the Baum-Welch algorithm has been used for model learning. In contrast to the above mentioned topological approaches, [17] proposes a topological approach, which rely heavily on odometry in order to better handle environment dynamics. All these approaches are robust, but have the drawback of losing in precision with respect to the fully metric ones: The robot pose is represented by a location without precise metric information. To face this, Markov localization [11] has been proposed: A fine-grained grid guarantees both precision and multimodality. However, this approach remains computationally intractable for current embedded systems. Monte Carlo localization has recently been proposed [10]. However, it has not been extended for simultaneous localization and mapping. Metric and topological approaches are converging, like [6, 10, 11], to hybrid solutions by adding advantageous characteristics of the opposite world. Going in this direction, in [22], the approach consists of extracting a topological map from a grid map by means of a Voronoi based method, while [23] proposes to use the Baum-Welch algorithm as in [15], but to build a topologically consistent global map which permits closing the loop for the global metric map too.

In contrast to the above mentioned approaches, for this system a natural integration of the metric and topological paradigm is proposed. The approaches are completely separated into two levels of abstraction. Metric maps are used only locally for structures (rooms) that are naturally defined by the environment. There, a fully metric method is adopted. As it has been shown in [7], for such small environments, where the drift in the odometry remains uncritical, the stochastic map allows for precise and consistent automatic mapping. The topological approach is used to connect the local metric maps that can be far away from each other. With this the robot can take advantage of the precision of a fully metric EKF navigation added to the robustness in the large of the POMDP approach. All this is achieved by maintaining a compactness of the environment representation and a low complexity, which allows an efficient implementation of the method on a fully autonomous system. This hybrid approach shows also its practicability for environments with loops. In this case the loop is closed in the global topological map based on the information from the topological localization, while the metric information remains local and does therefore not require further processing, contrasting to [23], where the topological information is used for mapping only, to close the loop in the metric map correctly.

This work is one of the first proposing such hybrid integration with [25] for localization. Since then, the hybrid navigation community is growing with for ex-

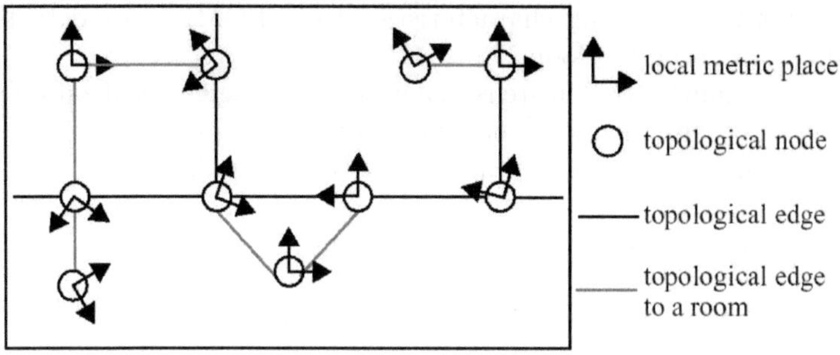

Fig. 2.9. This is an extension of the model presented in Fig. 2.1. The space is represented by places given by their metric maps, which corresponds in this case also to topological locations.

ample [1], where a discrete Markov Model is used to generate hypotheses, which are then tracked by multiple Kalman trackers and [13] with the combination of Markov Localization and Kalman Filter.

Due to the cross-fertilization of ideas from cognitive science and robotics in the last few years, it can be seen that these approaches have more and more in common with what is called *cognitive mapping*, a term first used in [24]. A remarkable similarity can be found in [8] and [14], showing that cognitive and robot mapping are converging toward a common solution.

2.7 Conclusion and Outlook

This chapter has presented a hybrid approach for both localization and map building. The metric and topological parts are completely separated into two levels of abstraction. Together they allow a very compact and computationally efficient representation of the environment for mobile robot navigation. Furthermore this combination permits both precision with the non-discrete metric estimator and robustness by means of the multimodal topological method.

The approach is validated empirically by extensive experimentation for a total of more than 1.5km. Map building is tested by performing five complete explorations of the environment and comparing the resulting maps. This comparison demonstrates that the maps are consistent with respect to the environment and that perception allows the extraction of important information. For localization, the success rate over the 0.9km of the 25 tests missions is 100%. Nevertheless a precise analysis of the state transitions shows that, between neighbor states, false state estimate occurs (1.4%) and sometimes are even treated as confident (0.5%). The relocation performance of the topological method has been shown with 10 successful experiments where the belief state converges within 1 to 4 state transitions depending on the distinctiveness of the part of the environment where the robot is navigating. It has been shown how loops can be closed on the localization level instead of the perception level. This is easily done by using the multi-hypothesis tracking characteristic of the POMDP for detection and backtracking for closing the loop.

A logical extension for the future is to have the topological localization running permanently and to use the metric navigation only when precision is needed. This requires an extension of the model as it is shown in Fig. 2.9.

Acknowledgements

I would like to thank my advisors: Roland Siegwart, Autonomous Systems Lab, EPFL, Lausanne, Switzerland and Illah Nourbakhsh, The Robotics Institute, Carnegie Mellon University, Pittsburgh, USA. Many thanks to all the friends at the Autonomous Systems Lab, EPFL for the fruitful discussions. This research has been partially supported the BIBA European Project (Bayesian Inspired Brains & Artefacts, IST-2001-32115).

References

1. Arras, K.O., Castellanos, J.A., Siegwart, R.: Feature-based multi-hypothesis localization and tracking for mobile robots using geometric constraints. In: Proceeding of IEEE International Conference on Robotics and Automation, Washington DC, USA, vol. 2, pp. 1371–1377. IEEE Computer Society Press, Los Alamitos (2002)
2. Arras, K.O., Siegwart, R.Y.: Feature extraction and scene interpretation for map-based navigation and map building. In: Proceeding of SPIE, Mobile Robotics XII, vol. 3210, pp. 42–53 (1997)
3. Arras, K.O., Tomatis, N., Jensen, B., Siegwart, R.: Multisensor On-the-Fly Localization: Precision and Reliability for Applications. Robotics and Autonomous Systems 24(2-3), 131–143 (2001)
4. Brega, R., Tomatis, N., Arras, K.: The need for autonomy and real-time in mobile robotics: A case study of xo/2 and pygmalion. In: Proceeding of IEEE International Conference on Intelligent Robots and Systems, Takamatsu, Japan, vol. 2, pp. 1422–1427. IEEE Computer Society Press, Los Alamitos (2000)
5. Cassandra, A.R., Kaelbling, L.P., Kurien, J.A.: Acting under uncertainty: Discrete bayesian models for mobile-robot navigation. In: Proceeding of IEEE International Conference on Robotics and Automation, Osaka, Japan, vol. 2, pp. 963–972. IEEE Computer Society Press, Los Alamitos (1996)
6. Castellanos, J.A., Devy, M., Tardós, J.D.: Simultaneous localization and map building for mobile robots: A landmark-based approach. In: ICRA. Proceeding of IEEE International Conference on Robotics and Automation, San Francisco, USA, IEEE Computer Society Press, Los Alamitos (2000)
7. Castellanos, J.A., Tardós, J.D., Schmidt, G.: Building a global map of the environment of a mobile robot: The importance of correlations. In: Proceeding of IEEE International Conference on Robotics and Automation, Albuquerque, USA, vol. 2, pp. 1053–1059. IEEE Computer Society Press, Los Alamitos (1997)
8. Chown, E.: Making predictions in an uncertain world: Environmental structure and cognitive maps. Adaptive Behavior 7(1), 17–33 (1999)
9. Crowley, J.L.: World modeling and position estimation for a mobile robot using ultrasonic ranging. In: Proceeding of IEEE International Conference on Robotics and Automation, Scottsdale, USA, vol. 2, pp. 674–680. IEEE Computer Society Press, Los Alamitos (1989)

10. Dellaert, F., Fox, D., Thrun, S.: Monte carlo localization for mobile robots. In: Proceeding of IEEE International Conference on Robotics and Automation, Detroit Michigan, vol. 2, pp. 1322–1328. IEEE Computer Society Press, Los Alamitos (1999)
11. Fox, D.: Markov Localization: A Probabilistic Framework for Mobile Robot Localization and Navigation. PhD thesis, Institute of Computer Science III, University of Bonn, Bonn, Germany (1998)
12. Gutierrez-Osuna, R., Luo, R.C.: LOLA: probabilistic navigation for topological maps. AI Magazine 17(1), 55–62 (1996)
13. Gutmann, J.S.: Markov-kalman localization for mobile robots. In: Proceeding of International Conference on Pattern Recognition, Quebec City, Canada, vol. 2, pp. 601–604 (1999)
14. Jefferies, M., Yeap, W., Smith, L., Ferguson, D.: Computing individual spaces for a mobile robot: An initial algorithm. In: Proceeding of International Conference on Cognitive Science, Beijing, China, pp. 541–545 (2001)
15. Koenig, S., Goodwin, R., Simmons, R.: Robot navigation with markov models: A framework for path planning and learning with limited computational resources. In: Dorst, L., Voorbraak, F., van Lambalgen, M. (eds.) RUR 1995. LNCS, vol. 1093, Springer, Heidelberg (1996)
16. Kuipers, B.J., Byun, Y.T.: A qualitative approach to robot exploration and map-learning. In: Proceeding of Workshop on Spatial Reasoning and Multi-Sensor Fusion, Los Altos, USA (1987)
17. Kunz, C., Willeke, T., Nourbakhsh, I.: Automatic Mapping of Dynamic Office Environments. Autonomous Robots 7, 131–142 (1999)
18. Leonard, J.J., Durrant-Whyte, H.F.: Dynamic Map Building for an Autonomous Mobile Robot. The International Journal of Robotics Research 11(4), 286–298 (1992)
19. Martinelli, A., Tomatis, N.N., Siegwart, R.: Open challenges in slam: An optimal solution based on shift and rotation invariants. In: Proceeding of IEEE International Conference on Robotics and Automation, New Orleans, USA, vol. 2, pp. 1327–1332. IEEE Computer Society Press, Los Alamitos (2004)
20. Nourbakhsh, I.: Dervish: An office-navigating robot. artificial intelligence and mobile robots. In: Bonasso, R.P., Kortenkamp, D., Murphy, R. (eds.) Human spatial memory: Remembering where, pp. 73–90. The AAAI Press/The MIT Press (1998)
21. Smith, R.C., Self, M., Cheeseman, P.: Estimating uncertain spatial relationships in robotics. In: Lemmer, J.F., Kanal, L.N. (eds.) Uncertainty in Artificial Intelligence 2, pp. 435–461. Elsevier Science Pub., Amsterdam (1988)
22. Thrun, S., Bücken, A.: Integrating grid-based and topological maps for mobile robot navigation. In: Proceeding of National Conference on Artificial Intelligence, Portland, USA, vol. 2, pp. 944–951 (1996)
23. Thrun, S., Gutmann, J.S., Fox, D., Burgard, W., Kuipers, B.: Integrating topological and metric maps for mobile robot navigation: A statistical approach. In: Proceeding of Tenth Conference on Innovative Applications of Artificial Intelligence, Madison, USA (1998)
24. Tolman, E.C.: Cognitive Maps in Rats and Men. Psychological Review 55, 189–208 (1948)
25. Tomatis, N., Nourbakhsh, I., Arras, K., Siegwart, R.: A hybrid approach for robust and precise mobile robot navigation with compact environment modeling. In: Proceeding of IEEE International Conference on Robotics and Automation, Seoul, Korea, vol. 2, pp. 1111–1116. IEEE Computer Society Press, Los Alamitos (2001)

3

Machine Perception in Unstructured and Unknown Environments

Steven Scheding, Richard Grover, and Hugh Durrant-Whyte

Summary. This chapter discusses the issue of machine perception from the perspective of a system design process. The three issues of information gathering, data representation and reasoning are discussed, leading to a general high-level model of the problem. The model is intended to be generic enough to allow a wide variety of tasks to be performed using a single set of sensory data. It is argued that the model has a direct correspondence with some recent biological models. Finally, an application is presented showing how the model may be applied to solving real-world problems, specifically an autonomous system operating in outdoor unstructured environments.

3.1 Introduction

Mapping unstructured and unknown terrain is an extremely important competency that any autonomous system must possess in order to be considered truly useful. The term mapping in this context will be considered to mean the process whereby complex information is gathered and subsequently abstractly and concisely represented. Additionally, the representation should readily support reasoning on the mapped information so that appropriate actions may be taken.

Therefore, when designing a perception system for an artificial agent, the following must be considered:

- Information gathering – Are the information sources (usually sensors) capable of measuring the quantities of interest? For example, water is extremely hard to classify using only video data, however during daylight, water has a fairly unique polarisation signature.
- Representation – Does the representation model the underlying uncertainties present in the system? Does the model possess desirable computational characteristics? How compact is the representation? It is conceivable that for many missions the data rates from sensors will be in the order of Gigabytes per minute.
- Reasoning – How does the representation facilitate reasoning? What mechanisms allow decisions to be made? Can you, as the designer, be confident that the *correct* decisions will be made? Can many parallel tasks be carried out using the *same* data sources?

This document will examine each of these issues in turn. The issues will be explored in the context of an autonomous ground vehicle. The vehicle must be capable of reconstructing the geometry of its environment, as well as extracting

M.E. Jefferies et al. (Eds.): Robot. & Cogn. Approach. to Spat. Map., STAR 38, pp. 65–81, 2008.
springerlink.com

other relevant information such as the identification of vegetation versus rocks and other high density obstacles.

3.2 Information Gathering

Historically, autonomous systems have relied on tightly constraining the complexity and structure of the operating environment through the use of artificial lighting, targets and structured environments. Natural outdoor environments, however, are characterised as structurally complex and require significant subtlety of interpretation. The tendency to underestimate the effects of this is a result of the (normally) highly-tuned abilities of the observer. A lack of precision in interpretation leads inexorably to fragility and increased likelihood of failure. The goal of a sensor system is to reduce the effects of ambiguity through the appropriate selection of sensors and task-specific processing. Two examples particularly relevant to an autonomous ground vehicle are identifying: the difference between a flat plain (such as a salt-pan) and a lake surface; or between a moss-encrusted rock and a similarly sized bush. Without appropriate sensing and interpretative capabilities these examples remain indistinguishable, with potential effects ranging from excessive caution to reckless confidence. Similar biological examples such as the difference between a leaf-shadow pattern and a predator can be readily imagined.

Resolving such ambiguity is possible if an appropriate set of sensory stimuli are available: polarisation information can reveal the surface of water as distinct from earth; while radar can reveal the lower density of the shrub directly. Figure 3.1 shows a schematic example where two objects are sensed in such a way that the depth and colour variations are not captured; the observational space is insufficient to clearly resolve the two objects. Conversely, this figure also suggests the sensory stimuli required to improve this discrimination: colour and/or range. Intuitively, the best solution would be to use both additional stimuli to reduce the ambiguities by expanding the dimensionality of the sensory space.

It is important to note that it is not necessary that each dimension introduced by a sensory stimuli be strictly orthogonal to one another. Rather, provided that the stimuli themselves span the decision space (that is, the subspace over which the important criteria for making a decision are supported) the system will be *capable* of interpreting the data successfully.

Further consider the example of determining a shrub from a similarly sized rock. They may have different colours or textural properties, but this is not guaranteed. Perhaps the only property that can uniquely identify the rock from the shrub is their density. For various reasons, this property is almost impossible to determine using cameras (single or stereo), lasers, or most other common imaging sensors. Radar is perhaps the only technology capable of determining the density of foliage, albeit indirectly.

To demonstrate these different properties, a small section of data is considered in more detail here. This portion of data contains a tree with some flat ground surrounding it. This data set is selected to highlight the subtlety required in

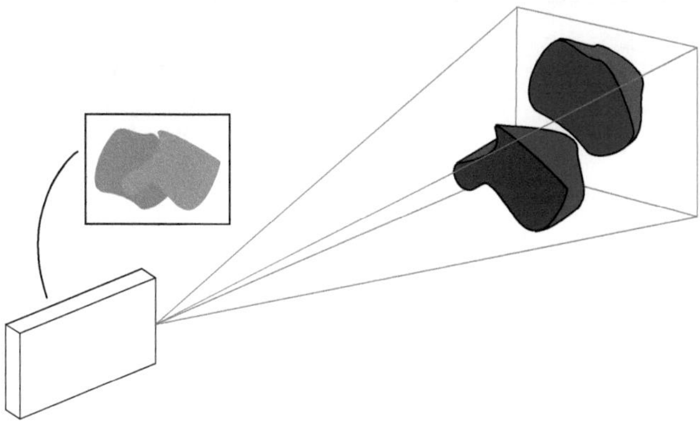

Fig. 3.1. Two objects are projected into a greyscale image subspace which makes their separation difficult. If colour or range, or even better both, measurements are available then their discrimination will be improved.

interpreting sensory data of natural scenes. Different parts of the scene often require quite different interpretations: the ground can be considered a surface, but a tree can be considered volumetric in structure.

Figure 3.2 shows the 3D view of the tree and ground that geometry measuring devices such as lasers and depth-of-field imaging sensors would generate. In this image it is difficult to discern the expected differences between the tree and the ground. To this data, colour and texture information can be added relatively simply, as seen in Fig. 3.3. This representation appears to be very rich in information, using both laser and visual data, however it remains difficult to identify the distinction between the ground and the tree using this extra information.

Figure 3.4 shows the same scene, this time imaged with a millimeter wave (mmWave) imaging radar capable of measuring the density of the elements of the scene. This figure is still geometric (as with Fig. 3.2), but the differences are now immediately apparent.

This example demonstrates the notion that the choice of sensors is very closely linked to the application. Without the appropriate sensor the task cannot be carried out, because the information necessary for task completion is simply not available. This idea has many precedents in the natural world, with many animals possessing senses and transducers that are uniquely suited to their own particular environment. Examples include:

- *Dolphin Sonar* [11] – In the underwater domain, sound propagation is generally far superior to the propagation of light. In fact, red light is absorbed by water which reduces the available spectra usable by photo-receptors. Similarly turbidity and depth decrease the available light. In [11] it is shown that dolphin sonar is (in most cases) able to discriminate between four different liquids contained in closed cylinders, a result simply not feasible with photo-receptor based sensing.

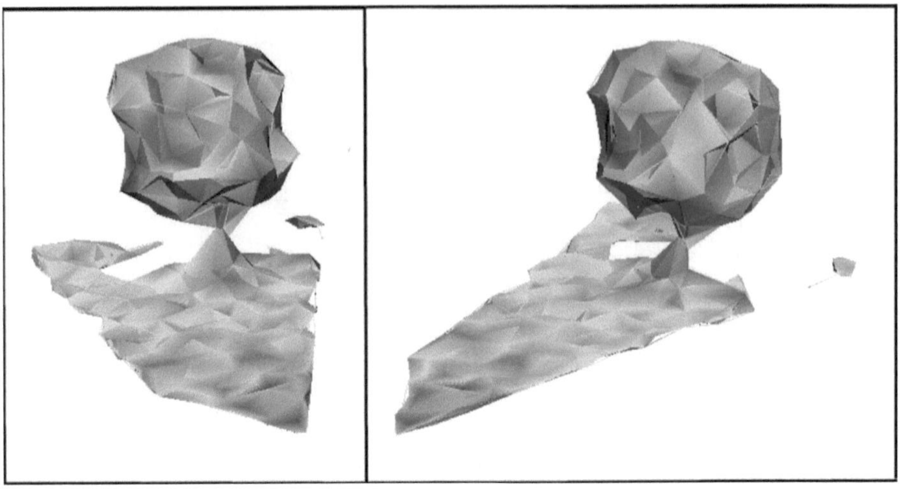

Fig. 3.2. 3D surface visulisation of the tree data set, from two view points

Fig. 3.3. 3D Delaunay surface visulisation of the tree data set, with texture mapping, from two view points

- *Invertebrate Polarisation Sensitivity* [19] – Many invertebrates are able to sense the polarisation of light directly. In low-light situations this ability would appear to aid in the discrimination of objects which may have similar 'colour', but different polarisation characteristics.

Fig. 3.4. Combined volumetric and surface visulisation of the tree data set, from two view points

3.3 Representation

The representation problem for autonomous artificial agents has been addressed at length in relatively structured internal, and some external, environments. Readily discernable structural features such as points, lines and other parametric elements have been utilised, and algorithms such as the Kalman filter [22] and particle filter [4] implemented to incrementally update such representations. As environments become more complex, however, they fail to conform to such simple parametric models and these methods often break-down in a fragile manner. This occurs in more dynamic indoor environments and, most importantly, in outdoor environments. In such cases, generalised non-parametric models become essential. One example of such a representation is the probability of occupancy which can be described using elevation maps, tesselated surface models [15] and structured [5] or unstructured [15] occupancy sampling. While such models are very general, they can also be unweildy [16, 12] and, in particular, basic learning operations such as incrementally combining new information or combining two separate representations can be very complicated.

Figure 3.5 shows graphically the dramatic effect the choice of representation can have on memory requirements and representation fidelity. In Fig. 3.5(a) the 'real' object can be seen as a circle or disc. Figure 3.5(b) shows the disc represented on an 8×8 grid whilst Fig. 3.5(c) shows the same disc represented by 64 point samples. The representations displayed in Figs. 3.5(b) and 3.5(c) therefore have the same memory requirements, however the representational fidelity is

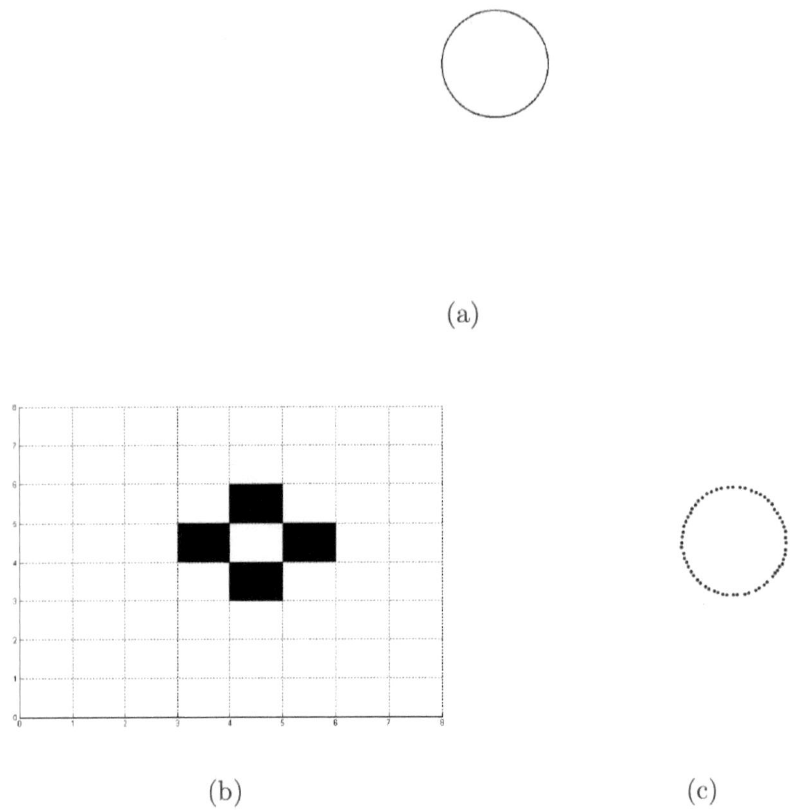

Fig. 3.5. (a) Real object (disk); (b) Grid representation of object with an 8×8 array; (c) particle representation of the object with the same storage requirements as the grid, i.e., 64 samples

very different. The grid representation shown in Fig. 3.5(b) would require vastly more memory to achieve the same fidelity as the particle representation.

Coupled to the memory issue is the need to apply a probabilistic approach, due to inherent uncertainty within and without the system. Two major operations are needed for such an approach: prediction of the future state of the probability given dynamic models of the agent and the environment; and updating the probability distribution given an observation of the environment. Respectively, these operations correspond to convolution (from the total probability theorem) and multiplication (from Bayes' rule) operations performed on probability distributions. These provide for a recursive prediction-correction style learning of a 'map' of the environment.

Thus, a practical representation should display two simultaneous characteristics: it must provide a reasonable level of compactness and must also support efficient manipulation by multiplication or convolution. The first criteria enables

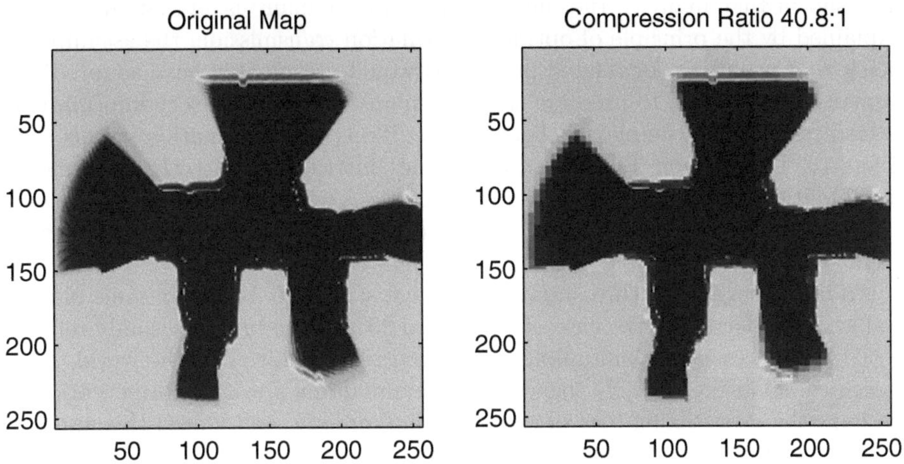

Fig. 3.6. 40:1 Compression of a Wavelet Represented Occupancy Grid

efficient storage and communication of the map information and the second ensures that the resulting data can be practically processed using consistent, standard probability formulas. There are many representations which achieve excellent compression or compactness, but many turn out to be extremely difficult to manipulate efficiently. Two of these previously considered are Gaussian Mixtures (or Sums of Gaussians) [4, 1] and Support Vector Machines (SVM) [17]. The Gaussian Mixtures approach suffers from an exponential increase in model complexity as the system evolves, requiring regular re-sampling, whilst the SVM approach yields representations which are extremely difficult to manipulate and combine mathematically.

A representation using Haar Wavelets and the Discrete Wavelet Transform (DWT) [7] is another interesting possibility. Analytical solutions to both the incorporation of positional and sensor uncertainty into the representation, and the multiplication of two wavelet-based representations (wholly within the wavelet form) have been derived. Their computational complexity is much improved over traditional occupancy grid implementations. Additionally, the Haar Representation for occupancy data is inherently compact, as demonstrated in Fig. 3.6. The figure shows an occupancy grid map originally built using laser data for which a compression ratio of over 40:1 has been achieved with marginal loss of information.

This section has shown that probabilistic representations are extremely important, as they allow the explicit encoding of uncertainty or even complete ignorance of a particular event or state of the world. Several methods were examined against the criteria of memory requirements, representational fidelity and computational complexity. Recently, there has been much work aimed at applying these types of probabilistic representations to aid in the understanding of the phenomenology of biological systems. In an interesting example [25], it is shown that many of the unique features of human visual systems can be explained by the principle of optimal information transmission, the assumption being that evolution (or related processes) would presumably have adapted the human visual system to preserve the maximum amount of sensory information, regardless of scene complexity, lighting conditions etc. Information in this case refers to 'Information Theory' which defines information directly in terms of probabilities. Storing probabilistic representations which preserve the maximum amount of sensory stimuli would therefore appear to have a strong biological analogue.

An interesting conjecture, implied by recent studies [6] is that in some biological systems (ferrets in the case of [6]), up to 80% of the brain's visually related activity is devoted to maintaining its learned representation of the world. This biological analogue strongly suggests that maintaining a world representation is perhaps the most taxing part of the 'mapping' process, and indeed this appears to be true of robotic systems also.

3.4 Reasoning

Reasoning, like information gathering and representation, may be thought of as being application specific and is intimately linked to the choice of sensors and representations. Whilst these three topics have been separated for clarity of discussion, in reality the three must be considered *in toto* before any real problem can be solved.

For the purposes of this discussion, reasoning will be defined as follows:

- The *transformation* of the sensory stimuli to another, more abstract representation, to aid the decision making process. In Fig. 3.7 this process is referred to as 'Data Condensation'.

This 'transformation' should have the following properties:

- It must be able to incorporate prior knowledge, should it exist.
- The transformation should be considered *optimal* if it retains only the information needed for a particular task, and no other.

These conditions refer to the ability of a reasoning algorithm or process to highlight the aspects of the data that are important to a particular task. For example if it were necessary to extract faces from an image sequence, it would be a reasonable *ad-hoc* assumption that the algorithm should be able to use knowledge of skin colour, facial geometry, or any other aspect deemed to be important. In a probabilistic framework, combining the incoming sensory data with a prior model of that data is usually achieved through convolution of the two distributions.

The criteria of optimality can only be measured with respect to the task being performed, in the sense that no task-specific information is lost during the transformation process. The 'prior' knowledge needed to complete a particular task could therefore, in principle, be learnt by posing it as an optimisation problem which maximises task performance, rather than as an *ad-hoc* process such as that described above.

In fact, in [25] it is shown that this single principle of maximising information (optimisation) over both short and long time-scales accounts for several well-known psychophysical phenomena in the visual systems of human-beings. In robotic systems, learning the correct (or optimal) transformation for reasoning tasks is typically computationally expensive, whilst using the transformation 'on-line' is usually comparatively inexpensive. It could be conjectured that the computationally expensive 'task optimisation' of the perceptual systems of biological systems is driven by standard evolutionary processes, with the resultant 'transformations' requiring very little 'brain power'.

3.5 Architecture

The proposed model is shown in Fig. 3.7. The model has three significant characterisitics:

- The data is modelled in a *sensor centric* manner, representing the perceptual information in terms of the observed sensory responses. Separate stimuli are described and quantified as separate degrees of freedom in a high-dimensional 'sensor space' which captures, in a single structure, the data from all available sensors. No attempt is made, at this stage, to infer the existence of important characteristics or features. The sensory data itself is treated as the best available model of the external world.
- *Uncertainty and ambiguity* in sensing is captured in a probabilistic form as a likelihood function. The explicit use of a probabilistic model allows operations of temporal propagation of data and temporal fusion of information to occur through the use of the Chapman-Kolmogorov and Bayes equations respectively. The actual form used to encode these likelihoods is an important computational issue and many different techniques are appropriate under different circumstances, including: sets of particles, kernel approximations and functional representations.
- The tasks of *perception and reasoning* are interpreted as processes which abstract and compress the stored information (data condensation). In this context, we focus on re-interpretation of the data for the purpose of increasing the contrast in the data relevant to some specific task. Entropy measures can be used to explicitly estimate the changes to the information content as a result of this process.

What distinguishes this approach is the emphasis on managing data in its 'raw' sensory form and delaying any interpretative tasks until the time and place where they are required to achieve some goal. This is motivated by the belief that the sensory stimuli themselves represent a robust, appropriate, and the most complete, summary of the available data.

Of course, simple recording of sensory information is not inherently valuable, for the information must be fused and propagated temporally to generate a

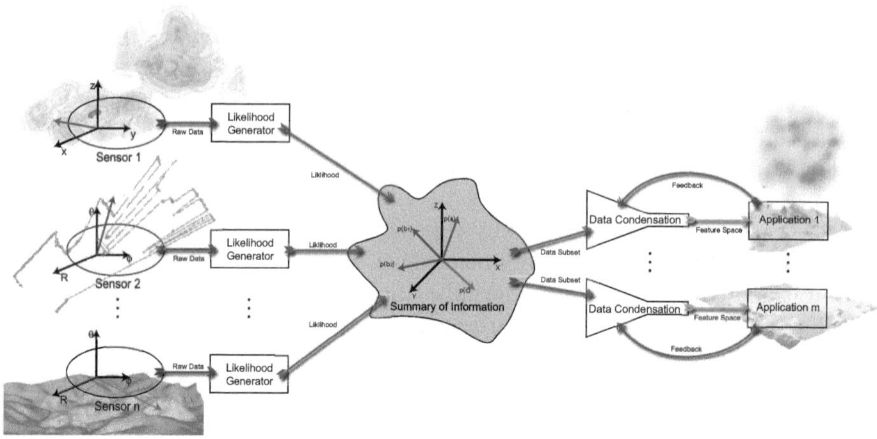

Fig. 3.7. A schematic view of the proposed perception model highlighting the three main characteristics: a sensor-centric information summary; explicit use of likelihood models; and delayed, application-specific interpretative stages

concise, composite and useful global reconstruction of the environment. Following the approaches of [22, 4] a probabilistic model of the sensory stimuli, in the form of observation likelihoods, is used to construct the sensor space. Practically, the responses of a sensor (amplitude returns, regions of irradiance, etc.) are used to construct a likelihood distribution around a region of true response, itself the underlying state to be inferred. As examples, consider the inference of occupancy measures according to laser scans, or interpreting a radar sensor as a source of information about the reflectivity of the scene.

The utility of delaying the interpretation is evidenced from the fact that the physical interactions between any single sensing modality and the environment is ill-posed with a many-to-one relationship between the true environment and data which makes it essentially impossible to model. If it were possible to construct these models, then it would also be possible to determine, in advance, the exact combination of sensory measurements which would enable unambiguous reconstruction of an environment model from the data. In the absence of complete models, performance can only be assessed using an appropriate metric which combines the characteristics of the data and the overall performance of the required task. For example, the direct interpretation of a given object's properties from radar and infra-red sensors will not necessarily capture its essential characteristics in a concise manner, whereas some combination of subsets of the measured stimuli of both sensors may provide a salient description *with respect to some given task.* Independently measurable components of this new description can be considered to form 'feature descriptors'. This interpretation has much in common with multi-sensor stimulus models described for mammalian data fusion of the Superior Colliculus [21].

3.6 Practical Implementation

3.6.1 The ARGO Demonstrator

In this section we consider the mapping of this approach to the autonomous ground vehicle (AGV) shown in Fig. 3.8. The vehicle subsystems have been developed with emphasis on two main goals: to ensure reliable operation over extended periods (greater than 24 hours continuous) and to provide a modular, scalable test platform for deployment of technologies resulting from research programmes. The vehicle has already been demonstrated to operate reliably for longer than 8 hours and has performed autonomous path-following over distances in excess of 7.5km and up to 2.5km between individual waypoints. The deployment arena is unstructured, expansive outdoor environments including desert, rural farmland and wooded areas.

3.6.2 Reasoning Tasks and Sensing Requirements

Reliable long-term navigation and control for this vehicle under mission scenarios including continuous day/night transitions, all-weather operation and with speeds of up to 9ms^{-1} requires at least the capabilities to navigate, plan and infer properties of potential obstacles. Most importantly the vehicle must be capable of reliable navigation under all design conditions. There are two obvious candidates: GPS/INS systems [10] and an implementation of the Simultaneous Localisation and Map Building (SLAM) algorithm [20]. It has been shown, however, that reliability and fault detectability considerations suggest that neither system is suitable in isolation, however an appropriate combination can be developed. Recent results from the DARPA Grand Challenge [24] have demonstrated the dramatic effects of undetectable faults in the navigation systems.

Determining the desired path based on available information is also critical for developing a truly autonomous platform. A reliable and efficient planner will necessarily interpret sensory data with respect to a vehicle specific model. In addition to estimating the local surface geometry, characteristics such as ground type (gravel, grass, tarmac, etc.) and condition will be required. Furthermore, the traversability of different regions must be estimated at a larger scale; identifying lakes, salt-pans, rivers, cliffs and known fence-lines among the necessary tasks. Even with reliable navigation and a global plan, it is well known that local variations will prohibit blind path-following. This suggests that a vehicle will require the ability to detect and respond to obstructions (both volumetrically positive and negative) in order to modify the trajectory appropriately.

Candidate sensing modalities for these reasoning capabilities include vehicle-based GPS/INS system, near-field laser and radar scanners, and depth augmented and non-augmented cameras for navigation and near-field sensing. Efficient global planning suggests the utilisation of satellite (or aerial) height data, visible and hyper-spectral imagery and surveyed meta-data (fence-lines, waypoints, beacons etc.).

This method is advocated as each of the separate reasoning tasks are dependent on different subsets of the *same* sensory data. For example, geometry may

Fig. 3.8. The Argo vehicle at a recent field trial

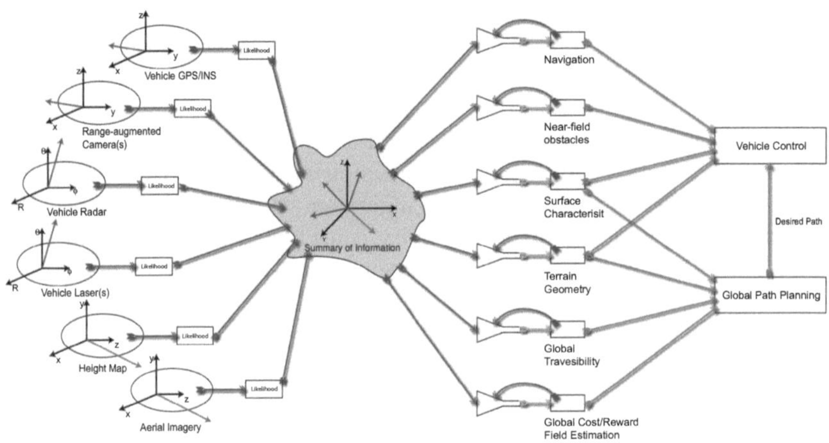

Fig. 3.9. High-level view of a simple implementation of the proposed approach to an Autonomous Ground Vehicle (AGV). For clarity the feedback between the Navigation solution and the likelihood generators is omitted.

depend on the direct measurement of the radar and laser reflectivities of the environment but robust feature association for navigation may require some highly abstracted combination of all the sensors. Furthermore, the most complete data is available for each task and assumptions required for a particular task do not result in data losses for other parallel operations. Finally, the system is able to handle asynchronous data gathering, providing the most complete available data at any point in time and is readily extensible to scenarios with multiple heterogeneous platforms. Figure 3.9 shows a customised version of Fig. 3.7 with these sensors and reasoning capabilities shown.

3.6.3 Surface Reconstruction

This section describes one possible application that can be applied to the information summary described in the previous section. It demonstrates the utility of the model, by allowing many application specific surfaces to be built using the same data sources. These surfaces or maps can then be used for planning subsequent actions.

Surface reconstruction can be described as a geometric methodology for representing an approximation of the objects or features in an environment. The applications for these surfaces are manifold; most commonly, in terrain-like structures, a generated surface (map) is used for navigation, localization, path planning, etc. [13, 23]. Another application for surface reconstruction is to generate a meshed model in order to perform differential computation on localized nodes of that object [3]; further employment of such techniques are for visualization purposes [9]. There is a large amount of literature on the subject of mesh generation, and surveys on various surface reconstruction and meshing techniques can be found in [2, 8, 18]. Reference [14] contains a more complete description of these approaches and techniques.

Most algorithms that generate surfaces often don't quantify the errors in the observed information and accept the generated representation as a 'true' model of the surface. This makes performing decisions (e.g., mesh decimation, localization, navigation, etc.) a difficult or even unfeasible task. Sensors are not perfect and usually provide incomplete information about the features of interest. Consequently, this information is subject to errors. If these are not accounted for, the result of making decisions under the assumption of the representation being correct may result in equipment loss and mission failure. Unfortunately, it is not possible to identify and quantify the exact sources of all these errors, which consequently results in the need to be able to perform decisions under uncertainty. The approach presented here introduces a summary of the uncertainties involved in the reconstruction and information gain processes of map-building, resulting in a representation that enables the quantification of bounds on the uncertainty of a surface region. Therefore, decisions can be performed under the knowledge of the amount of uncertainty in the surface. The algorithm also allows for fusing data from multiple viewpoints and different types of ranging sensors (e.g., Laser, Radar, Stereo vision, etc.).

Results from using a novel surface reconstruction algorithm with the data obtained from the experimental platform from multiple viewpoints are shown in Figs. 3.10 and 3.11. Figure 3.10 depicts the reconstructed terrain from various viewpoints. Figures 3.10(a) and (b), were obtained at a low resolution from a distant viewpoint. Figures 3.10(c) and (d) were obtained after the first scan by adding information from another sensor in a closer location providing higher resolution data. Additionally, the surface management layer was designed to maximize the resolution of the reconstructed surface, by limiting the normal error of every triangle to zero variance. This resulted in a highly triangulated terrain. On the other hand, in Fig. 3.11, the utilities were set to obtain a lower

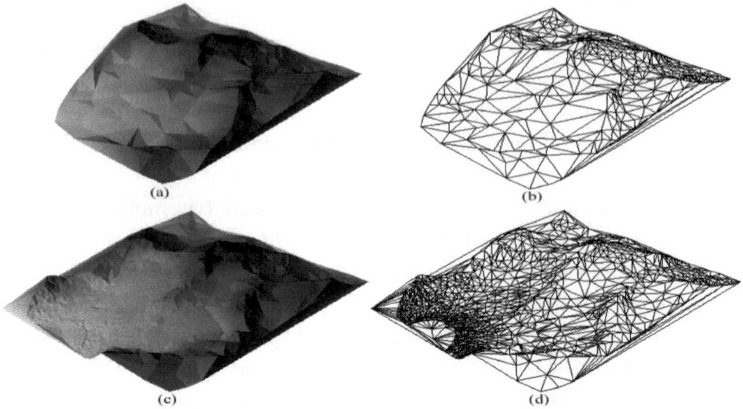

Fig. 3.10. The triangulated reconstructed surface from multiple viewpoints, and with multiple scanning resolutions; (a, b) after a scan from first viewpoint with a two degree scanning resolution in pan and 0.5 degree in the tilt axis; (c, d) after an additional scan from a second viewpoint with a 1.5 degree scanning resolution in both axis. The MMSE threshold of the local particle representation over triangles was set to $\sigma^2 = 0m^2$.

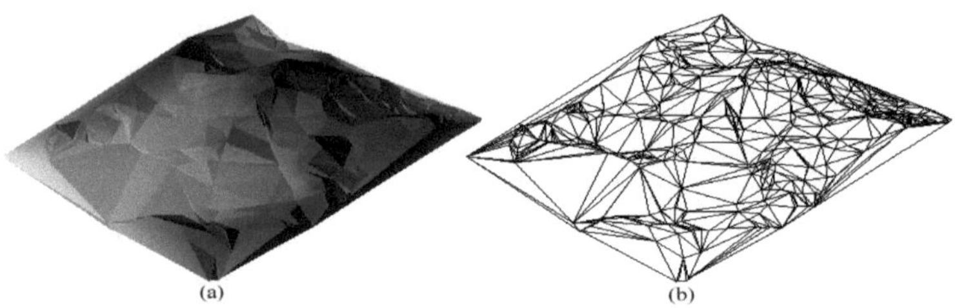

Fig. 3.11. Triangulated reconstruction of the surface from multiple viewpoints, and with multiple scanning resolutions. Obtained from the same particle representation as in Fig. 3.10. In this case, the MMSE threshold of the local particle representation over triangles was set to $\sigma^2 = 0.0005m^2$.

resolution triangulation, by limiting the normal error of every triang le to a 0.0005 variance. This small deviation from the first example allowed the result of a less triangulated surface without loss of the main features of the map. This is due to a more triangulated terrain where surface variations are greater and lower triangulated regions, where there is less variation (flatter regions).

Figure 3.12 depicts the two levels of representation in the knowledge base. One is the geometric model (a Delaunay Triangulation DT) that is designed to fit in a Least Mean Square (LMS) sense to the other, the particle representation (information summary). The particle representation is simply another probabilistic representation similar to those discussed in Sect. 3.3.

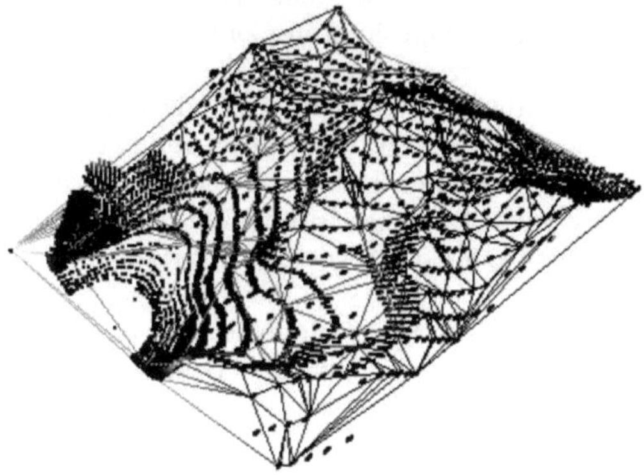

Fig. 3.12. The representation of two levels of knowledge about the surface: The particle representation, and the triangulated surface (DT)

3.7 Conclusion

This chapter has presented a discussion of the various aspects of 'perceptual systems' from a practical system designers viewpoint. The task of perception was broken into three main, but interrelated parts; those of information gathering, representation and reasoning. It was argued that reasoning is the process of taking abstract sensory data and transforming it (in combination with any prior information) into a more abstract representation that contains only the information relevant to a particular task. Further, it was argued that in order for this to work effectively, that parameters relevant to the task must be sensed, and then stored in a representation which may be manipulated efficiently and is compact. Several representations were examined for use in this task. An optimal implementation would be one in which the sensors used were uniquely suited to the task being performed, the representation was computationally efficient and compact, and the reasoning process discarded only the information not relevant to the current task. Interestingly, it would appear that evolutionary processes have shaped many biological systems against these exact criteria, and that quite simple principles of Information Theory can be used to describe both robotic and biological systems against the criteria. Finally, an architecture and an implementation were presented to illustrate the concepts as applied to a large outdoor robotic system.

References

1. Alspach, D.L., Sorenson, H.W.: Nonlinear bayesian estimation using gaussian sum approximations. IEEE Transactions on Automatic Control AC-17(4), 439–447 (1972)
2. Bern, M., Eppstein, D.: Mesh generation and optimal triangulation. Csl-92-1, Xerox PARC (March 1992)

3. Chew, P.L.: Guaranteed-quality mesh generation for curved surfaces. In: Proceedings of the Ninth Annual Symposium on Computational Geometry, pp. 274–280 (1993)
4. Durrant-Whyte, H., Majumder, S., Thrun, S., de Battista, M., Scheding, S.: A bayesian algorithm for simultaneous localisation and map building. In: ISSR. Proc. 10th International Symposium on Robotics Research, Springer, Heidelberg (2002)
5. Elfes, A.: Occupancy grids: A stochastic representation for active robot perception. In: Proceedings of the Sixth Conference on Uncertainty in AI, pp. 249–265 (July 1987)
6. Fiser, J., Chiu, C., Weliky, M.: Small modulation of ongoing cortical dynamics by sensory input during natural vision. Nature 431, 573–578 (2004)
7. Graps, A.: An introduction to wavelets. IEEE Computational Science and Engineering 2(2), 1–18 (1995)
8. Heckbert, P.S., Garland, M.: Survey of surface approximation algorithms. Technical Report CMU-CS-95-194, CS Department, Carnegie Mellon University (1995)
9. Hoppe, H., DeRose, T., Duchamp, T., McDonald, J., Stuetzle, W.: Surface reconstruction from unorganized points. Computer Graphics (SIGGRAPH) 26(2), 71–78 (1992)
10. Kim, J.H., Sukkarieh, S.: A baro-altimeter augmented INS/GPS navigation system for an uninhabited aerial vehicle. In: SATNAV 2003. Proc. 6th Int. Conf. on Satellite Navigation Technology (July 2003)
11. Lainscsek, C., Gorodnitsky, I.: Characterization of various fluids in cylinders from dolphin sonar data in the interval domain. In: Proceedings of Oceans 2003, pp. 629–632 (2003)
12. Leal, J.: Stochastic Environment Representation. Phd thesis, The University of Sydney (2003), `http://www.acfr.usyd.edu.au`
13. Leal, J., Scheding, S., Dissanayake, G.: Probabilistic 2D mapping in unstructured environments. In: Proceedings of the Australian Conference on Robotics and Automation, Melbourne, Australia, pp. 19–24 (August 2000)
14. Leal, J., Scheding, S., Dissanayake, G.: 3D terrain mapping: A stochastic approach. In: Proceedings of the Australian Conference on Robotics and Automation, pp. 135–140 (2001)
15. Leal, J., Scheding, S., Dissanayake, G.: Stochastic simulation in surface reconstruction and application to 3D mapping. In: Proceedings of the IEEE International Conference on Robotics and Automation, pp. 1765–1770. IEEE Computer Society Press, Los Alamitos (2002)
16. Leonard, J., Durrant-Whyte, H.F.: Directed Sonar Sensing for Mobile Robot Navigation. Kluwer Academic Press, Boston (1992)
17. Müller, K., Mika, S., Rätsch, G., Tsuda, K., Schölkopf, B.: An introduction to kernel-based learning algorithms. IEEE Trans. Nerual Networks 12(2), 181–202 (2001)
18. Owen, S.J.: A survey of unstructured mesh generation technology. In: 7th Proceedings of the International Meshing Roundtable, pp. 239–267 (1998)
19. Rowe, M.P., Engheta, N., Easter, S.S., Pugh Jr., E.N.: Graded index model of a fish double cone exhibits differential polarization sensitivity. Optical Society of America 11, 55–70 (1994)
20. Smith, R., Self, M., Cheeseman, P.: Estimating Uncertain Spatial Relationships in Robotics. Autonomous Robot Vehicles. Springer, Heidelberg (1990)

21. Stein, Merideth: Merging of the Senses. MIT Press, Cambridge (1992)
22. Thrun, S.: Probabilistic algorithms in robotics. AI Magazine 21(4) (2000)
23. Thrun, S., Burgard, W., Fox, D.: A real-time algorithm for mobile robot mapping with applications to multi-robot and 3d mapping. In: IEEE International Conference on Robotics and Automation, San Feancisco, CA, pp. 321–328. IEEE Computer Society Press, Los Alamitos (2000)
24. Urmson, C., Anhalt, J., Clark, M., Galatali, T., Gonzalez, J.P., Gowdy, J., Gutierrez, A., Harbaugh, S., Johnson-Roberson, M., Kato, Y.H., Koon, P., Peterson, K., Smith, B., Spiker, S., Tryzelaar, E., Whittaker, W.R.: High speed navigation of unrehearsed terrain: Red team technology for grand challenge 2004. Technical Report CMU-RI-TR-04-37, Carnegie Mellon University: Robotics Institute (June 2004)
25. Wainwright, M.J.: Visual adaptation as optimal information transmission. Vision Research 39, 3960–3974 (1999)

4

Emergent Cognitive Mappings in Mobile Robots Through Self-organisation

Ulrich Nehmzow

Summary. This chapter presents examples of emergent cognitive mappings in mobile robotics, in application areas as diverse as self-localisation, route learning, novelty detection or action selection.

In all cases, be it navigational or non-navigational tasks, the robot control mechanisms share fundamental properties:

- they exploit an internal mapping between "world space" and "map space" (the cognitive map),
- they acquire maps through learning, rather than using pre-installed maps, and
- they use these maps for a broad range of tasks, rather than narrowly defined ones.

In other words, this chapter presents evidence for a convergence towards one common underlying mechanism over a broad spectrum of mobile robot control architectures.

4.1 Introduction

Capabilities such as the ability to learn, to adapt continuously to changing circumstances during operation, the ability to interpret contradictory, inconsistent or noisy data or the ability to structure perceptual space autonomously are all desirable for controllers of autonomous mobile robots. In fact, it is widely accepted that for reliable robot operation in the real world, over extended periods of time, they are indispensable.

Early robotics research focused on control architectures specifically designed to suit the specific niche (in terms of task and environment) that the robot was to operate in. Few common control principles emerged between such disparate tasks as acquisition of sensor-motor couplings (a sensor-motor competence) or self-localisation, route learning or planning of entire paths (navigational competences). Increasingly, however, control architectures for widely different tasks concerning the control of autonomous mobile robots show similarities that indicate an emerging, fundamental low-level cognitive architecture. This chapter presents examples from robot navigation (localisation, route learning and path planning), novelty detection (detecting uncommon sensor stimuli, without human intervention, and without using pre-installed models) and action planning (i.e. the autonomous determination of task-achieving sequences of actions) that all share a common, emerging cognitive mapping mechanism.

M.E. Jefferies et al. (Eds.): Robot. & Cogn. Approach. to Spat. Map., STAR 38, pp. 83–104, 2008.
springerlink.com

4.1.1 "Map" and "Cognitive Map"

Map

Commonly, the term "map" is of course used to refer to a navigational map, the kind one can buy in a stationer's shop. In the context of autonomous agents, such as for instance mobile robots, it is however useful to expand the definition of "map". For the purposes of this chapter, the term "map" therefore is used to describe a bijection[1] or an injection[2] from a "world space" A onto a "map space" B, so that B is a representation of A, internal to the robot. If this bijection represents a non-navigational map, the term "mapping" is often used.

A "classical map" therefore is indeed a map in the sense of this chapter, in that every location in the physical world is represented by a unique entry on the map, and vice versa. But so are telephone books or family trees: they respectively map the space of physically existing telephones or people to an entry in the phone book or a graph in the family tree, and are therefore "maps" in the sense used in this chapter.

The reason that such mappings are useful in robotics is that they allow the robot to reason about its world — concerning navigation and otherwise, and that for many high level robotic tasks they are indispensable. In this chapter, we will therefore present some examples of cognitive mappings in autonomous mobile robots, used both for navigation and for other tasks.

Cognitive Map

The term "cognitive map" goes back to Tolman's 1948 paper [21], in which he introduces the term to contrast "narrow, strip-like" maps with "broad and comprehensive" ones. "Map" here refers to classical maps used for navigation. However, even in the original 1948 article Tolman expanded the scope of cognitive maps to include behaviour other than navigation, concluding his article with the role of human cognitive maps concerning regression ("the return to earlier, more childish ways of behaving"), fixation ("undue persistence of early maps") and displacement of aggression onto outgroups. He concludes by equating "cognitive map" with "reason", praising "the virtues of reason — of, that is, broad cognitive maps."

The term "cognitive map" in Tolman's sense, therefore, implies a representation of knowledge that goes beyond simple storage of information, obtained along a strip-like path (be it in physical or some other cognitive space).

Specifically regarding navigation, two criteria are commonly used to identify such cognitive map-like organisation of knowledge [14]:

1. the ability to make spatial inferences without direct experience of the location in question, and
2. the ability to take mentally a different perspective of a spatial layout.

[1] A one-to-one mapping from A to B, with A and B having the same number of elements.

[2] A one-to-one mapping from A to B, where B has more elements than A.

Especially the first criterion — the ability to make inferences without direct experience — can be demonstrated in internal representations other than navigational maps, and this chapter gives some examples of this in autonomous mobile robotics.

4.1.2 Mobile Robot Control and Cognitive Mapping

In mobile robot control, be it for real world ("industrial") applications or experiments in cognitive science, some control structure is needed that provides a mapping between the robot's perceptual space and its action space. This mapping might be provided through designed control algorithms (e.g. PID control), or through evolved control strategies (e.g. artificial neural networks, fuzzy logic etc.). By whatever means the control strategy is obtained, it forms the core of the robot controller.

The Problem of Perceptual Discrepancy

The reason why roboticists are often interested in *evolving* robot control algorithms, for example by using machine learning mechanisms, rather than using fixed, pre-installed controllers, lies in the "problem of perceptual discrepancy".

It is inevitable that humans perceive their environment through their own senses, thus interpreting the world in certain ways. For example, we are unable to perceive infrared light, which therefore does not feature in our reasoning about the environment. If the sensitivity of our eyes was in a different wave length, windows would not be made of glass, but of some other, "opaque" material (only, of course, it would not be called "opaque"). This unavoidable influence of our everyday experience can have two negative effects when designing robot control code:

1. Sensory perceptions are postulated to exist, based on human perception of the world, and therefore used as part of the control program, while in reality they do not exist for the robot (for instance, "chair", or "food").
2. Sensory perceptions may exist for the robot, but are not exploited by the human, because he is unaware of them (for instance, "electromagnetic radiation", or "infrared light").

Figure 4.1 gives an example. It shows the signals obtained from a side-looking sonar sensor (bottom graph) and a sonar sensor looking 45° ahead. The robot's task is to detect doors.

Based on human experience, one would expect that the side-looking sonar would be best to detect the slight recess of a door, and indeed the graph of the side-looking sonar shows a peak of about 20 cm above baseline when the robot passes a door.

However, it turns out that the door jambs actually act like beacons in the darkness, returning a very strong peak on the sonar scan (top graph in Fig. 4.1). This is an example of perceptual discrepancy: because humans do not identify the presence of a door by the presence of a door jamb, this possibility would not normally enter the mind of a robot programmer.

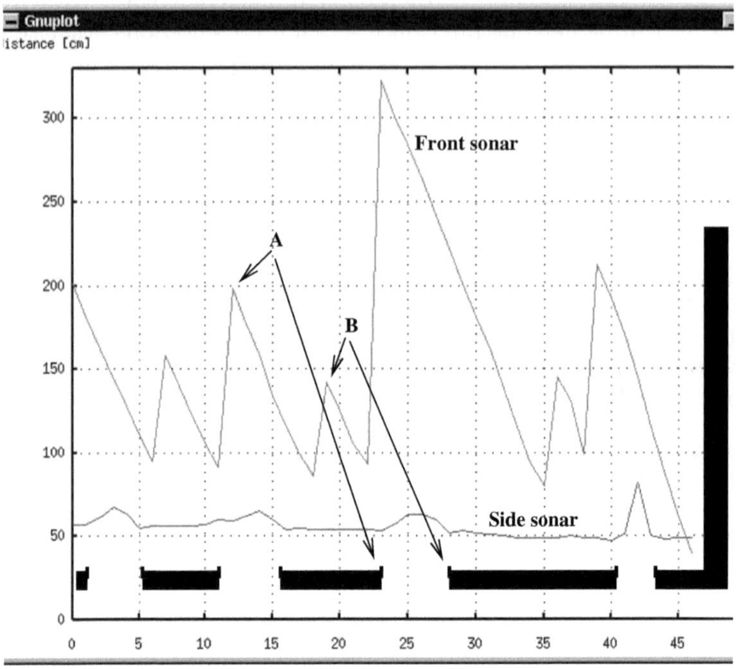

Fig. 4.1. Sonar sensor signals obtained when following a corridor. The small "humps" in the side-looking sonar measurement can be used to detect doors, but peaks A and B, which indicate detections of door jambs, are far more prominent. The abscissa indicates range measured in centimetres, the ordinate position along the corridor.

If, however, the control code was *evolved* through the robot's interaction with its environment, that is, if the robot's own perception of its environment was used to create the mapping between perception and action through self-organisation, then postulated, but non-existent perceptions would never be used, and prominent, but never expected perceptions would be exploited automatically. For this reason it is often desirable to use mechanisms of self-organisation and emergence to construct the mappings central to a mobile robot's controller.

4.2 Experiments with Autonomous Mobile Robots

4.2.1 Navigational Tasks

Location Identification

The simple scenario shown in Fig. 4.2 shows an experimental setup that is regularly used at Essex University as part of student assignments. The robot, moving along the straight line indicated in the figure, has to stop at a previously visited goal location (marked by the cross in Fig. 4.2).

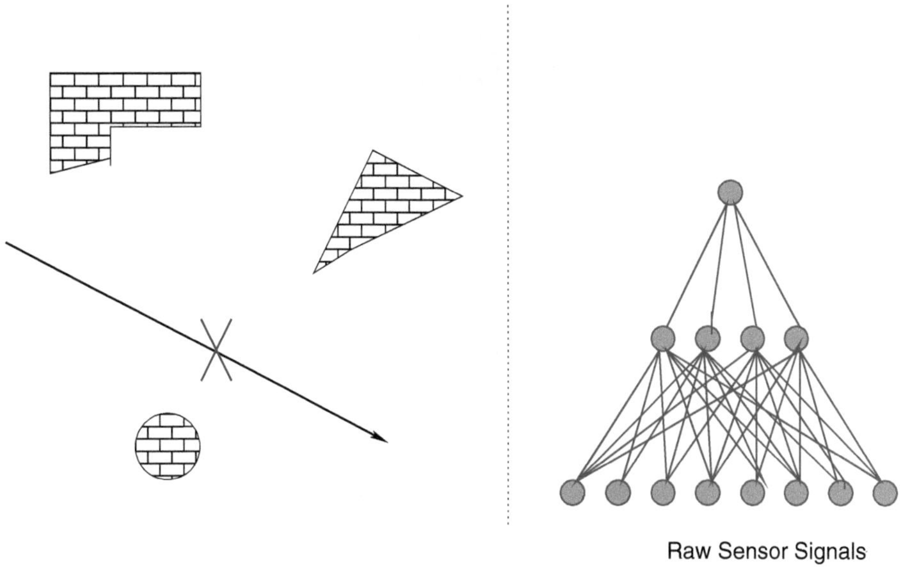

Fig. 4.2. A target location can be identified using perceptual landmarks alone

The simplest method to achieve this might be to just store the robot's sensory perception at the goal location, and to use this to determine when the goal has been reached — a simple stimulus-response system. However, the alternative of using a cognitive mapping between physical space and the robot's perceptual space also achieves the goal efficiently, and offers the advantage that other, different mapped locations can be defined and identified as goal positions as well.

Our robot therefore uses an abstracted representation of its sensory perception, for example an artificial neural network that has been trained with the sensor signals observed at the goal location. Reaching the goal location is then indicated by a peak output of the network.

Why could such a mapping be called a cognitive map? One manifestation of the presence of a cognitive map given earlier was that inferences can be made based on the map alone, without experiencing the physical counterpart of a selected map state. In the simple example shown in Fig. 4.2, the output of the network changes in a characteristic manner as the robot travels along the straight line, and can therefore be used to identify other, new target locations, even if the network has been trained on the target location indicated by the cross! The network weight space and the resulting network response as a whole therefore constitute a cognitive mapping of the robot's perceptual space.

That arbitrary, previously visited locations can be identified using such a cognitive map, is demonstrated by experiments we conducted in 1991. Here, a simple mobile robot, equipped with just two front whiskers and a wheel revolution counter for crude distance measuring (see Fig. 4.4) was given the task to localise in the environment shown in Fig. 4.5 [8].

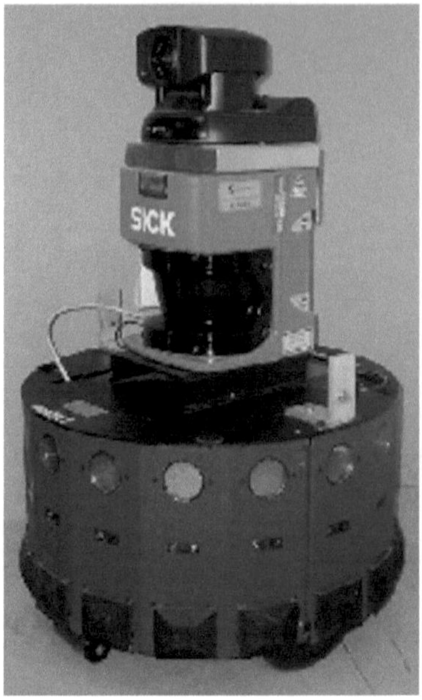

Fig. 4.3. The Magellan Pro mobile robot *Radix*, used in the experiments discussed in this chapter

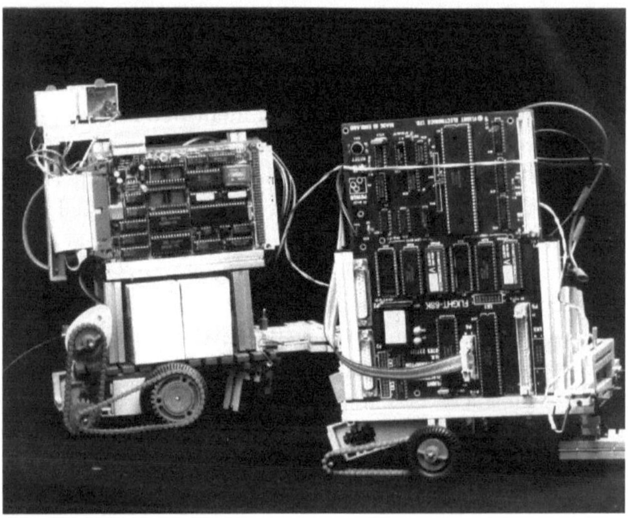

Fig. 4.4. The mobile robots *Alder* (left) and *Cairngorm*

To achieve this, the robot's whisker signals were used to detect concave and convex corners, and information about the current as well as previously visited

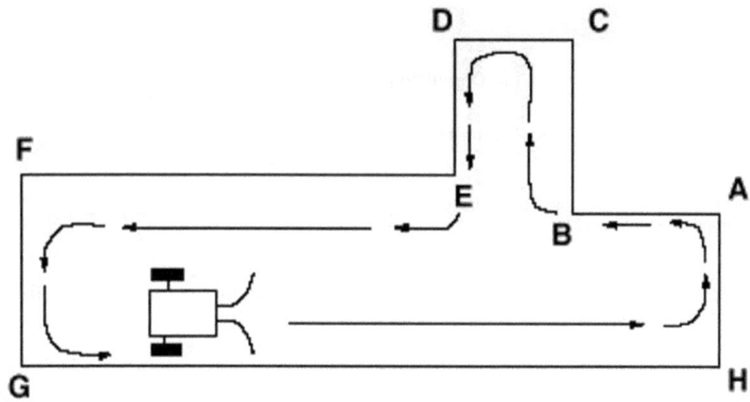

Fig. 4.5. The environment in which localisation experiments with *Alder* were carried out

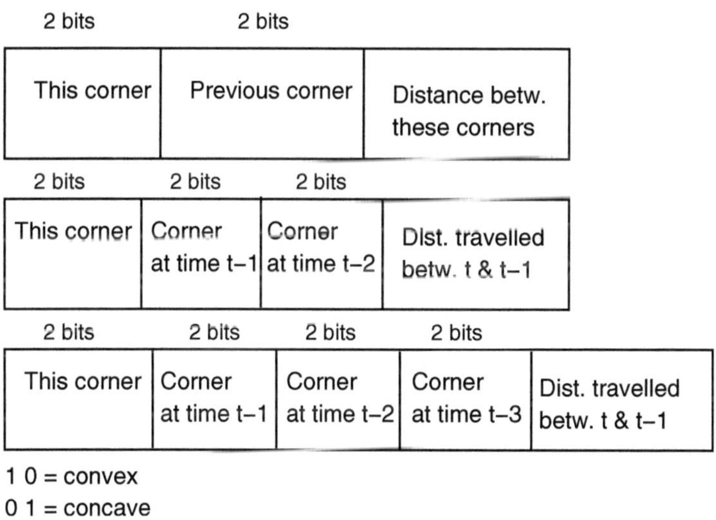

Fig. 4.6. Input vectors used for the localisation experiment with *Alder*

corners was used to construct an input vector to a self-organising feature map (SOFM) [5]. Three different types of input vectors are shown in Fig. 4.6.

These input vectors were used to train a self-organising feature map (SOFM) (Fig. 4.7), whose excitation patterns can be used to identify the robot's location. There are functional similarities here to rat self localisation [1]. When the rat is at a certain locations, place cells in the rat's hippocampus fire, and thus indicate the rat's position. Likewise, the robot's SOFM firing pattern changes with changing location, and can therefore be used to identify the robot's current

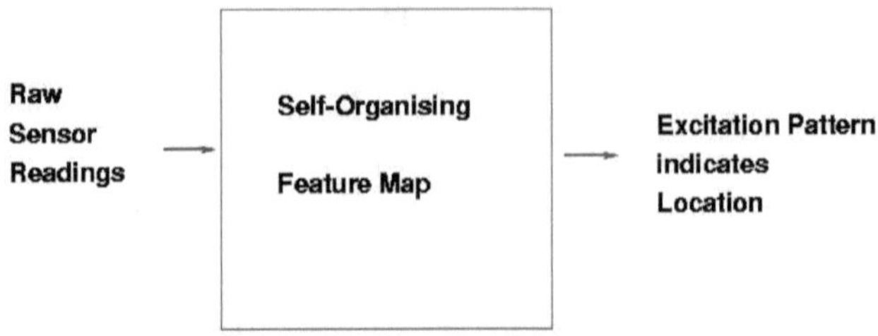

Fig. 4.7. Input vectors used for the localisation experiment with *Alder*

place in the world. In both cases, rat and robot, the cognitive map is learnt through agent-environment interaction.

These experimental results demonstrate that provided sufficient information is available from the input vector (i.e. the input vector shown at the bottom of Fig. 4.6 is used), localisation is possible and reliable. The important point for the purpose of this chapter is that the robot's cognitive map, the SOFM shown in Fig. 4.7, can be used to identify a range of different locations, without having been trained specifically for each location.

Self-Localisation in the Real World

The examples given in the previous section described laboratory experiments, that is, experiments under conditions in which the occurrence of perceptual aliasing (ambiguity in the robot's sensory perception, which makes location identification through perception unreliable) can be controlled. In the real world, perceptual aliasing occurs frequently, and has to be dealt with. One of the most reliable methods currently in use is to use Bayesian methods [20]; the following experiment is an example of this [4, 9].

In this experiment, the Nomad 200 mobile robot *FortyTwo* (Fig. 4.8) was to localise in the environment shown in Fig. 4.9.

In this environment, the robot obtains identical sensory perceptions at different locations, due to the limited resolution of its sensors. In other words, the "perceptual signatures" obtained in this environment are not unique (perceptual aliasing is indicated by shaded areas in Fig. 4.9), so that localisation by perception alone is impossible.

To address this problem, we used a combination of cognitive navigational map, evolved through unsupervised exploration of the environment and self-organisation, and metric information obtained from the robot's odometry (path integration) system (Fig. 4.10). Neither odometry nor perception alone would have been successful in this case, the former would fail due to the inevitable accumulation of drift error, the latter due to the encountered perceptual aliasing. The combination of the two, however, is capable of eliminating implausible location candidates.

In this experiment, the cognitive mapping between sensory perception in the real world and the robot's internal representation of these sensory perceptions

Fig. 4.8. The Nomad 200 mobile robot *FortyTwo*

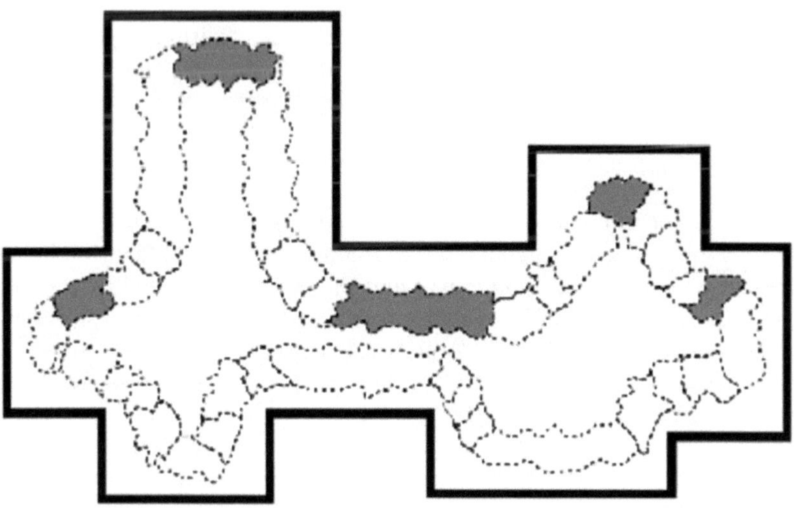

Fig. 4.9. The environment in which *FortyTwo* was to self-localise. Areas where perceptual aliasing occurred are shown in grey.

was established by training an adaptive resonance theory network (ART 2, [2]). This clustered the robot's sensory perceptions (infrared and sonar signals), and generated a cognitive mapping between perceptual and map space. In parallel,

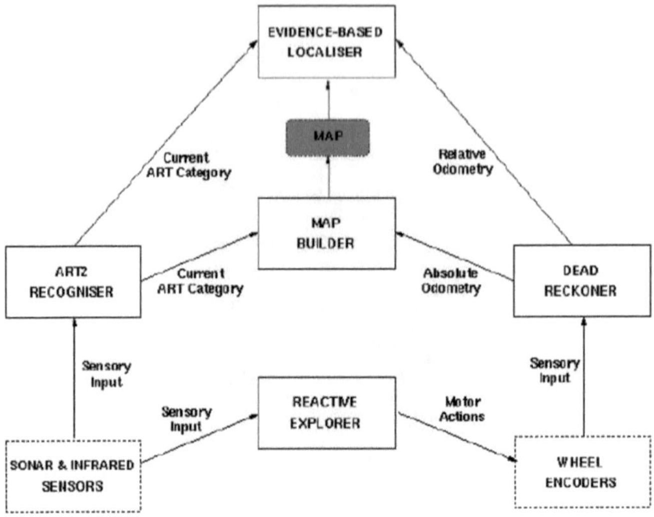

Fig. 4.10. The localisation mechanism used by *FortyTwo*, combining a cognitive navigational map with metric information. The cognitive mapping between sensory perception and the robot's internal representation of it is encapsulated in the ART2 recogniser. "Map" denotes the combined representation of perceptual and odometry information.

the robot's relative motion, that is, not the motion within a *global* reference frame, but merely motion relative to a previously identified location, was used to disambiguate between locations that appeared perceptually identical.

Quantitative Assessment of Localisation Performance

The performance of a robot localisation system can easily be measured quantitatively by analysing contingency tables that indicate the correlation between a robot's actual position (ground truth) and the position assumed by the localisation system. A useful measure of this correlation is the uncertainty coefficient U, which lies between 0 and 1 [3, 9]. A perfect localisation system will produce perfect correlation between the robot's actual and perceived position (in this case, the uncertainty coefficient U of the contingency table would be 1 [17, 11]). If, at the other extreme, a robot localisation system would merely guess the robot's location, using a random process, the uncertainty coefficient U would become 0.

Figure 4.11 shows the result of this quantitative assessment: a localisation mechanism purely based on odometry initially provides perfect localisation (uncertainty coefficient $U = 1$), and deteriorates with distance travelled. Using perceptual information alone for localisation results in a localisation capability that is independent from distance travelled, obviously. Furthermore, perception-based localisation is initially less accurate than odometry-based localisation, because of perceptual ambiguities in the environment. Combining the two mechanisms, however, gradually removes perceptual ambiguities, so that the localisation

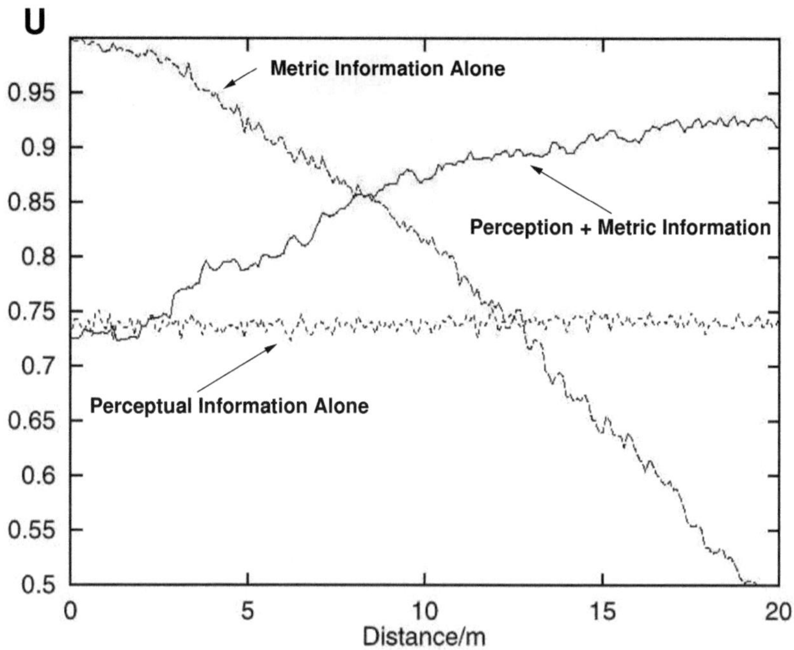

Fig. 4.11. Localisation capability of *FortyTwo*, using metric information alone, perceptual information alone, and the two combined. The uncertainty coefficient U shown here is 1 for perfect localisation, and 0 for localisation no better than random guessing.

performance improves with distance travelled, because more and more information is gathered *en route*, until localisation performance is so constant that U reaches a high, constant level.

In the example shown in Fig. 4.11, one can see that localisation by odometry alone is the most accurate localisation method for travel distances below 7m, after that taking perceptual information into account produces better results.

Route Learning

Mechanisms of self-organisation and topological clustering of sensory perceptions — the cognitive mapping mechanism used in the previous example — can not only be used for self-localisation, but also for path planning or the planning of entire sequences of actions (Sect. 4.2.2).

The following experiment demonstrates, how *both* perception and action can be stored together in one cognitive map, which is obtained through self-organisation [12]. The objective in these experiments was for a Nomad 200 mobile robot (Fig. 4.8) to learn to follow a route in an unmodified environment. In other words, no specific information concerning the desired route was given to the robot *a priori*; instead, the robot had to acquire the necessary perception-response mapping in the target environment, through learning.

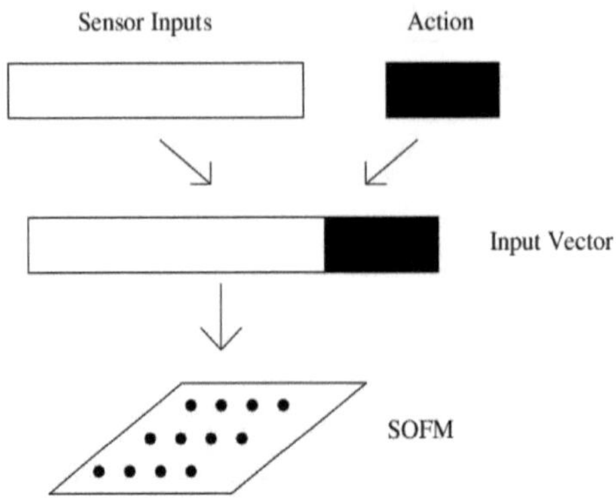

Fig. 4.12. Route learning (learning phase): Perception-action pairs are stored in a self-organising feature map for subsequent autonomous route following

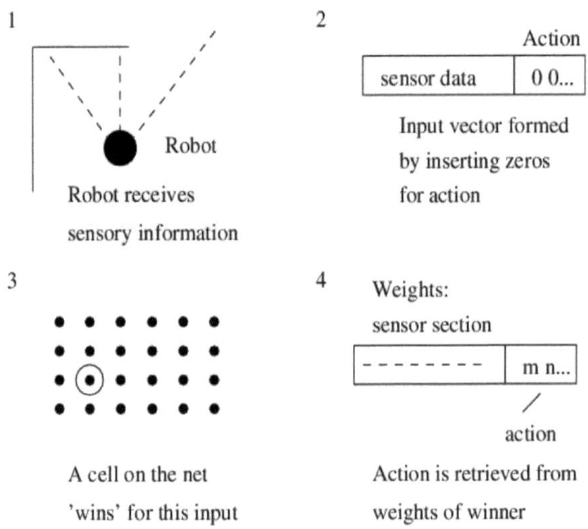

Fig. 4.13. Retrieval of action information stored in the route-learning self-organising feature map

During an initial learning phase, the robot is driven along the desired route by a human operator. During this learning phase, a self-organising feature map (SOFM [5]) is trained, using input vectors that combine sensory perception and the desired motor response at that location (see Fig. 4.12). This training results in an internal representation (cognitive map) of perception-action space that can later be used to follow the desired route autonomously.

The SOFM, a self-organising structure that learns without external teaching feedback, through unsupervised learning, thus produces an internal

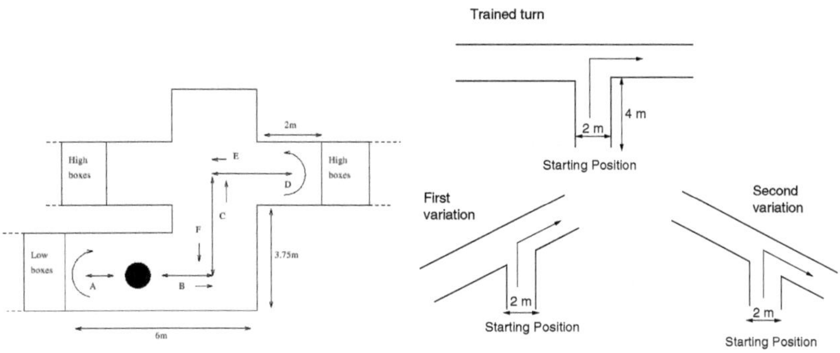

Fig. 4.14. Example of a route learned by *FortyTwo* (left), and generalisation ability of the algorithm (right)

representation of the desired path which emerges through the robot's interaction with its environment.

When the robot follows the route autonomously after the learning is completed, the required motor information is retrieved from the SOFM by replacing the "action part" of the input vector with zeros (step 2 of Fig. 4.13).

If such an input vector is presented to the trained SOFM, the network will find the closest matching neuron purely on the basis of sensory perception (step 3 of Fig. 4.13). The "action part" of the winning unit's weight vector then contains the required action information (step 4).

Cognitive Mapping Independent from Underlying Mechanism

Note that it is not the actual artificial neural network used that is particularly interesting here, but the underlying mechanism it supports: without the user supplying information about location, landmarks etc., the robot generates an internal representation of its perceptions, the "cognitive map", and uses this to perform navigational tasks.

In [13], for example, we used a different kind or emergent mapping, the RCE network described in [18], to achieve robot navigation. Landmarks, as they were perceived by the robot's sensors, were stored in the RCE network (Fig. 4.15), their topological and geometrical relationship, obtained from the robot's wheel encoders, were also stored to allow route planning and route following.

Robustness with Respect to Noise

FortyTwo was able to learn various different routes, using this mechanism, even if the environment was altered between training and autonomous route following. In real world robotics, all sensory perception is subject to noise, so that the mechanisms ability to deal with this is particularly important. For instance, the changes shown in Fig. 4.14 generated different sensory perceptions at the junctions, yet the robot was nevertheless able to select the correct path.

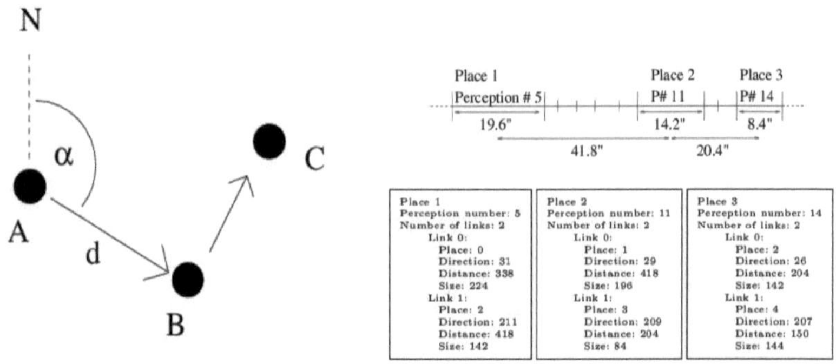

Fig. 4.15. Vectormap for robot navigation. Left: A, B and C are perceptual landmarks, stored in the robot's RCE map. Right: Distances d_i and angles α_i between landmarks are stored elsewhere to allow navigation. "Place" denotes a perceptual cluster in the RCE map, "N" gives the reference direction for angles.

4.2.2 Non-navigational Tasks

The examples given so far show how emergent structures — for instance those established by self-organising feature maps — can be used in navigational tasks. The following example shows how they can be applied to other, non-navigational cognitive tasks.

The specific examples this section will look at are action planning [15], i.e. the concatenation of individual, known actions that achieve a global goal that has not been achieved before by the agent, and novelty detection [23, 10], i.e. the detection of novel sensory perceptions, without using pre-installed knowledge or representations of novelty.

Unsupervised Planning of Action Sequences

The objective of the experiments presented in this section was for the mobile robot *Radix* (Fig. 4.3) to determine a task-achieving sequence of individual actions, each of which had been learned earlier, but which have never been brought into the sequence required to achieve the specific task given to the robot. This is the "unlocking and opening a door" scenario, which involves i) obtaining a key, ii) inserting it in the lock, iii) turning it, iv) turning the door handle, and v) opening the door. All of these actions may have been encountered individually beforehand, but not in one complete, task-achieving sequence.

In the specific experiments with *Radix*, the robot initially explored its environment randomly. During this exploration, the robot used a self-organising feature map (SOFM) to develop a cognitive mapping between physical sensory perception and an internal representation of this sensor space (the SOFM). During exploration, the robot also recorded which physical motion linked individual perceptual clusters on the SOFM.

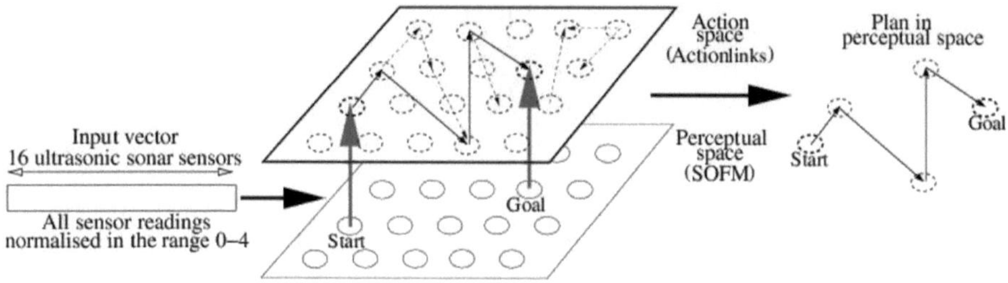

Fig. 4.16. Architecture used for autonomous, unsupervised action planning. A self-organising feature map (SOFM) "contains" a cognitive map of perceptual space, an additional layer (Action Layer) records motion of the robot between perceptual clusters.

Once this representation of perception-action-perception triples was acquired, it could be used to determine novel sequences of actions that would take the robot from some arbitrary, but previously encountered starting location to a user-specified goal location in perceptual space: the robot was first taken to the goal location, so that the sensory perception at the goal location was known to the action selection mechanism. The robot was then taken away from the goal location to some arbitrary start location, and given the task to move to the goal.

Mechanism

To determine which motor actions will take the robot closer to the goal from its current position, an imaginary marker is placed at the goal location's position on the SOFM, and propagated at constant speed along all action links. If a connection exists at all between goal and start, the marker (Fig. 4.17) will eventually reach the robot's current location in perceptual space, encoded on the SOFM, thereby indicating a complete path of actions that will take the robot from the current position to the goal. In our experiments *Radix* would then execute that motor action that would take it closer to the goal (in perceptual space), then repeat the entire action selection mechanism, until the goal was reached. It is important to realise that all this reasoning happens in perceptual space, not physical space: only sensory perceptions are encoded in the robot's cognitive map, the robot has no notion of Cartesian space in these experiments [16]!

This mechanism of determining links between locations on a "map" was first introduced by Alan Turing [22], and in Artificial Intelligence is sometimes referred to as Reaction-Diffusion Dynamics [19] (i.e. a marker diffuses along known links across the cognitive map, and produces a reaction). Figure 4.18 shows a scenario, in which the mechanism depicted in Fig. 4.16 was successfully used by the robot to move from the start to the goal.

Novelty Detection

The final non-navigational example of applications of "cognitive maps" in robotics was inspired by an inspection scenario, in which the robot has to detect

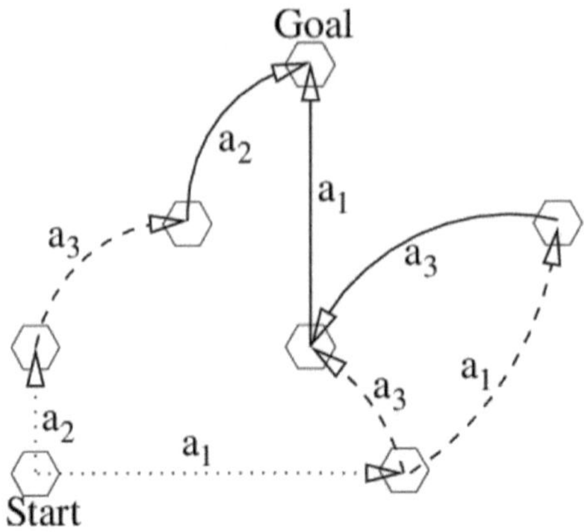

Fig. 4.17. Action selection using a cognitive map. Hexagons indicate perceptual landmarks (the cognitive map of the robot's perceptual space), a_i denote motor actions. To determine a link from the goal perception to the start perception, a marker diffuses from the goal location at equal pace along all perception-action links, until it reaches the current position in perceptual space, thus indicating a complete path between current and goal location. Action sequences $[a_1, a_3, a_1]$ and $[a_2, a_3, a_2]$ would both lead to the goal in this example.

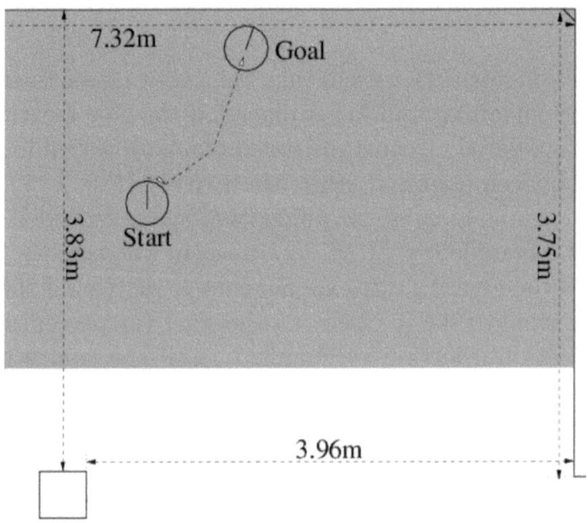

Fig. 4.18. Autonomous action planning: the robot's task was to determine a task-achieving sequence of actions that would take it from the "Start" to the "Goal". Three different actions are needed to reach the goal, resulting in movement along three straight segments.

a sensory stimulus that had not been encountered before. The scenario that inspired this research is the inspection of underground pipes. Such inspection is very costly and, more importantly, the manual evaluation of video coverage

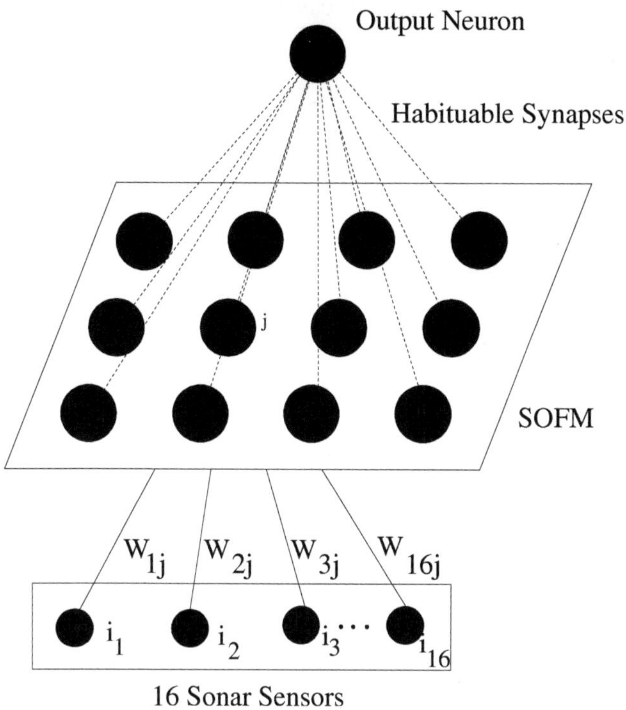

Fig. 4.19. The self-organising novelty filter. Input stimuli (sonar sensor perceptions) are clustered, without supervision, by the SOFM, providing a cognitive mapping of sensory perceptions. Links from SOFM neurons to the one output neuron habituate over time, so that infrequent input signals will generate a stronger response than frequent ones.

of underground pipes is so boring that human operators tend to miss genuine faults.

Detecting abnormality, therefore, is a task that promises high returns of investment. The difficulty of this task lies in the fact that it is *a priori* unknown what the autonomous robot should be looking for. "Novel" items therefore cannot be defined beforehand, and standard template-matching methods cannot be used.

Instead of detecting *abnormality*, therefore, we used an acquired model of *normality* to detect novelty in the robot's perception. Figure 4.19 shows the mechanism.

A self-organising structure — either a self-organising feature map (SOFM) or another self-organising network, the Grow-When-Required network [7] — clusters the robot's sensory perception topologically, without external supervision. The network is the cognitive map in this case, containing an internal representation of all the robot's perceptual experiences. Artificial neurons representing common stimuli will therefore fire frequently, those neurons responding to rare

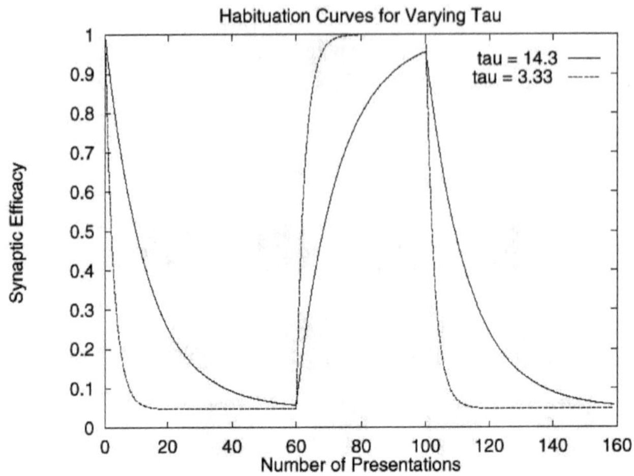

Fig. 4.20. Habituation: synaptic efficacy drops as the number of presentations of a stimulus increases

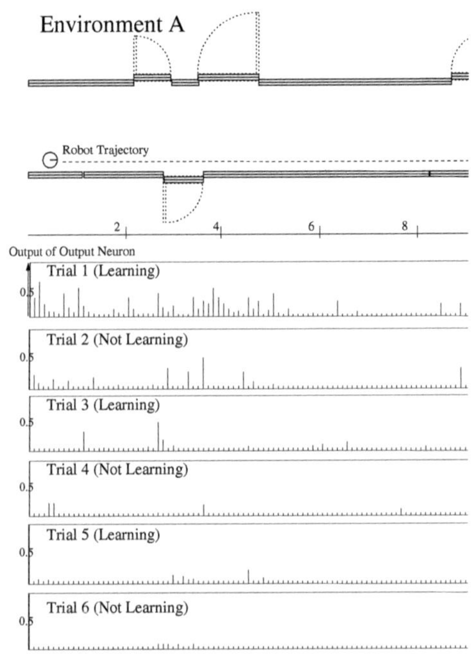

Fig. 4.21. Acquiring a model of normality. *FortyTwo* explores the test environment A, and after three exploration runs has habituated to all stimuli present in that environment.

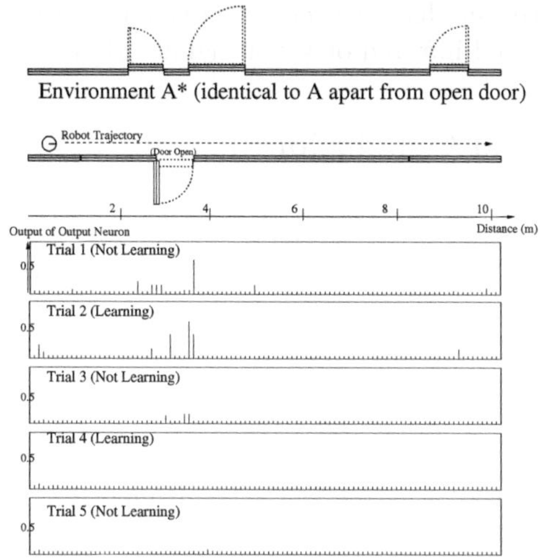

Fig. 4.22. Detecting novelty. The open door — not present during model acquisition (Fig. 4.21) — is clearly detected.

sensory perceptions will fire seldomly. This fact can be exploited for novelty detection, as follows.

Each artificial neuron of the SOFM is connected to one output neuron via a habituable "synapse". With each activation of a particular SOFM neuron, the link between this neuron and the output neuron is weakened, according to the graph shown in Fig. 4.20. This process is known as habituation.

In other words, the more common a sensory perception, the less activation the output neuron receives. The output neuron thus serves as a novelty detector, and can be used to differentiate between common and uncommon sensory perceptions.

In initial experiments [6] we used the sonar sensor perception of *FortyTwo* (Fig. 4.8), as the robot travelled along a corridor, to detect abnormalities in the robot's environment. Figure 4.21 (left) shows the output of the novelty filter during successive traversals of the corridor during the cognitive map acquisition phase. As can be seen, the robot habituated fully to all perceptions after three traversals of the corridor.

After the representation of normality had been acquired, we introduced a novel perception by opening a door (Fig. 4.22). As can be seen, the novelty filter immediately highlights the area around the door. If the robot is presented with an open door repeatedly, it will habituate to that perception, too (Fig. 4.22).

One concluding remark: in the novelty-detection scheme presented here common and rare perceptions are detected at the cognitive map level, independent from the input stimuli used. Input stimuli of any kind are clustered by the SOFM or GWR network, the cognitive map, and classified there as novel or common.

The example discussed in this chapter detected novel stimuli in sonar sensor signals, but we have used infrared or vision signals [23] as well.

4.3 Summary and Conclusion

4.3.1 Summary

Reference [21] introduced the term "cognitive map" to mean a "broad and comprehensive" map, rather than a "narrow, strip-like" one, used for the purpose of navigation. We argue that the term "map" can be expanded to include not only maps for the purpose of navigation, but any bijection (one-to-one mapping) between the physical world and a representation inside the agent. "Broad and comprehensive" we take to mean "usable in circumstances other than those under which the map was acquired". A second important aspect to the term "cognitive map" is that the map is *acquired* through the agent's interaction with its environment, not pre-installed.

This chapter illustrates these points with robotics examples concerning navigational tasks — location identification, self-localisation and route learning — and non-navigational ones — planning of action sequences and detection of novelty.

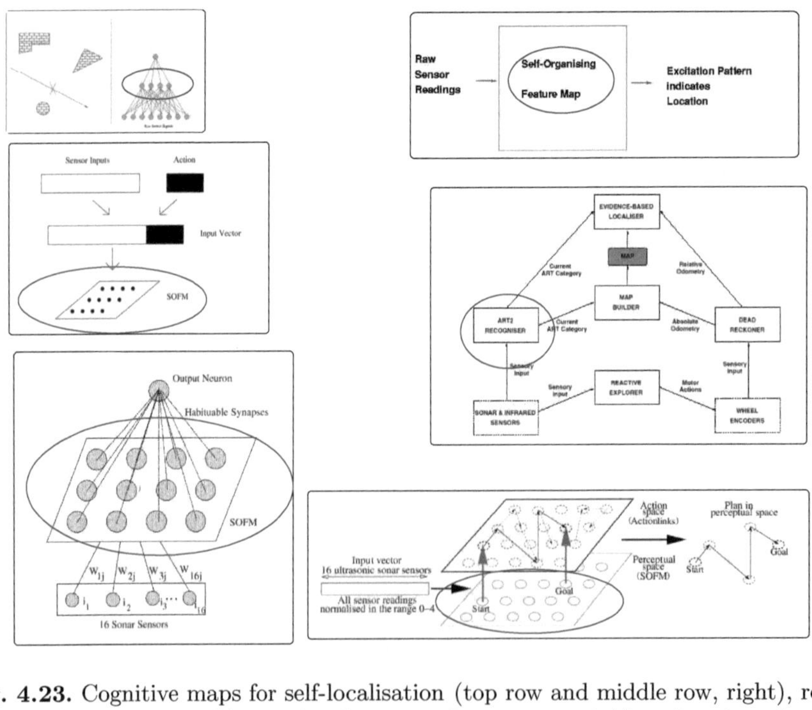

Fig. 4.23. Cognitive maps for self-localisation (top row and middle row, right), route learning (middle row, left), novelty detection (bottom row, left) and action planning (bottom row, right). Cognitive maps are circled.

4.3.2 Conclusion

This paper presents a number of examples of how such acquired cognitive maps can be used in mobile robotics. The examples in Sect. 4.2.1 all relate to Tolman's original application, navigation. Self-localisation and route learning are the specific examples given.

But cognitive maps, emerging through agent-environment interaction, can be used for other applications, too. Action planning (Sect. 4.2.2) and novelty detection (Sect. 4.2.2) are given in this paper as examples; others are conceivable.

There are commonalities to all approaches:

- A core component of the mechanism used is a mapping between physical space (the "world") and the agent's perceptual space, the cognitive map,
- the map is not pre-installed, but *acquired* through agent-environment interaction, and
- the map can be used by the agent to perform more than one narrowly defined task.

Figure 4.23 once more shows all examples discussed in this chapter, with the cognitive map circled in each case.

The observation is that *one* mechanism, one of self-organisation, unsupervised learning and one showing emergent functionality, is employed to establish the robot's cognitive map, be it for localisation, path planning, action planning or the detection of novelty. This result is surprising, in that classical approaches to robot control are usually algorithms dedicated to one specific application. The convergence demonstrated here is one step away from such "insular" solutions.

References

1. Burgess, N., Donnett, J., Jeffery, K., O'Keefe, J.: Robotic and neuronal simulation of the hippocampus and rat navigation. Philosophical Transactions of the Royal Society B 352 (1997)
2. Carpenter, G., Grossberg, S.: Adaptive Resonance Theory. In: Arbib, M. (ed.) The Handbook of Brain Theory and Neural Networks, MIT Press, Cambridge (1995)
3. Duckett, T., Nehmzow, U.: Quantitative analysis of mobile robot localisation systems. In: Proc. Towards Intelligent Mobile Robots 97, Manchester (1997)
4. Duckett, T., Nehmzow, U.: Mobile Robot Self-Localization and Measurement of Performance in Middle Scale Environments. J. Robotics and Autonomous Systems 1-2(24) (1998)
5. Kohonen, T.: Self-Organization and Associative Memory. Springer, Heidelberg (1984)
6. Marsland, S., Nehmzow, U., Shapiro, J.: On-line Novelty Detection for Autonomous Mobile Robots. J. Robotics and Autonomous Systems 51, 191–206 (2005)
7. Marsland, S., Shapiro, J., Nehmzow, U.: A Self-organising Network that Grows When Required. Neural Networks 15(8-9), 1041–1058 (2002)
8. Nehmzow, U., Smithers, T.: Mapbuilding using Self-Organising Networks in Really Useful Robots. In: Meyer, J.A., Wilson, S. (eds.) From Animals to Animats, Proc. 1st conference on Simulation of Adaptive Behaviour, MIT Press, Cambridge (1991)

9. Nehmzow, U.: Mobile Robotics: A Practical Introduction, 2nd edn. Springer, Heidelberg (2003)
10. Nehmzow, U., Neto, H.V.: Novelty-Based Visual Inspection Using Mobile Robots. In: TAROS 2004. Proc. "Towards Autonomous Robotic Systems", Essex (2004)
11. Nehmzow, U.: Scientific Methods in Mobile Robotics - Quantitative Analysis of Agent Behaviour. Springer, Heidelberg (2006)
12. Owen, C., Nehmzow, U.: Route Learning in Mobile Robots through Self-Organisation. In: Proc. Eurobot 1996, IEEE Computer Society, Los Alamitos (1996)
13. Owen, C., Nehmzow, U.: Landmark-Based Navigation for a Mobile Robot. In: Proc. Simulation of Adaptive Behaviour, Zurich, MIT Press, Cambridge (1998)
14. Pick, H.: Cognitive Maps. In: Pick, H., Wilson, R., Keil, F. (eds.) The MIT Encyclopedia of the Cognitive Sciences, MIT Press, Cambridge (1999)
15. Pisokas, J., Nehmzow, U.: Robot Action Selection through Subsymbolic Planning, Department of Computer Science, University of Essex, Technical Report CSM-374 (October 2002)
16. Pisokas, J., Nehmzow, U.: Performance Comparison of Three Subsymbolic Action Planners for Mobile Robots. J. Robotics and Autonomous Systems 51(1) (2005)
17. Press, W., Teukolsky, S., Vetterling, W., Flannery, B.: Numerical recipes in C. Cambridge University Press, Cambridge (1992)
18. Reilly, D.L., Cooper, L.N., Erlbaum, C.: A neural model for category learning. Biological Cybernetics 45, 35–41 (1982)
19. Steels, L.: Steps towards common sense. In: Proc. European Conference on Artificial Intelligence (1988)
20. Thrun, S.: Simultaneous Localization and Mapping. In: Jefferies, M.E., et al. (eds.) Robot and Cognitive Approaches to Spatial Mapping. STAR, vol. 38, pp. 13–42. Springer, Heidelberg (2008)
21. Tolman, E.C.: Cognitive Maps in Rats and Men. Psychological Review 55(4), 189–208 (1948)
22. Turing, A.: The chemical basis for morphogenesis. Phil. Trans. Royal Soc. 37, 129–152 (1952)
23. Neto, H.V., Nehmzow, U.: Visual Novelty Detection for Inspection Tasks using Mobile Robots. In: SBRN. Proc. Brasilian Symposium on Neural Networks, IEEE Computer Society, Los Alamitos (2004)

5

Towards a Generalization of Self-localization

Diedrich Wolter, Christian Freksa, and Longin Jan Latecki

Summary. Self-localization is an important task for humans and autonomous robots as it is the basis for orientation and navigation in a spatial environment and for performing mapping tasks. In robotics, self-localization on the basis of monomodal perceptual information has been investigated intensively. The present chapter looks at self-localization in a more general setting where the reference information may be provided by different types of sensors or by descriptions of locations under a variety of conditions. We introduce some of these conditions and discuss general approaches to identifying locations in perceived environments. Taking into account cognitive considerations, we propose an approach to identify locations on a high, abstract level of representation. The approach combines qualitative and quantitative information to recognize locations described as configurations of shape features. We evaluate this approach in comparison to other approaches in a self-localization task and a generalized localization task based on a schematic map.

5.1 Introduction

Humans and autonomous robots need to know where they are located to successfully orientate themselves, to navigate in a spatial environment, and to perform mapping tasks. The notion of "self-localization" (SL) refers to an agent's procedure of determining where it is located. SL procedures require spatial reference systems, for example a coordinate system or a map.

In autonomous robotics, approaches to SL have been developed that determine a robot's position and orientation (jointly referred to as "pose") based on sensor readings. To accomplish this, the robots relate the sensor information about their environments with their internal knowledge about these environments. Detected correspondences between the two knowledge sources are used to infer the presumed location of the robot.

The approach to SL outlined above even enables wheeled robots to identify locations not visited before. Henceforth, it also enables robots to incrementally build up spatial knowledge about initially unknown environments by determining their location and registering new observations in relation to this location. Coping with a-priori unknown environments is an important ingredient to intelligent autonomous navigation and consequently has been studied intensively.

However, in many situations, agents (humans, robots, software agents) have extensive a priori knowledge about the spatial environment, for example in the form of maps, sketches, natural language descriptions, or (precise or vague) memories of previous observations or descriptions. In such cases it may be desirable

M.E. Jefferies et al. (Eds.): Robot. & Cogn. Approach. to Spat. Map., STAR 38, pp. 105–134, 2008.
springerlink.com

to make use of this knowledge to enable robots to localize themselves more efficiently or in ways that are similar to human self-localization.

For certain tasks the utilization of a priori knowledge is not only desirable but indispensable, for example when a robot is expected to visit places which are described by reference to this a priori knowledge; this may frequently be the case in natural instructions by a human instructor. Furthermore, it may be necessary that a robot specifies its position not in terms of its internal reference system but in terms of a reference system that is available to its human instructor and can be understood by him or by her.

From a technical point of view, this is a different task than conventional SL, as the knowledge employed exhibits different structures and characteristics than conventional sensor readings. In particular, this knowledge may not have an immediate geometric interpretation and it may lack details. Different types of reference systems will require different ways of self-localization; this does not imply, however, that localization will be less precise.

From a more abstract point of view, both tasks — sensor-based and knowledge-based self-localization — can be viewed as belonging to the same class of tasks, as both answer the question of the robot's pose with respect to a given spatial reference system. Therefore we will call this class of tasks "generalized self-localization" (GSL).

In the present chapter we explore several variations of the SL problem and investigate how we can extend existing SL approaches in such a way that they can solve the GSL problem. To this end we propose to employ more abstract forms of knowledge in order to integrate the dissimilitude of potential information sources for common treatment. We illustrate this approach using a specific robot task: spatial orientation by means of schematic maps. Schematic maps (e.g. public transportation maps or emergency evacuation maps) are successfully employed by humans due to their fast and efficient use. We will show how GSL can be used for human-robot communication on the basis of schematic maps.

5.2 The Generalized Localization Task

In robotics, the notion of self-localization has been used in a rather restricted sense: in its most elementary form it is used to denote the task of identifying the robots' locations on the basis of the same type of sensor information that has been retrieved from the location previously. More specifically, self-localization in so-called view-based robot navigation (see for example [13]) is performed with the *same* sensors and the *same* spatial resolution by an agent with more or less the *same* perspective as before. Thus, the robot can use characteristic features to identify a specific place in a finite set of places.

However, we may have situations in which a robot has to localize itself from perspectives it never has encountered before under comparable conditions, possibly not even with the same sensors, or even never encountered before at all. A human, another robot, or a data base may have provided information about the environment; this information is now to be used by the robot for its self-localization task. To cope with such situations, we will adopt a more general notion of self-localization.

5.2.1 Generalizing the Self-localization Task

Starting with the aforementioned case of SL, an agent recognizes a location from an observation obtained with the same sensors, with the same spatial resolution, and from the same perspective — a simple task provided the agent receives the same percept as obtained in a reference cognition event. In realistic situations, however, the sameness of all these parameters is never given — let alone guaranteed; therefore it is not a trivial task to solve this self-localization problem. Successful approaches must deal with the unavoidable deviations of parameter values. However, this problem can be solved with little effort purely on the level of sensor data. We refer to this type of SL (not varying any parameters) as the *elementary case* of SL. It is utilized in view-based robot navigation (for an example, see [13]).

Which abilities does an agent need to recognize places under even less favorable conditions: from different locations with different spatial orientation, with different sensors, at different sensor resolution, or under different environmental conditions? In the following, we will consider incremental abilities required with respect to the restricted case of self-localization. We will consider three strands of generalizing the SL problem: (1) different perspective; (2) different spatial resolution; and (3) different kinds of sensors. Fig. 5.1 presents an overview of these generalization strands and indicates specific classes of SL tasks.

Different Perspective

Variations of view poses can be considered a first step of generalizing SL. Robots identify their location using sensor readings taken at different view poses. Perceiving objects from varying locations, their appearance or visibility can change — for example, due to occlusion; such changes are reflected in the generalization axis 'perspective'. Spatial reasoning allows inferring how a physical phenomenon observed from one perspective appears when observed from another perspective.

The elementary case of SL permits place recognition in an agent-centered reference system. To recognize locations independently of the agent's perspective we must transform sensory information into a location-independent, absolute reference system, e.g. a geographic map. Transformation from agent-centered observations to an absolute map is an abstraction process that abstracts from individual sensor readings and mediates between differences in multiple observations of the same physical phenomenon. This step is particularly easy for sensor data obtained from range sensors. It is still an unsolved problem if relying on camera images, though.

Using elementary spatial reasoning on an absolute spatial representation allows us to partially infer the expected view caused by a different pose. Most approaches to robot navigation or robot mapping utilize some kind of absolute representation, typically a coordinate-based map (see, e.g. [30, 41]). In this representation, perspective generalization can easily be handled.

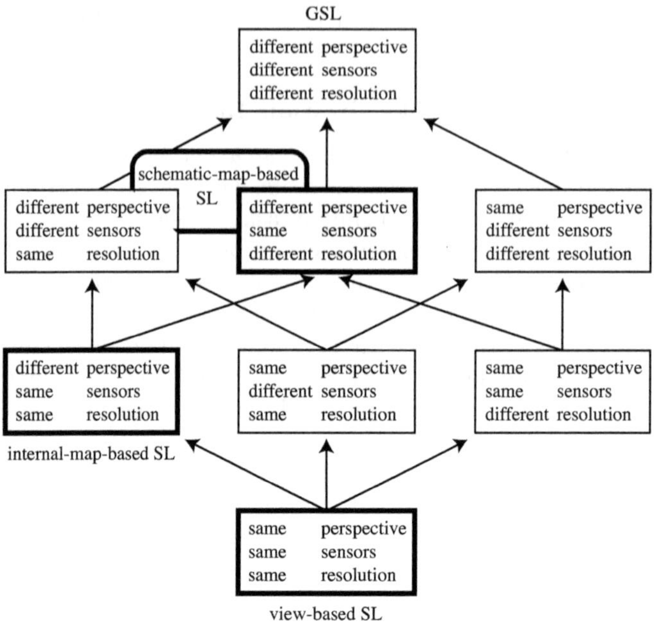

Fig. 5.1. Generalization in self-localization: The elementary case in SL (sameness in all apsects) is the localization problem faced in view-based robot navigation (bottom). Three strands of generalizing the elementary situation are depicted: perspective (left); sensors (center); and resolution (right). SL using an absolute spatial representation can cope with varying perspectives and handles SL based on a robot's internal map. The approach presented in this article uses a schematic map as reference; it is depicted at the generalization strand from map-based localization to GSL (upper left).

Different Kinds of Sensors and Knowledge Sources

To describe different kinds of sensors and knowledge sources with a single label, we employ the notion of abstract sensor readings. For example, a map can provide abstract sensor readings by retrieving sensor information available at a given pose. An agent that has to recognize a place through perception with a different kind of sensor than initially will not be able to successfully match the corresponding percepts, in general; rather it will require a representation that relates different perceptions in terms of common traits.

For example, the boundary of a physical object may be perceived visually in terms of a transition between different brightness or hue values, through tactile perception in terms of a transition of physical resistance values, and through distance sensors in terms of an abrupt transition between distance values, while the object surface appearance may exhibit differentiated readings on some sensors and stable readings on others. Therefore, object boundaries are suitable concepts of a spatial scene that support multimodal recognition while object surfaces may be less suitable. Especially object boundaries which are boundaries to passable space are of importance to navigation as they constrain possible movements. We find these boundaries registered in maps, including schematic maps; boundaries are easily accessible to a robot utilizing range sensors.

To enable multimodal recognition on the basis of different abstract sensor readings, we may develop a representation that features the notion of an object boundary while it abstracts from object surfaces, for example. Such a representation also can be used to relate sensory information to conceptual knowledge that has been conveyed through object descriptions in terms of natural language or by graphical means. In other words, to make cross-modal use of a variety of knowledge sources we can abstract from the specifics of individual modalities and identify modality-independent features or concepts. We then must provide mappings between the modality-specific percepts and those concepts.

Different Spatial Sensor Resolution

Even if we stay within the same modality, we will get problems with matching abstract sensor readings from a given place if the sensors provide spatial data at different levels of spatial resolution, as they will identify different sets of sensory features. A suitable abstraction from low-level perceptual features also will be helpful in this case: a resolution-adaptive representation will enable the comparison of sensor data obtained at different levels of spatial resolution.

We point out that a change of resolution (granularity) does not necessarily happen uniformly, as in the case of smoothening filter application. Rather, coarsening can occur selectively like in schematization processes (see [4]). Here information characteristic for a spatial configuration or relevant to a considered task may remain on a high level of detail whereas irrelevant information may be discarded completely. To interrelate different levels of granularity it is advantageous to define a notion of saliency for features; only salient features remain represented when the resolution is reduced. Moreover, it is essential to estimate whether a feature at hand will be represented on a specific level of granularity or not.

5.2.2 High-Level Knowledge for GSL

In the elementary case of SL, sensory information obtained by independently sensing the same physical phenomenon can be correlated in a rather straightforward manner. Moving along one of the three strands of generalization, adequately abstracted information and abstraction processes are required to enable correlation of sensor readings, i.e. matching, by focusing on essential features. When the perspective of observation changes, sensory information is abstracted to yield view independent images by employing an absolute representation, e.g. a map. To mediate between different abstract sensor readings, information can be abstracted to cross-modal concepts. Features present on different levels of resolution can be related using an abstraction process to reduce spatial resolution or handling information in a granularity-adaptive or granularity-insensitive manner. Qualitative spatial representations provide an anchor to handle varying

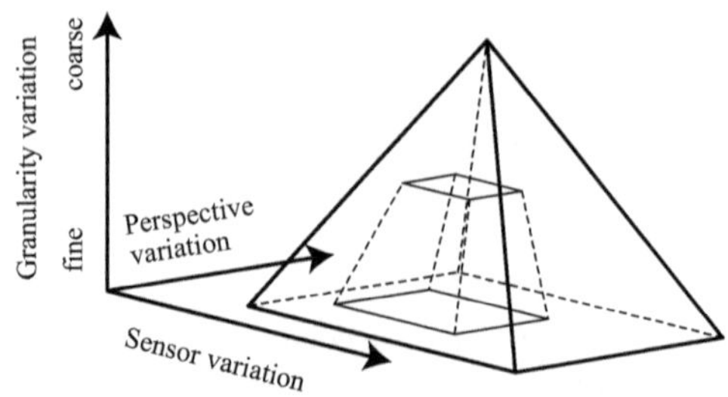

Fig. 5.2. The pyramid of generalized localization problems from the perspective of spatial information processing. Possible variations due to change of perspective or change of source for (abstract) sensor readings decrease when the level of spatial resolution is decreased. A reduction of granularity abstracts from metric details and gives rise to the importance of characteristic qualitative information.

levels of granularity as only the most relevant relations — which are not subject to change of resolution — are made explicit. We point out that in all strands of generalization abstraction is the key to master generalized localization tasks.

SL is primarily a problem of spatial information processing and we are especially interested in understanding spatial abstraction. Reconsidering the generalization strands in SL from the point of spatial abstraction, it can be organized as the pyramid presented in Fig. 5.2. On the finest level of granularity, fine-grained metric information is available; sensor and perspective variations cover a wide range. On a coarser level of granularity the multitude of possible variations decreases as the expressiveness in spatial information is reduced.

Coarse spatial information is available through language or through rough or schematic overview maps; it is typically qualitative information that classifies spatial information into distinct categories [23]. Notably, qualitative representations are not restricted to representations of coarse knowledge. Qualitative information can also be retrieved from fine-grained representations and, for example, can be exploited in reasoning process. We conclude that it is advantageous to explicitly address abstract qualitative information when interrelating spatial information on significantly varying levels of granularity or to bridge cross-modal variations.

5.2.3 Localization Using Schematic Maps

In this paper, we use the term "map" to denote a representation that relates landmarks and features to spatial locations. This subsumes internal spatial representations of a robot and external maps, e.g. floor plans. A map typically preserves spatial information metrically (on a certain level of spatial granularity). If it abstracts from metric properties and represents qualitative aspects of spatial information (specifically topological and ordering information), we refer to it as "schematic map". Schematic maps as characterized by Klippel et al. [23] abstract from information irrelevant to a specific task considered. For example, a schematic floor plan giving directions to visitors typically abstracts from furniture, doors, etc. One may even abstract from the shape of a room. Indeed, a schematic floor plan represents salient boundaries to free space, for example the outline of rooms, corridors, etc. Schematic maps can however be designed for arbitrary environments featuring a great variety of objects. Therefore, approaches to schematic map interpretation by means of recognizing specific objects or specific spatial properties are restricted to specific environments. We are interested in the fundamental principles of recognizing spatial environments and, henceforth, do not aim at recognizing hallways, doors, etc. This allows responding to arbitrary environments and maps.

By considering abstract sensor readings retrieved from a schematic floor plan in relation to a robot capable of scanning the borderline of free space (e.g. by means of a range sensor) we characterize the localization task using a schematic map. First, as with any absolute representation, schematic maps provide information free of a specific perspective. Second, the granularity of schematic maps is coarser than of sensor information. However, not all differences between metric maps and schematic maps are due to reduced resolution. When transforming a metric map into a schematic map, it undergoes a selection process that only retains salient, characteristic information (compare [4]). For a robot not capable of distinguishing different kinds of obstacles the relation between sensor information and schematic map information presents a small modality change, as the set of objects that can be perceived by the robot differs from the set of objects registered in the schematic map. Therefore, we classify the localization using a schematic map on the generalization strand different perspective, different resolution, same sensor towards GSL (see Fig. 5.1).

5.3 Localization Strategies

In this section we will analyze approaches to SL that employ some kind of map representation. We will evaluate their applicability to more generalized localization tasks. First, the SL problem is decomposed into distinct subtasks and aspects; this decomposition provides a classification scheme for individual approaches to SL.

5.3.1 Computing the Pose

The challenge in SL is to find a sensible transformation from an agent-centered perspective to a specific reference system, typically an absolute one. Therefore, SL primarily is a question of spatial reasoning. On a closer look, additional aspects emerge, though.

A robot can localize itself by determining the correspondence between its sensory input and the map. In other words, we compute the pose which — according to its map — explains the sensory input. The problem of determining this correspondence is termed the *correspondence problem* or the task of *data association*; a good solution to the correspondence problem is among the hardest problems in mobile robot navigation [41, 21]. Once a correspondence between perceived features in their local frame of reference and map features in the absolute frame of reference is established, simple trigonometric computation yields the robot's absolute pose. Important criteria of the applicability of specific approaches are the robot's perceptual features. The ability to uniquely identify landmarks, for example, would make the correspondence problem trivial. Industrial applications sometimes use unique artificial tags to simplify recognition in a robot's working environment [20, 16]. In the present chapter, however, we will consider unaltered environments, though.

To approach the correspondence problem if a — possibly vague — pose estimate is available, matching algorithms are employed. These algorithms calculate the most likely correspondence between the sensory input and the expected perception on the basis of the pose estimate and the internal map. On the basis of this correspondence they infer the expected percept. In the context of statistical frameworks for robot localization the role of matching algorithms is providing a solid perceptual model to infer the probability of each individual pose hypothesis (compare [40]). The more robust a correspondence can be determined, even in absence of precise pose estimates, the fewer hypotheses need to be considered; this improves efficiency. Differences between true and estimated robot perspective result in differences between actual and expected percept. Robustness of matching algorithms is important, especially in the context of GSL. Here, variations may also appear due to shifts of modality or granularity.

To sum up, the key challenge in map-based localization is to find a good solution to the correspondence problem. There are four essential factors that shape approaches to localization:

Feature representation: Which features are made explicit in the map? (sensor reflection points, extracted feature points, ...)

Representation of configurations: Which spatial relations are made explicit in the map? (qualitative knowledge, metric data, ...)

Spatial reasoning/configuration matching: Which matching algorithm is used? (Iterative Closest Point, shape matching, ...)

Temporal reasoning: How is history information handled? (stochastic estimators, conceptual neighborhoods, ...)

In the following sections we will discuss these factors in some detail.

5.3.2 Feature Representation

Sensor data is interpreted in terms of environmental features. Features can range from hardly interpreted sensor patterns to complex objects and their properties. The manifold of features possible can be classified into spatial properties (e.g. position, size, shape) and non-spatial properties (e.g. color, object category). In the following we will focus on spatial features in unprepared environments that can be perceived by robots as well as by humans. Though exploitation of non-spatial properties would support the recognition processes and would complement spatial information, intelligent processing of spatial information is one indispensable ingredient to successful localization.

The choice of features to be used for localization depends on the type of sensors; applicability to GSL adds further requirements. In external representations such as schematic maps a coarse level of granularity entails a complete lack of unimportant features whereas other features may be schematized, i.e. they are coarsely represented. To successfully match information on different levels of granularity, means for determining the saliency of a feature and means for shifting the level of granularity are required. Determination of saliency allows to estimate whether a feature at hand will be represented on a specific level of granularity or not; means of shifting granularity levels are required to identify correspondences. Proceeding from simple to more complex features we examine these properties as well as the contribution of a specific feature to robust localization.

Raw Sensor Patterns

A prominent approach relying on matching sensor data is the 'view-based approach'. It matches raw sensor images and does not extract features from sensory input. Typically, sensor snapshots are obtained and stored for different discrete view points. For example, Franz et al. [13] handle linear panoramic camera images taken at specific locations in the environment. Similar to the view-based map representation, the lowest level of Kuipers' spatial semantic hierarchy ([25, 26]) associates the robot's action patterns at decision points with the corresponding locations.

Uninterpreted data does not allow for granularity shifts and cannot be integrated with external information. Furthermore, uninterpreted data provides no information about the local spatial configuration; data can only related to the view point.

Landmarks

Landmarks are objects in space that are easy to identify; for localization purposes, they can be represented by their position. Landmarks are typical environmental features for localization in human navigation (see e.g. [9]). Landmarks are well-researched in the context of human navigation, but the detection of landmarks that are commonly used in human communication (e.g. "the gas station") is not yet possible in computer implementations. Landmarks that can be used in robotics still must be comparatively simple. For example, Forsman [11] developed a tree detection approach on the basis of range data; it was tailored to an outdoor park scenario. Similarly, corners detected in the environment can be used as landmarks [1]. In human-robot communication it is desirable to identify entities in the environment that provides both species a spatial reference for their interaction.

Specific landmark identification approaches restrict applicability to environments that contain those landmarks. It is however possible to derive additional information from landmarks which can be used, for example, to estimate their appearance in a representation at a specific level of granularity. The utilization of landmarks in human-robot interaction is still a challenge; its solution depends on sophisticated object recognition which is still beyond reach.

Free Space

The boundary of free space is of special importance to robots and humans since it limits the accessible environment and it constrains possible actions. Consequently, many approaches represent free space, its boundary, or geometric features derived from it. Information about free space also can be obtained from maps that are used by humans. Sensors like laser range finders or sonars measure the boundary of free space directly. We will now review the most important features for representing boundaries of free space.

Cell Occupancy

In cell occupancy representations, spatial cells are classified as occupied or free. The spatial domain is partitioned into square-shaped cells of fixed size (e.g., 10cm x 10cm). The typical map representation employed is the so-called occupancy grid [35]. This technique is particularly popular when using range sensors like laser range finders (LRF); sensor output can be used directly without processing (other than noise filtering). A clear advantage is the universality of the approach, as it can be used in arbitrary environments. [2]. The simplicity entails severe limitations, though. Occupancy grids are basically bitmap images that, if related to externally provided maps, would require sophisticated image processing techniques for matching. As of today, communication on the basis of occupancy grids is limited to strongly constrained settings like multi-robot mapping involving identical robots and known start poses of all robots (see e.g. [24]).

Free Space Boundary

Reflection points measured by a range finder represent the boundary of free space. To capture a wider context than single points and to reduce the amount of data, points can be grouped to geometric primitives. For indoor environments grouping into line segments is especially popular (e.g. [33, 37, 8, 10]). In connection with communication tasks it may be desirable to identify salient boundary configurations. A starting-point for defining saliency is given by considering the size of configurations, e.g. the length of a line.

Existing grouping approaches are limited to environments whose boundaries present mostly straight lines. To achieve more universal applicability, Wolter & Latecki [46, 47] propose to use polygonal lines to approximate arbitrarily shaped boundaries. In this way, the universality of point-based representations and the compactness of abstract geometric features can be retained. Feature saliency based on shape complexity and an approach to schematization complex shapes have been proposed by Barkowsky et al. [4].

Routes

A prominent geometric feature derived from free space is the Generalized Voronoi diagram (GVD) [31]. The GVD represents the medial axis of free space ("skeleton"), the set of all points equally and maximally apart from the nearest boundaries. Each point of the GVD is the center of a circle inscribed in the free space that touches at least two points of obstacle boundaries. A graph, the so-called Generalized Voronoi Graph (GVG), is then derived from the GVD; meet points and end points of the GVD constitute the nodes in the GVG. Nodes belonging to a GVG are identified by their degree. GVD point touches the boundary). Roughly speaking, the degree corresponds to the number of Voronoi paths emanating from a given point on the GVD. GVGs offer abstract and compact means for representation [39]. Furthermore, routes that follow the GVD are maximally safe as they maintain maximum distance to obstacles. However, the graph structure of GVGs is susceptible to noise in input data; the problem of robust recognition on the basis of GVGs has not yet been solved. It is not yet possible to handle the absence of environmental features in external maps when matching them to perceived information, as the graph structure changes fundamentally when objects disappear. The applicability of GVGs to place recognition depends on improvements in handling multiple levels of granularity and in skeleton-based recognition. These topics are currently under investigation (see [45]).

5.3.3 Representation of Configurations

A configuration describes the spatial arrangement of features that can be perceived in the environment. Frequently coordinate systems are used to represent the position of objects, but qualitative spatial relations describing relative positions (e.g., "A is north of B") or topology information may also be used.

Qualitative Representations

Qualitative representations employ a finite, typically small set of relations to model spatial information. Relations usually describe by means of relative information as obtained by comparison; for example, "north of" and "south of" can serve as qualitative relations acquired by comparing the geographic location of two objects.

Some authors confide the set of potential relations to a single connectivity relation, topology (among others, see [7, 26, 49]). Topological information captures connectivity information of distinctive places and can be represented by an (attributed) graph structure. For example, Yeap & Jefferies [49] represent connectivity of local maps. Graph labeling is required to enable agents to identify individual edges that meet in a single node of the graph. Yeap & Jefferies associate edges with exits of the local maps. Kuipers [25] labels directed edges by robot commands. The execution of an action associated with an edge takes the robot from one node to the other. In contrast, Franz et al. [13] use directional information to label edges. Hereby, directions are determined by the relative positions of the two nodes connected. The kind of information used to attribute the graph structure influences the matching process in important ways so that general statements about the properties of relational representations cannot be made.

Ordering information is another important representative of qualitative information in navigation. Schlieder [38], for example, represents the cyclic order of point-like landmarks and Barkowsky et al. [3] utilize cyclic order of extended landmarks in non-cyclic environments. Cyclic order of perceivable objects has also been used to instruct a mobile robot by means of a schematic map [48]. The self-localization approach proposed in the present chapter utilizes cyclic ordering, as well.

Qualitative representations have been claimed to provide adequate means for communicating spatial information; Moratz & Tenbrink [34] utilize projective relations between objects in a robot instruction setting. A robot is instructed to move to a position described by qualitative relations. This task is strongly connected to the localization problem.

Qualitative Calculi

Qualitative calculi extend qualitative relations by introducing means to "calculate with relations", e.g. to infer, if the relations holding between A and B & B and C are known, which relation holds between A and C (relation composition). To relate spatial relations, reasoning — often based on relation composition & constraint propagation — is applied. With respect to correspondence determination, constraint-based reasoning could be exploited to prune the search space. A mapping of objects is only admissible, if it is consistent with qualitative constraints posed on the objects. Thus, qualitative calculi can be employed to introduce hard constraints in correspondence computation (compare [43]). Additionally, conceptual neighborhood structures (see Sect. "Spatio-temporal reasoning", p. 119) have been introduced for qualitative reasoning. Conceptual neighborhoods are in particular valuable to resolve conflicts on the symbolic level by defining an interrelation on the level of relations. However, the application of qualitative reasoning to the correspondence problem, e.g. by means of constraint propagation (see Sect. 5.3.4) has not yet been thoroughly investigated.

Quantitative Representations

Quantitative formalisms describe the world by means of absolute, often fine-grained, uniform scales. Quantitative representations employ no abstraction besides reduction of resolution. Henceforth, sensor data, e.g. distance information sensed by a range finder, can be mapped directly to a quantitative representation. The most prominent form of quantitative representation is coordinate-based geometry; landmark positions, for example, are represented as points in the Euclidean plane. Most approaches in robotics represent positions as coordinates in the absolute frame of reference given by the global map (see Thrun [41] for an overview).

Generally speaking, in quantitative representations all available information is maintained while in qualitative approaches some details may be intentionally discarded. In quantitative approaches all values are treated equally and no aspects are made explicit. This can hamper recognition, as a small example on coordinate-based geometry shows. Consider an agent that observes two landmarks that are located close to one another. By measuring their position the agent determines two similar coordinates that are both subject to measurement errors. By evaluating the measurements and taking into account the error margins, we may not be able to decide which of the landmarks is located on the left and which is located on the right. The agent can, however, observe with certainty which of the two landmarks is left of the other. In a quantitative approach, this knowledge is shadowed by a representation that relates observations to an external scale rather than to one another. Notably, there are situations where we cannot decide in advance which spatial relations will be required later on. In such cases, quantitative approaches are more economical as it is impossible to record all potentially relevant spatial relations in an environment.

5.3.4 Matching

Matching establishes the correspondence between observed features and features represented in the robot's internal map.[1]. A transformation from an agent-centered to the absolute frame of reference can then be computed on the basis of correspondences between observed features and map features. In other words, by establishing the correspondence the agent is localized.

The correspondence problem is challenging in three regards: obtaining a feasible solution, handling uncertainty, and integrating spatio-temporal knowledge. In the following we will review strategies addressing these problems and we will analyze how these strategies meet the requirements of GSL.

Achieving Feasibility in Data Association

Considering a map containing n features and an observation comprising m features, there are

$$\sum_{i=0}^{n} \binom{n}{i} \cdot \binom{m}{i} \cdot i! \tag{5.1}$$

potential correspondences if observed features are not necessarily represented in the map and only correspondences of type 1-to-1 are taken into account. Even this restricted case is infeasibly complex, so additional knowledge must be exploited to reduce the search space and computation time. Confident knowledge, for example, can be exploited in terms of hard constraints restricting the search space. If a pose estimate is available, the *projection filter* [33] can be employed to disregard map features that are estimated to be hidden to the robot. Likewise, observed features are filtered. The pose estimate must be of high quality in order not to disregard features erroneously classified as invisible; this would affect the

[1] In the case of SL by means of feature tracking (e.g. [37, 32]), the agent's previous observation assumes the role of the internal map in map-based SL.

matching result. In many robot applications pose estimates are provided by odometry.

Computational complexity can be further reduced, if distinguishable features are exploited. For example, using extremes in range scans, Lingemann and Hertzberg [32] restrict consideration of correspondences to features of the same type (minimum or maximum). If uncertainty in feature classification is an issue, a feature similarity measure is used as heuristic, i.e. similarity provides a soft constraint, and matching is transformed into a discrete optimization task that assigns the most similar features to one another. In the case of using occupancy as feature, feature similarity considers difference of cell occupancy; this difference is typically represented as probability value [19, 41]. Utilization of complex features allows for fine distinctions in the similarity measure and yields both efficient matching and robustness. In our approach we will argue for shape features that represent the boundary of free space to exploit distinctive shape similarity in the matching procedure.

An alternative approach to increase the efficiency of the matching procedure is to respect the spatial configuration of observed features in relation to the configuration of map features. Admissible mappings from perception to the map preserve the configuration of the features. This can, in principle, be achieved similarly as in constraint propagation (compare [43]), treating relative position of features as constraints. If, for example, feature A is observed north of feature B, then by assigning A to some map feature, the set of candidates for B can be pruned. Unfortunately, uncertainty inherent in map and observation requires a careful selection of hard constraints that model confident knowledge. In our approach, we utilize circular order of visibility as a source of certain information (compare Sect. 5.5.2). Notably, the application of the Mahalanobis distance for pruning potential candidates can be interpreted as an application to constraint propagation. Here, correlations of distances are exploited for gating in a statistical framework (compare [36]). To our knowledge, constraint propagation has not been further utilized in this context and remains an open research issue. Instead, correspondences are sometimes pruned in a successive step; correspondences which entail a transformation from an agent-centered to an absolute frame of reference that deviate significantly from the transformation obtained by averaging the individually obtained transformations can be removed [17].

To avoid costly computation of robust matching, some SL approaches handle the correspondence problem indirectly. They seek to directly determine the robot pose which explains the percepts (e.g. [41, 18, 33, 6, 8]). In this family of approaches, the robot pose is no longer derived from the discrete correspondence problem; instead, it is obtained by a continuous optimization search for an optimal pose. A pose estimate is required as a start value. Within each step, a simple but fast matching procedure relates perceived features already transformed to the absolute frame of reference to map features. Typically, nearest neighbor algorithms are applied to perform the matching [18, 33, 6, 8]. Embedded in an iterative optimization framework, erroneous results of the matching algorithm can be recovered in successive steps. Notably, all optimization algorithms are susceptible to local minima and erroneous matching can further affect the overall performance. Therefore, this family of approaches relies on a high quality pose estimate as start value.

Handling Uncertain Information

Inescapable uncertainty in real-world data inhibits perfectly congruent correspondences. Therefore, the goal must be to find those correspondences which explain the agent's observations *best*. This requires integrating differences of assigned features on the level of feature appearance and configuration. The most successful approaches today use statistical methods to "explain" and correct for these differences (see Thrun [40, 41] for an extensive overview). The role of matching algorithms in a statistical framework is to determine the degree of belief in a specific hypothesis of observation, robot pose, and map appearance [19].

Statistical models also are helpful to handle uncertainty beyond sensor noise, e.g. sporadical errors in feature detection — given that a stochastic distribution can be found to model this phenomenon. Hähnel et al. [19] regard a uniform distribution as sufficient to handle erroneous measurement of individual laser beams by a laser range finder. However, in cross-modality, granularity, or perspective shifts of GSL it is unclear if and how differing appearances for a specific source of abstract sensor readings can be adequately modeled by means of a probability distribution. For example, it appears impractical to model which perceived objects are registered in a schematic map. Therefore, we argue for an additional utilization of qualitative knowledge in GSL which, by advancing to a more abstract representation, allows disregarding deviations on a fine level of granularity.

Spatio-Temporal Reasoning

Spatio-temporal reasoning ties spatial and temporal information together. The possible sequences of physical robot locations and orientations constrain hypotheses about its actual and future pose; therefore spatio-temporal reasoning is an important ingredient to determining the pose of a robot.

In robotics, spatio-temporal reasoning often is tightly coupled with stochastic models to represent uncertainty. Therefore, robot movements are modeled stochastically. SL can then, for example, be approached by means of Markov processes [22] or Monte Carlo methods [42, 12]. This is advantageous in a stochastic framework of SL, but likewise to the aforementioned considerations it is questionable how to express spatio-temporal constraints when information on different levels of granularity needs to be interrelated. In qualitative representations, changes on the level of qualitative information can be represented by discrete *conceptual neighborhood* [15, 14] structures of qualitative spatial relations. Conceptual neighborhoods denote transitions between qualitative relations. Two relations are neighbored, if and only if they can be directly transformed into each other by steady motion. For example, when distinguishing four cardinal directions, "north" and "west" are conceptual neighbors, but "east" and "west" are not. If, for example, a landmark is expected in direction "north" but cannot be observed, this conflict may be resolved most easily by searching in the conceptually neighboring directions "west" or "east". Conceptual neighborhoods allow expressing spatio-temporal constraints in terms of admissible transitions on the qualitative level.

5.3.5 Conclusions

In reviewing the variety of map features, we identify three main categories of information sources used in localization: features represented in absolute maps, sensor patterns in egocentric observations, and approaches completely abstaining from map representations (see Fig. 5.3 for an overview). Completely abstaining from maps either requires employing position sensors like GPS — this requires external information to establish a frame of reference — or to incrementally determine the robot's movement by tracking static features and updating the pose estimate. In principle, any approach to feature tracking can be related to a map-based approach, if considering a map that exclusively represents the last observation. However, we are interested in approaches that allow expressing the robot's pose in an externally supplied reference system, i.e. a schematic map. Thus, approaches handling allocentric map information are most adequate.

There are two principle alternatives in map features to choose from, namely features that represent landmark positions and features that describe free space (either directly, e.g. occupancy grids, it's boundary, e.g. line-based maps, or derived geometric information, e.g. Voronoi diagrams). Considering maps commonly used by humans we conclude that landmarks and representation of free space are both suitable choices, boundary of free space being the more fundamental feature, though. Moreover, landmarks typically used by humans are difficult to identify for a robot. Therefore we suggest anchoring map representations on a representation of free space boundaries.

With regards to representing configurations, we reviewed relational approaches which link features by means of a graph, qualitative approaches describing the relative position of objects, and quantitative approaches employing coordinate systems. An overview of this classification is presented in Fig. 5.4. Quantitative approaches support expressive and precise pose representation. Relational and qualitative approaches, on the other hand, are valuable for handling spatial information on a coarser level of granularity. They abstain from metrics and, by doing so, avoid inescapable differences on the metrical level, e.g., if two configurations on a different level of granularity are related. In localization tasks employing coarse external maps we therefore propose to explicitly integrate qualitative or relational information.

Many matching algorithms employed in robotics align perception and map by means of continuous optimization which searches for the pose value which best aligns perception and map. The correspondence problem is eclipsed. This approach has two major drawbacks. First, they require a pose estimate to start the search. In the case of using an external map no good start estimate may

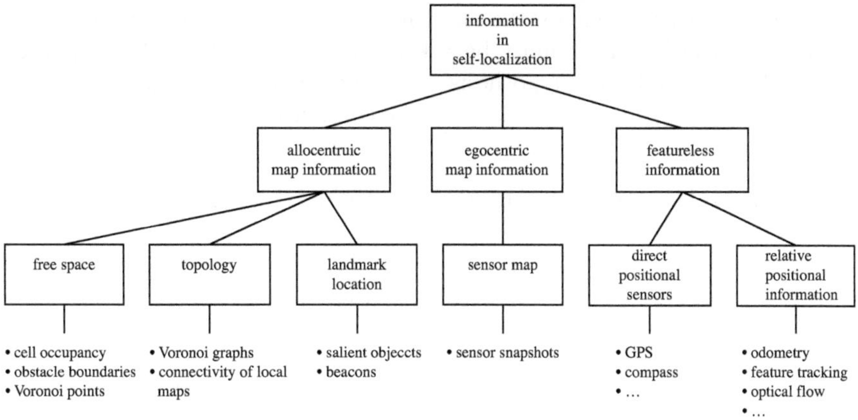

Fig. 5.3. Categories of information employed in localization

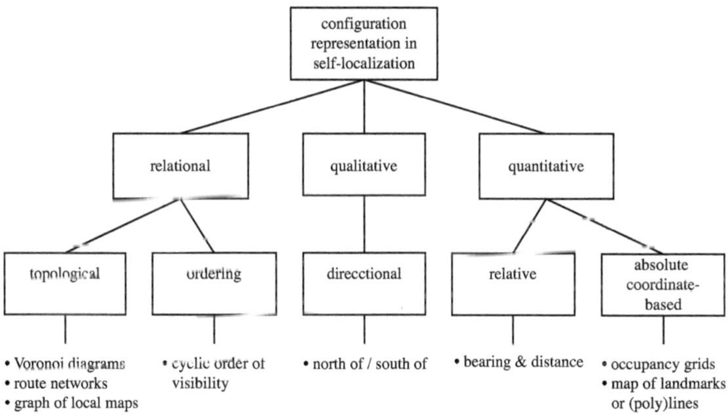

Fig. 5.4. Categorization of representing configurations

be available. Second, optimization algorithms are susceptible to getting stuck in local minima. This can easily happen when the optimal alignment of perception and map is of poor quality, i.e. when features identified in the sensor information are not registered in the map, or vice versa. This may be the case, for example, if we use a schematic map.

We suggest focusing on the correspondence problem in order to find an optimal correspondence between perception and map. The problem may then be formulated as a discrete optimization problem that can be solved analytically, i.e. without the risk of getting stuck in a local minimum. Certain information can be explicitly introduced by means of qualitative information, its exploitation allows for an efficient approach. In summary, by incorporating qualitative constraints on spatio-temporal processes, on one hand, and by relaxing requirements on insignificant distinctions, on the other hand, we can considerably reduce the number of alternatives that must be taken into consideration. This approach resembles knowledge-based hypothesis matching in natural cognitive systems more closely and considerably cuts down the computational complexity.

5.4 Spatial Representation Based on Shape Information

In our approach, the spatial representation utilizes shape features that describe the boundary of free space as basic map entities. Shapes are represented by configurations of polygonal lines. In these configurations, scene features are simultaneously related by qualitative ordering information and by quantitative position information. In the following we refer to this approach as shape-based localization or, shortly, SBL. This section presents details on the construction of its underlying representation.

From the sensor readings of a range sensor we extract shape information as polygonal lines, termed *polylines*. Polylines resemble the discrete structure of sensor data; they allow us to approximate arbitrary contours with arbitrary precision. SBL differs from other approaches to extracting complex features in that it is parameter-free and does not require a noise model of the sensor[2]. All control-values are determined adaptively, but preset values reduce computational cost. In the following, we will present a brief description of the OA algorithm; for an extensive description refer to [46, 47]; intermediate stages of the shape extraction process are shown in Fig. 5.5.

5.4.1 Extracting Shape Features from Range Information

Shape extraction starts by grouping sensor reading points. The maximum distance between neighboring points within a single polyline is controlled by a threshold. Ideally, each polyline represents a single object in view and each object is represented by a single polyline. As different view points and noise can cause different groupings, we need to account for differences in later processing stages. When we match perceived shapes against the map, we allow for re-joining and splitting polylines. The threshold that controls the grouping is chosen to resemble an assumed minimum object distance of 10 cm.

To obtain a compact representation of shapes without loss of important shape information and to cancel the effects of sensor noise, we apply Latecki's & Lakämper's Discrete Curve Evolution method (DCE) [27]. DCE describes a context-sensitive process of evolving polylines by iterative vertex removal. A vertex relevance measure is defined to determine individual vertices' contribution to the shape information; the measure can be computed locally. It is defined for neighboring vertices u, v, w (see Figure 5.5 (d)) as

[2] Veeck & Burgard [44] also suggest to use polygonal lines. Their approach requires an accurately aligned set of scans as starting point of their computation. In contrast, we pursue incremental map construction [47, 46].

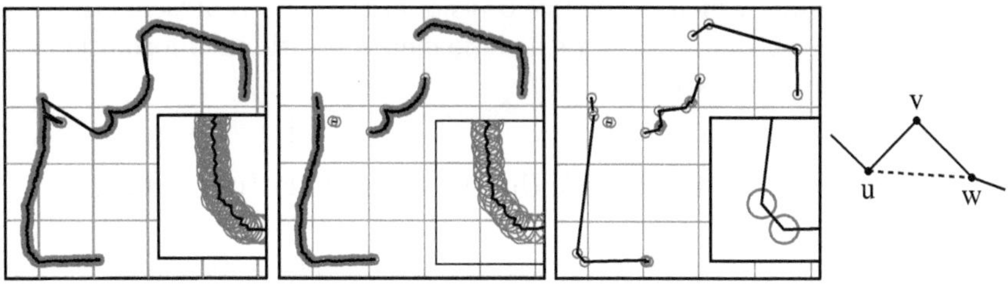

(a) (b) (c) (d)

Fig. 5.5. States in extracting shape features; grid size is 1m x 1m. (a) input data obtained from a range sensor in an indoor environment, (b) grouping, (c) application of DCE, (d) vertex labels in the relevance computation. Framed boxes in (a), (b), and (c) show enlargements.

$$r(u, v, w) = d(u, v) + d(v, w) - d(u, w) \tag{5.2}$$

where d denotes Euclidean distance.

After vertex removal, the relevance measures of neighboring vertices get updated. Hence, DCE is a fast process (complexity $O(n \log n)$ for polylines with n vertices). In practical use, DCE can process laser range scans consisting of 361 measurements in just a few milliseconds. Besides for noise cancellation, DCE can be used in schematization processes to coarsen the granularity level [4], it simplifies the contour but maintains the overall appearance.

DCE selects vertices to be removed in the context of a single polyline. The identification of relevant vertices can be improved by extending the context to sets of corresponding polylines. To this end, we do not stop the evolution of polylines on the basis of a fixed threshold; rather, we terminate the evolution process on the basis of shape similarity (see Sect. 5.5.1). Efficiency is improved by first performing DCE without consideration of shape similarity until an intermediate, fixed stop threshold is reached. Then, DCE is continued under consideration of shape similarity. The evolution process for exemplary polylines obtained from a simulated laser range finder is depicted in Fig. 5.5 (a) – (c).

5.4.2 Representing Configurations

Configurations describe the relative positions of shape features, i.e. polygonal lines. Most importantly, the cyclic order of visibility is represented. As laser range data is ordered cyclically to begin with — i.e. ordered by angle of perception — we simply need to retain the sequence of shape features. We can consider ordering information as reliable information, i.e., there is no uncertainty about the ordering of perceived features. Respecting ordering as a hard constraint in the matching process greatly improves efficiency and robustness. However, if we were to restrict the representation of configurations to cyclic ordering, we would face some limitations. For example, if the map were to contain two objects of identical shape, but only one similar object was found in the sensor data, it would not be possible to determine which of the two objects in the map represents the sensor data. To overcome this limitation, we include metric positional information along with the ordering information.

5.5 Matching Based on Ordered Shape Information

Matching integrates the recognition of individual shape features and the recognition of configurations. We first describe the recognition of polylines which is based on shape similarity. Thereafter, recognition of configurations is described.

5.5.1 Shape Similarity

SBL examines shape similarity to determine potentially corresponding shape features. Shape similarity is modeled by a shape distance measure — the minimum distance of 0 refers to identical shapes.

Shape distance measures play an important role in computer vision, particularly in object recognition. They measure the difference between two shapes and aim at mirroring human intuition. There is a strong connection between object recognition in vision and recognition processes in localization, although the connection between computer vision and robot mapping has not yet been sufficiently exploited according to Thrun [41]. We derived a shape distance measure from state-of-the-art shape matching used in computer vision [29, 27]. To tailor the approach to the domain of range data, some adaptations have been made (for details see [29, 28]).

The idea of measuring the distance between a polyline p and a model q is to disregard irrelevant features that make polylines dissimilar from one and another; in other words, we focus on the subset of vertices that exhibit maximal similarity. Therefore, the measure has been termed *partial optimal similarity* [29]. Here, p corresponds to a polyline extracted from LRF data (which may still contain some noise), whereas q will be a matching candidate extracted from the map. The map is typically derived from multiple observations; we consider map data as absolute reference. The algorithm proceeds as follows: Evolution by means of DCE is continued for polyline p while a simplification of p improves the similarity to q.

The basic similarity measure for comparing simplified ps and qs as detailed in [46, 28] establishes an optimal correspondence of maximal arcs and accumulates differences in relative angular directions. Optimal correspondence of arcs is computed by means of Dynamic Programming — see [27] for details.

5.5.2 Matching Configurations

Provided we have two configurations of features; the task of the matching algorithm is to determine a sensible correspondence relation on the level of polylines. In SBL the currently observed configuration is related to the configuration extracted from the map using the robot's last location as view point. In other words, we do not make use of odometry information to achieve a pose estimate. Due to the distinctiveness of the shape information and the sensibility of the shape distance measure we do not require such pose estimate [46]. Differences of perceived configurations are small on the qualitative level of ordered shape features if the robot has not traveled too far (e.g. less than 1m). These differences can easily be handled by configuration matching.

Matching is formulated here as a discrete optimization problem. We seek to determine the optimal correspondence of shape features. When matching two configurations, changes in the environment, variations of perspective, or noise can cause differences. Constraints and observations that must be considered are as follows:

- Only polylines showing similar shape may correspond.
- The cyclic order of shape features must not be violated. For example, when finding corresponding counterparts for polylines p and q, where p proceeds q, p's counterpart must also proceed q's counterpart.
- An object's visibility can change. Therefore, some polylines may need to be disregarded.
- Correspondences are not necessarily of the type 1-to-1 due to different outcomes of the segmentation process. Instead, 1-to-n, n-to-1, and n-to-m types of correspondence must also be considered.
- Each potential correspondence of two polylines induces an alignment that would adjust the complete shapes involved. We demand that all alignments induced by corresponding polylines are consistent.

We now formulate the discrete minimization problem. Let $S^\star$: polyline $\times$ polyline $\rightarrow \mathbb{R}^+ \cup \{0\}$ denote the shape distance measure described earlier. We will denote configurations, i.e. cyclic ordered vectors of polylines by $\mathbf{P} = (p_1, p_2, \ldots, p_n)$ and $\mathbf{Q}=(q_1, q_2, \ldots, q_m)$ respectively; a sub-vector $(p_i, p_{i+1}, \ldots, p_j)$ will be denoted $P_{i,j}$. $P_{i,i}$ will be abbreviated P_i. Sub-vectors represent a single polyline composed by concatenating a sequence of polylines; they are introduced to correct segmentation differences. Furthermore, let $\sim$ denote the relation of correspondence which pins polylines from two configurations together. Our aim is to compute the *optimal* correspondence relation $\sim$.

The quality of a match $\sim$ is determined as the sum of corresponding polylines' shape distances. To compute the optimal match as an optimization process, a penalty for not finding a polyline's counterpart is introduced; otherwise, the empty correspondence relation would yield 0, the lowest possible value, i.e. the optimal choice. The counterweight used is a penalty function R : polyline $\rightarrow \mathbb{R}^+ \cup \{0\}$ that grows linearly with the polyline's angular size in the field of view. A linearly growing penalty reflects the observation that the shape distance of two polylines that differ only by independent noise grows linearly, too ([46, 47]). This penalty function also addresses feature saliency by consideration of size. Preferring larger features over smaller ones is advantageous in matching a perceived configuration with many details against a schematic map which only presents salient shape features.

The observation that an object is to the left (or right, respectively) of another object is not affected by noise in sensor data. Cyclic order of visibility can therefore be considered certain knowledge. This allows to exploit order as hard constraint and reduce the search space. Observe that the task of determining the optimal correspondence relation of polylines restricted to only correspondences of type 1-to-1 which respect the cyclic order, i.e. $(p_i \sim q_{i'} \wedge p_j \sim q_{j'} \wedge i < j) \rightarrow i' < j'$, is a standard application of Dynamic Programming [5]. Therefore, the unconstrained search space declared in (5.1) is reduced to

$$n \cdot m \tag{5.3}$$

We now formulate the matching which respects the constraints and observations listed above as a minimization problem and we show how it can be solved by an extended Dynamic Programming scheme.

We require that an estimate for the alignment induced by any pair of corresponding polylines exists. This estimate can either be derived from odometry or it can be computed purely based on shape information (see Sect. 5.5.3). Let us now assume that such an estimate, i.e. a translation vector $\mathbf{t}$ and a rotation by Φ exists. We denote the alignment induced by corresponding polylines P and Q by $A(P,Q)$. The difference of the induced alignment $A(P,Q)$ and the estimated alignment is denoted as $\Delta A(P,Q)$. To measure $\Delta A(P,Q)$ our experimental system utilizes

$$D(\mathbf{dt}, d\Phi) = ||dt|| + 10 d\Phi \tag{5.4}$$

Denoting the set of polylines $\{p_i, p_{i'}, \ldots, q_j, q_{j'}, \ldots\}$ not belonging to any correspondence by $\overline{PQ}$, determination of the optimal correspondence relation $\sim^\star$ is formulated as follows:

$$\sim^\star = \operatorname{argmin} \sum_{(\mathbf{P}_{i,j}, \mathbf{Q}_{i',j'}) \in \sim} \left(\overbrace{S(\mathbf{P}_{i,j}, \mathbf{Q}_{i',j'})}^{\text{shape distance}} + \overbrace{\Delta A(\mathbf{P}_{i,j}, \mathbf{Q}_{i',j'})}^{\text{robot pose consistency}} \right) + \overbrace{\sum_{r \in \overline{PQ}} R(r)}^{\text{penalty}} \tag{5.5}$$

To solve the equation, the Dynamic Programming scheme is extended. To enable detection of correspondences of types 1-to-n, n-to-1, or n-to-m, we introduce an updating step that reconsiders the correspondence determined in the previous step. This overcomes the prefix requirement of Dynamic Programming: Suppose a polyline p shall be matched against two polylines q_1, q_2 which are created by splitting p. In classical DP, the result of comparing (the prefix) q_1 to p cannot be altered in subsequent computation. Thus, if p and q_1 are significantly dissimilar, q_1 is disregarded once and for all. Consequently, q_2 would not be matched either. In our extension to DP, we reconsider q_1 when comparing q_2 and p; this gives us the correct correspondence of p and the concatenation of q_1 and q_2.

5.5.3 Shape Complexity and Correspondence Quality

Matching correlates two sets of shape features which are expected to have a correspondence relation. In the case of relating significantly different configurations (e.g. relating robot perception with schematic map information or perceptions from significantly different view points) metric information about position of objects is of little help; yet if considered, different metric information can even hinder correspondence association. To overcome this limitation we introduced a shape complexity measure that allows us to perform the matching restricted to salient shape features (compare [46] for details). Matching the subset of the most salient shape features in a configuration is more robust than matching nearly featureless, small shapes. Hence we perform matching as a two-step process. In a first step we only consider the most similar and most complex pairs of the corresponding shape features; we can estimate the metric displacement required for the robot pose consistency measure ΔA. In a second matching step, this knowledge can be taken into account; it allows to robustly associate simple shape features even in significantly different configurations [46].

5.6 Experimental Comparison

To evaluate different approaches to localization, we set up a simulated environment using a virtual robot equipped with a laser range finder. Simulation allows us to easily measure the performance of individual simulation methods, as the ground truth is known. Additionally, we can systematically alter the environment and other parameters like sampling rate or sensor quality to gain a better understanding of the capabilities of individual methods.

To maintain the focus on spatial aspects we do not incorporate stochastic models; we only determine the most plausible pose. In the context of a hypothetical stochastic framework this would mean that we focus on the development of individual hypotheses. The more reliably a single hypothesis can estimate the robot's true pose, the better a complete system including stochastics performs. Furthermore, incorporating comprehensive uncertainty handling would conceal the ability of judging the performance of spatial representation and reasoning techniques, to a certain extent.

5.6.1 Experiments and Discussion

We examined two experimental setups. The first setup is a typical map-based robot SL task. A simulated robot traveled a total distance of 43.03m in the environment depicted in Fig. 5.6 (a). The average travel distance of the robot between sensing the environment amounts to 11 mm and the average rotation between sensing amounts to 4.0°. The true map was accessible to the localization methods. Hence, the main challenge of this setup is to robustly extract features from noisy input data and to robustly handle the correspondence problem.

In the second setup we investigated into generalized SL using a schematic map as reference system. The robot traveled along the same route as in the first experiment, localizing once every 104 mm on the average; this entails an average

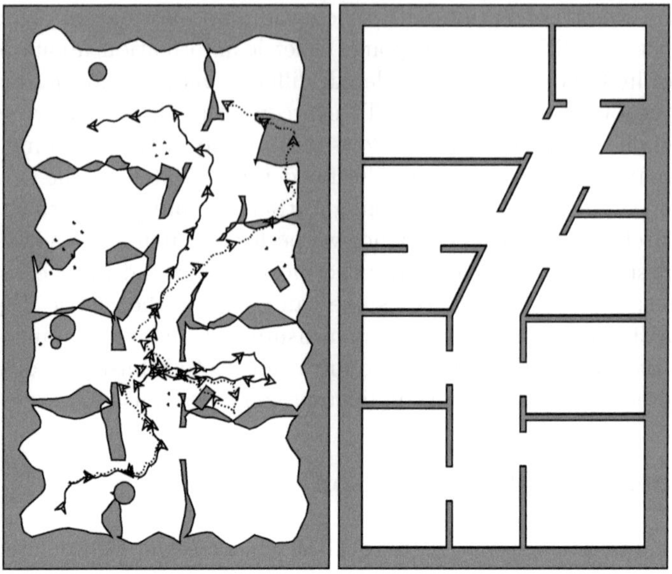

Fig. 5.6. Experimental setups to test robot self-localization performance. (a) depicts the test environment of aprox. 14×23 meters containing furniture, complex obstacles, etc. The path of the robot (dark line), and the path as reconstructed from the simulated odometry readings (dashed line); (b) shows a schematic map of the test environment.

rotation between sensing and SL of 30.6°. In this experiment, the schematic map presented in Fig. 5.6(b) was supplied to the localization methods. This added an extra challenge to mediating between information present in different levels of resolution and to robustly handle objects that were missing in the map (compare Sect. 5.2.3).

In the experiments we compare our approach with the following localization methods from different categories discussed in Sect. 5.3.

- Map-based localization by line matching [8, 33, 17, 18]
- Iterative Closest Point (ICP) used in connection with occupancy grids [6]
- SBL based on shape matching and ordering information [46, 47]

The methods listed above have been implemented according to the specifications given in the literature. Grid size for occupancy grids in ICP was 50mm x 50mm. We determined the quality of the localization by comparing the differences between true pose (ground truth) and localized pose. A proximity test was applied to compare the deviation between computed pose and ground truth against a threshold. In map-based localization, the test is passed if the position deviation is less than 100mm and the heading deviation is less than 45°. The proximity test for SL using a schematic map allows for a difference in position

Table 5.1. Tabular overview of localization results obtained. The proximity test evaluates, if a determined pose is close enough to ground truth. In the map-based localization, the test is passed if the difference is less than 100mm in position an less than 45° in heading. The proximity test for SL using a schematic map allows for a difference in position less than 500 mm and 45° in heading.

map-based SL:

method	average difference to true position [mm]	heading [°]	proximity test [%]
ICP	534	1.5	21
line-based	5167	65	0.4
shape-based	144	1.21	78

SL using schematized map:

method	average difference to true position [mm]	heading [°]	proximity test [%]
ICP	2234	25.9	50
line-based	1836	16.6	30
shape-based	553	3.3	86

less than 50cm and 45° in heading. Results presented in Table 5.1 and in the Fig. 5.7 and 5.8 will be discussed in the following.

Considering the map-based localization experiment, we observe that line-based localization relying on line detection in the LRF data quickly looses track of the correct path; it passed the proximity test in less than 1% of the cases. At first glance, ICP seems to resemble the robot's true trajectory (see Fig. 5.7 (a)). However, due to susceptibility for local minima in ICP's optimization process, pose estimates often get stuck in their local surroundings; however, ICP recovers when the robot moves on further. 21% of the estimated poses satisfied the proximity test. Shape-based localization passed the proximity test in 78% of the cases. This demonstrates that our approach is able to robustly perform standard SL tasks.

Using LRF data corresponding to the same test environment as before, but providing a simplified schematic map for localization instead of the true map simulates wayfinding using an overview map. In this setting we relaxed the proximity limit to a distance of 50cm, since sensed LRF data and schematic map differ significantly. We observe a decrease in localization performance which is caused by the large differences between map and perception. ICP met the proximity constraint in 50% of the computed poses, line-based localization succeeded in 30%, and shape-based localization in 86% of the cases. Of these methods, ICP first looses track of the robot's path, but coarsely resembles the true path (see Fig. 5.8 (a)). An interesting observation is that line-based localization performs better in the localization using the schematic map than in the classical localization task; a reason for this can be seen in the eased line extraction from the schematic map as compared to the true environment map containing mostly non-linear obstacles. However, as regards the average localization error, line-based localization

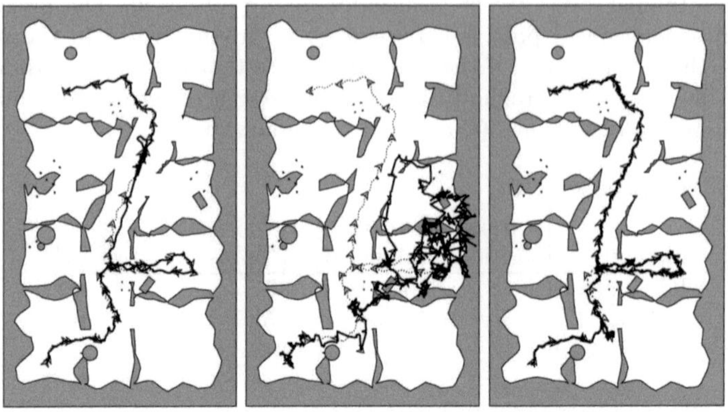

(a) (b) (c)

Fig. 5.7. Results obtained in the map-based localization experiment. Determined poses and true poses are plotted. (a) ICP, (b) line-based, and (c) shape-based.

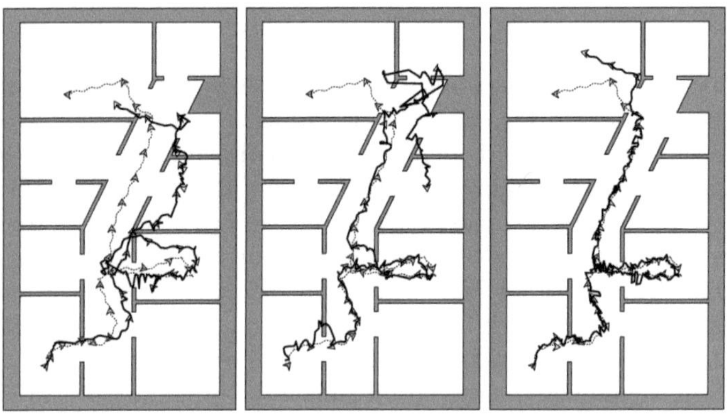

(a) (b) (c)

Fig. 5.8. Results obtained in the localization experiment involving a schematic map. Computed and true poses are plotted. (a) ICP, (b) line-based, and (c) shape-based.

is outperformed by ICP with an average error in line-based localization of about 2.2 m as compared to about 1.8 m in ICP. In contrast, SBL estimates poses with an average error of 0.55m, just about the proximity test threshold of 0.5m. SBL estimates the path closely until the robot enters the last room in the top-left corner. The failure when entering the room was caused by erroneously matching the perceived circular obstacle against the wall registered in the schematic map. Considering the average differences between true and estimated trajectory (see Table 5.1), it can be concluded that only SBL is able to master the generalized localization setting.

5.7 Conclusion

We proposed a generalization of the self-localization problem for robots to integrate a variety of localization tasks. These tasks include localization with respect to an externally supplied coarse or schematic map and localization based on route descriptions. We identified three strands of generalization: change of perspective, change of sensor, and change of resolution.

SL approaches are classified with respect to their choice of map features, their representation of configuration information, their approach to the correspondence problem, and their integration of spatio-temporal reasoning. For utilizing external floor plans in a generalized localization task, map features describing the boundary of free space are particularly valuable. The involvement of coarse maps requires an abstraction of configuration information from fine-grained metric details which are meaningless in schematic maps to a qualitative level. We argue in favor of improving matching algorithms to approach the correspondence problem analytically rather than by means of optimization. Analytical solutions do not get stuck in local minima; getting stuck in local minima i nevitably occurs when differing views are correlated, as in cross-modal setups or due to a granularity change.

We describe our approach to SL which makes use of expressive shape features. Configuration information makes qualitative knowledge explicit alongside metric information. Qualitative knowledge about cyclic ordering enables design of an efficient analytical approach t o the correspondence problem.

In an experimental evaluation we demonstrated the applicability of our approach to standard map-based localization and SL using a schematic map. The experiments highlighted that our approach performs comparably well as often-used ICP based localization in map-based localization. In the case of SL using a schematic map only the shape-based approach is able to robustly perform localization.

To sum up, several tasks exist that have a close relation to SL and can all be integrated into a more general task definition. For all dimensions of generalization, a sensible abstraction is the key to finding a solution. Sensible abstraction of spatial information can be achieved by including abstract qualitative knowledge and advancing to more expressive map features.

Acknowledgment

This work was carried out in the framework of the SFB/TR 8 Spatial Cognition, project R3 [Q-Shape]. Financial support by the Deutsche Forschungsgemeinschaft is gratefully acknowledged. This work was supported in part by the National Science Foundation under grants NSF INT-0331786 and NSF IIS-0534929. We thank the editors of this volume for their comprehensive review and for their valuable comments that helped us to improve the paper.

References

1. Altermatt, M., Martinelli, A., Tomatis, N., Siegwart, R.: SLAM with corner features based on a relative map. In: Proceedings of the IROS-2004 (2004)
2. Baker, C., Morris, A., Ferguson, D., Thayer, S., Whittaker, C., Omohundro, Z., Reverte, C., Whittaker, W., Hähnel, D., Thrun, S.: A campaign in autonomous mine mapping. In: ICRA. Proceedings of the IEEE International Conference on Robotics and Automation, IEEE Computer Society Press, Los Alamitos (2004)
3. Barkowsky, T., Berendt, B., Egner, S., Freksa, C., Krink, T., Röhrig, R., Wulf, A.: The REALATOR: How to construct reality. In: Proceedings of the ECAI 1994 Workshop of Spatial and Temporal Reasoning (1994)
4. Barkowsky, T., Jan Latecki, L., Richter, K.-F.: Schematizing maps: Simplification of geographic shape by discrete curve evolution. In: Freksa, C., Brauer, W., Habel, C., Wender, K.F. (eds.) Spatial Cognition II - Integrating Abstract Theories, Empirical Studies, Formal Methods, and Practical Applications, pp. 41–53. Springer, Berlin (2000)
5. Bellman, R.E.: Dynamic Programming. Princeton University Press, Princeton (1957)
6. Besl, P., McKay, N.: A method for registration of 3D shapes. IEEE Transactions on Pattern Analysis and Machine Intelligence 14(2), 239–256 (1992)
7. Choset, H., Walker, S., Eiamsa-Ard, K., Burdick, J.: Sensor-based exploration: Incremental construction of the hierarchical generalized voronoi graph. International Journal of Robotics Research 19(2), 126–148 (2000)
8. Cox, I.J.: Blanche: Position estimation for an autonomous robot vehicle. In: Cox, I.J., Wilfong, G.T. (eds.) Autonomous Robot Vehicles, pp. 221–228. Springer, Heidelberg (1990)
9. Denis, M.: The description of routes: A cognitive approach to the production of spatial discourse. Cahiers Psychologie Cognitive 16(4), 409–458 (1997)
10. Forsberg, J., Larsson, U., Wernersson, Å.: Mobile robot navigation using the range-weighted Hough transform. IEEE Robotics & Automation Magazine 21, 18–26 (1995)
11. Forsman, P.: Feature based registration of 3D perception data for indoor and outdoor map building. In: Int. Conference on Field and Service Robotics, Helsinki, Finland (2001)
12. Fox, D., Thrun, S., Dellaert, F., Burgard, W.: Particle filters for mobile robot localization. In: Doucet, A., de Freitas, N., Gordon, N. (eds.) Sequential Monte Carlo Methods in Practice, Springer, New York (2000)
13. Franz, M.O., Schölkopf, B., Mallot, H.A., Bülthoff, H.H.: Learning view graphs for robot navigation. Autonomous Robots 5, 111–125 (1998)
14. Freksa, C.: Conceptual neighborhood and its role in temporal and spatial reasoning. In: Singh, M., Travé-Massuyès, L. (eds.) Decision Support Systems and Qualitative Reasoning, pp. 181–187. North-Holland, Amsterdam (1991)
15. Freksa, C.: Temporal reasoning based on semi-intervals. Artificial Intelligence 54(1), 199–227 (1992)
16. Frese, U.: An O(log n) Algorithm for Simulateneous Localization and Mapping of Mobile Robots in Indoor Environments. PhD thesis, University of Erlangen-Nürnberg (2004)
17. Gutmann, J.-S.: Robuste Navigation autonomer mobiler Systeme. PhD thesis, University of Freiburg (2000) (in German)

18. Gutmann, J.-S., Weigel, T., Nebel, B.: A fast, accurate and robust method for self-localization in polygonal environments using laser range finders. Advanced Robotics 14(8), 651–667 (2001)
19. Hähnel, D., Schulz, D., Burgard, W.: Map building with mobile robots in populated environments. In: IROS 2002. Proceedings of International Conference on Intelligent Robots and Systems (2002)
20. Hähnel, D., Burgard, W., Fox, D., Fishkin, K., Philipose, M.: Mapping and localization with RFID technology. In: ICRA. Proceedings of the IEEE International Conference on Robotics and Automation (2004)
21. Jefferies, M.E., Cosgrove, M.C., Baker, J.T., Yeap, W.K.: The correspondence problem in topological metric mapping — using absolute metric maps to close cycles. In: Negoita, M.G., Howlett, R.J., Jain, L.C. (eds.) KES 2004. LNCS (LNAI), vol. 3213, pp. 232–239. Springer, Heidelberg (2004)
22. Kirchner, F., Hertzberg, J.: A prototype study of an autonomous robot platform for sewerage system maintenance. Autonomous Robots 4(4), 319–333 (1997)
23. Klippel, A., Richter, K.-F., Barkowsky, T., Freksa, C.: The cognitive reality of schematic maps. In: Meng, L., Zipf, A., Reichenbacher, T. (eds.) Map-based Mobile Services — Theories, Methods and Implementations, pp. 57–74. Springer, Berlin (2005)
24. Ko, J., Stewart, B., Fox, D., Konolige, K., Limketkai, B.: A practical, decision-theoretic approach to multi-robot mapping and exploration. In: IROS 2003. Proceedings. of the IEEE/RSJ International Conference on Intelligent Robots and Systems (2003)
25. Kuipers, B.: The spatial semantic hierarchy. Artificial Intelligence 119, 191–233 (2000)
26. Kuipers, B., Byun, Y.-T.: A robot exploration and mapping strategy based on a semantic hierarchy of spatial representations. Journal of Robotics and Autonomous Systems 8, 47–63 (1991)
27. Latecki, L.J., Lakämper, R.: Shape similarity measure based on correspondence of visual parts. IEEE Trans. Pattern Analysis and Machine Intelligence 22(10) (2000)
28. Latecki, L.J., Lakämper, R., Wolter, D.: Shape similarity and visual parts. In: Nyström, I., Sanniti di Baja, G., Svensson, S. (eds.) DGCI 2003. LNCS, vol. 2886, pp. 34–51. Springer, Heidelberg (2003)
29. Latecki, L.J., Lakämper, R., Wolter, D.: Partial optimal shape similarity. Image and Vision Computing Journal 23(2), 227–236 (2005)
30. Latombe, J.-C.: Robot Motion Planning. Kluwer Academic Publishers, Norwell (MA), USA (1991)
31. Lee, D.T., Drysdale, R.L.: Generalization of Voronoi diagrams in the plane. SIAM Journal on Computing 10(1), 73–87 (1981)
32. Lingemann, K., Surmann, H., Nüchter, A., Hertzberg, J.: Indoor and outdoor localizations for fast mobile robots. In: IROS. Proceedings of International Conference on Intelligent Robots and Systems (2004)
33. Lu, F., Milios, E.: Robot pose estimation in unknown environments by matching 2D range scans. Journal of Intelligent and Robotic Systems 18(3), 249–275 (1997)
34. Moratz, R., Tenbrink, T.: Spatial reference in linguistic human-robot interaction: Iterative, empirically supported development of a model of projective relations. Spatial Cognition and Computation, 2006 (in press)
35. Moravec, H.P., Elfes, A.E.: High resolution maps from wide angle sonar. In: ICRA. Proceedings of the IEEE International Conference on Robotics and Automation, IEEE Computer Society Press, Los Alamitos (1985)

36. Neira, J., Tardós, J.D.: Data association in stochastic mapping using the joint compatibility test. IEEE Transactions on robotics and automation 17(6), 890–897 (2001)
37. Röfer, T.: Using histogram correlation to create consistent laser scan maps. In: IROS-2002. Proceedings of the IEEE International Conference on Robotics Systems, IEEE Computer Society Press, Los Alamitos (2002)
38. Schlieder, C.: Reasoning about ordering. In: COSIT. Proceedings of the 3rd International Conference on Spatial Information Theory (1995)
39. Thrun, S.: Learning metric-topological maps for indoor mobile robot navigation. Artificial Intelligence 99(1), 21–71 (1998)
40. Thrun, S.: Probabilistic algorithms in robotics. AI Magazine 21(4), 93–109 (2000)
41. Thrun, S.: Robotic mapping: A survey. In: Lakemeyer, G., Nebel, B. (eds.) Exploring Artificial Intelligence in the New Millenium, Morgan Kaufmann, San Francisco (2002)
42. Thrun, S., Fox, D., Burgard, W., Dellaert, F.: Robust Monte Carlo localization for mobile robots. Artificial Intelligence 128(1-2), 99–141 (2001)
43. Tsang, E.: Foundations of Constraint Satisfaction. Academic Press, London (1993)
44. Veeck, M., Burgard, W.: Learning polyline maps from range scan data acquired with mobile robots. In: IROS. Proceedings of the IEEE/RSJ International Conference on Intelligent Robots and Systems (2004)
45. Wallgrün, J.O.: Autonomous construction of hierarchical Voronoi-based route graph representations. In: Freksa, C., Knauff, M., Krieg-Brückner, B., Nebel, B., Barkowsky, T. (eds.) Spatial Cognition IV. LNCS (LNAI), vol. 3343, pp. 413–433. Springer, Heidelberg (2005)
46. Wolter, D., Latecki, L.J.: Shape matching for robot mapping. In: Zhang, C., W. Guesgen, H., Yeap, W.-K. (eds.) PRICAI 2004. LNCS (LNAI), vol. 3157, pp. 693–702. Springer, Heidelberg (2004)
47. Wolter, D., Latecki, L., Lakämper, R., Sun, X.: Shape-based robot mapping. In: Biundo, S., Frühwirth, T., Palm, G. (eds.) KI 2004. LNCS (LNAI), vol. 3238, pp. 439–452. Springer, Heidelberg (2004)
48. Wolter, D., Richter, K.-F.: Schematized aspect maps for robot guidance. In: Doherty, P. (ed.) CogRob. Proceedings of the workshop cognitive robotics, pp. 71–76 (2004)
49. Yeap, W.K., Jefferies, M.E.: On early cognitive mapping. Spatial Cognition and Computation 2(2), 85–116 (2000)

Part II

Cognitive Mapping

Dead Reckoning, Cognitive Maps, Animal Navigation and the Representation of Space: An Introduction

Charles R. Gallistel

The chapters in this part deal with some aspects of current behavioral and neurobiological research on the construction of maps by animal brains. A map is an encoding of some or all of the geometric relations between locations. Map making is a fundamental part of navigation. It enables the navigator to set a course for a destination that is not currently perceived by reference to the navigator's currently perceptible surroundings.

Navigation and map making are fundamentally computational problems. The character of the problem and the relevant mathematical principles do not change when one shifts from the study of animal navigation to the creation of autonomous robots that navigate, any more than character of the imaging problem and the principles of optics change when one shifts from the study of eyes to the design of cameras. We can, therefore, answer with some confidence the question posed by Chown and Boots, whether there is any common ground between cognitive mapping and robot mapping. There is because in both cases workable solutions are strongly constrained by the character of the computational problem that must be solved. It is the distinctive geometric structure of the problems that leads to modularity in the mechanisms that have evolved or been designed to solve the problem (cf. Chap. "Geometry and Navigation", p. 145). The structure of the modules reflects the structure of the problem just as the structure of the eye reflects the principles of optics.

Navigation and map making have engaged mathematical and engineering minds for centuries. In reading these chapters, I am reminded of my long-standing conviction that fruitful interchange between psychologists, neuroscientists and roboticists–and simple, clear thinking–would be facilitated if we all relied more on the conceptual framework and terminology provided by standard texts on navigation and map making. If each discipline, or worse yet, each lab or working group, makes up its own terminology and creates its own conceptual framework for dealing with universal, long-understood aspects of map making and navigation, it leads to a great deal of reinventing of the wheel, and it creates barriers to mutual understanding.

In this introduction, I describe aspects of the common computational problems faced by robots and animals that make maps for navigational use.

Dead Reckoning. Roboticists find that it is surprisingly difficult for a robot to estimate where it is from processing sensor input so as to recognize landmarks.

M.E. Jefferies et al. (Eds.): Robot. & Cogn. Approach. to Spat. Map., STAR 38, pp. 137–143, 2008.
springerlink.com

They also find that in many environments, there are long stretches with no useable landmarks. Animals, including humans, find the same thing. Therefore, the taking of a fix–establishing where you are from sightings of recognized landmarks–is done only intermittently. Between fixes, a continuous estimate of position is kept by means of dead reckoning. This is true in traditional marine and aeronautical navigation and in animal navigation, including the routine navigation of familiar spaces, both large (city scale) and small (room scale), that humans carry out throughout a large portion of every day. That is why there is some discussion of dead reckoning in many of these chapters.

The navigation of familiar spaces is so routine and done with so little thought that many humans are under the erroneous impression that they cannot dead reckon and are amazed to learn that ants, bees and rats can. When put to the test, however, humans are also good at estimating how far they have gone and in what direction [11]. They are good at it even when walking blind, which deprives them of a major source of the velocity cues critical to dead reckoning.

The reliance on dead reckoning for the moment-to-moment estimation of position on the map becomes apparent in experiments with rats in which familiar routes are lengthened or shortened. The running rat collides with the walls at the ends of shortened corridors, turns into the wall when it reaches what used to be the correct turning point in a lengthened corridor, and runs off the end of a shortened path in an elevated maze (see [5], p. 91ff for review and citations).

Dead reckoning is the integration of velocity with respect to time to obtain position as a function of time. If viewed as a discrete process, it is the summation of successive displacements to obtain net displacement. The velocity vector is displacement per unit time. If displacements, $\langle \Delta x, \Delta y \rangle$, are estimated at regular intervals, $\Delta \tau$, the velocity vector is $\lim_{\Delta \tau \to 0} \langle \Delta x, \Delta y \rangle / \Delta \tau$. Displacement is the distances one has moved in different (preferably orthogonal) directions. A metric space is by definition a space in which a measure of the distance between any two points has been defined, so displacement and displacement per unit time are inherently metric.

Given the fundamental role of dead reckoning in navigation, a central computational problem is to make the reckoning of the distances moved as accurate as possible. In animals, and no doubt in robots also, a key to this is integrating the information from many different cues. Because all of the cues have errors associated with them, because the validity (accuracy and precision) of the information provided by any one cue varies from environment to environment, and because the integration must be done in real time, the integration should be Bayesian [18].

Making a metric map. Dead reckoning provides the navigator with moment-to-moment estimates of position and heading in a common allocentric coordinate framework. This always accessible estimate of current position and heading makes it possible to record in that same common coordinate framework the position of a landmark seen in one location and the position of a landmark seen in a different location (see [5], p. 106ff).

The map thus created inherits its metric structure from the metric structure in the dead reckoning that mediates its construction. The metric structure of the map is critical to many of its uses, in particular to the estimation of range and bearing (distance and direction), the fundamental metric measures. Because metric maps are the formally most powerful of spatial representations and because weaker geometric properties (e.g., topological properties) are sometimes erroneously equated with properties that are simpler to compute or, from a computational perspective more basic or primitive, there is a tendency to assume that cognitive maps in animals have a"simpler" non-metric structure. Given the role of inherently metric dead reckoning in the construction and utilization of cognitive maps and the fundamental role of distance and direction in navigational computations, this is, in my estimation, misguided. The topology is contained in a metric representation and may be computed from it. The reverse is not true; topological representations lack the metric information that is essential in navigation, and there is no way to extract this information from them.

A major computational problem in connection with map making is that the estimates for the positions and orientations of the surfaces recorded on the map are themselves subject to error, deriving as they do from error-prone dead-reckoning and error-prone distance and direction sensing. As a result of errors in the reckoning of the navigator's position and heading and errors in the sensing of his distance and direction from landmarks, the same surface will be estimated to have different positions on the map when approached from different directions and via different routes on different occasions. This raises two computational issues:

- how to integrate position and orientation estimates obtained on different occasions;
- how to avoid, insofar as possible, recording two different surfaces when there is in fact only one, that is, how to recognize that one has arrived at or is currently picking up sensor input from a surface whose position has already been recorded on the map, albeit in a location and with an orientation somewhat different from those currently imputed to it.

One key to solving these problems is again to treat the positions of surfaces on the map probabilistically and use a Bayesian (optimal estimation) approach to integrating in real time past and current information about them [16].

A second key is to focus on global structure, which, as several chapters in this book emphasize, naturally takes a hierarchical form: the environment has parts; and the parts themselves have parts. The success of a focus on the structure of the environment depends on a felicitous choice of a scheme for representing its shape. This latter issue-how best to represent the shape of the environment-has received rather little attention. It is one of the deepest issues. Roboticists may someday make a major contribution because of mathematical sophistication that they bring to the problem from their engineering training. At present, however, most robotic map making that I am familiar with uses grid-occupancy [17]. Although this simple representation may facilitate probabilistic integration, it is an infelicitous choice from the second perspective: using constraints from global structure to help solve the ambiguity problem (is this a surface already on the map or a new surface?). The grid-occupancy representation treats the world as an unstructured array, composed of cells that are independently either occupied or unoccupied. In fact, of course, the world has useful a priori predictable statistical structure (see Chap. "Cue and Goal Encoding in Rodents: A source of Inspiration for Robotics?", p. 163): the probability that a given cell is or is not occupied depends strongly on the occupancy pattern for neighboring cells and also, to some extent, on global occupancy patterns, the pattern of occupancy in remote cells. The challenge is to find computationally efficient algorithms that capture and exploit this probabilistic structure to help resolve the ambiguity problem.

The best approach to the shape representation problem would simultaneously achieve both compression (a compact encoding of the experienced environmental shape) while making explicit useful features of the shape, such as, for example, its part structure, axes of symmetry, efficient routes from part to part, and so on. Here, roboticists may profit from theoretical work on shape representation by psychologists. Leyton's [10] group-theoretic approach with its emphasis on the parameterized Euclidean operations of translation and rotation that could most parsimoniously have generated a shape seems particularly promising in that it suggests how the encoding of the shape of the environment, along whose surfaces the robot moves, could be based on a natural decomposition of the run (straight) and turn commands that effect its movements.

Leyton's work also emphasizes the importance of axes of symmetry in the encoding and manipulation of shape [9]. The use of medial-axis transforms to generate an abstract hierarchically structured shape skeleton has been popular in computer vision and image processing [6, 7, 14, 15] and in psychological and neurobiological theorizing [8] since Blum first suggested it [1, 2]. Enthusiasm for this representation has been tempered by the fact that existing approaches to computing the skeleton are hypersensitive to contour noise and give counterintuitive skeletons for several simple shapes (e.g., rectangles). Recently, Feldman and Singh ([4], in press) have found an ingenious solution to this problem. In their approach, Bayesian estimation is used to identify the skeleton most likely to have "produced" the shape, that is, the skeleton that best "explains" it in a sense made clear by Leyton's generative theory of shape.

The medial axis of a part is an example of what Cheng calls a global shape parameter. These parameters play an important role in the registration of images in computerized image processing [14]. Cheng (see Chap. "Geometry and Navigation", p. 145 and [3]) suggests that they also play an important role in establishing an animal's orientation within a familiar environment.

Using the map. The behavioral signature of a cognitive map is the ability to estimate the range and bearing of one arbitrary location within familiar (that is, mapped) terrain from another arbitrary location within that same terrain. This ability extends to insects [12, 13], emphasizing the fundamental role of a metric map in solving the problems inherent in navigation.

Another fundamental use is in route planning, because many environments discourage to varying degrees travel by the straight-line route. Among the attractive features of an axial skeleton representation of the shape of the experienced environment is that the skeleton itself is a plausible route map: it shows the midlines through the parts and where those midlines join one another (cf. Chap. "Learning Cognitive Maps: Finding Useful Structure in an Uncertain World", p. 215 and Chap. "These Maps are Made for Walking – Task Hierarchy of Spatial Cognition", p. 181). Route-finding algorithms that work on road maps will work on axial skeleton representations of environmental shape.

A less often emphasized use is in recognizing places and landmarks (see Chap. "These Maps are Made for Walking – Task Hierarchy of Spatial Cognition", p. 181 and Chap. "Landmarks for Navigation in Humans and Robots", p. 203). A precondition for using the perception of the distance and bearing of some part of the environment to improve one's estimate of one's position (that is, to take a fix) is recognizing it. To recognize a part of the environment is to identify it with a charted feature. As several chapters emphasize, place and landmark recognition is no small computational problem (unlike dead reckoning, which is, in its basics, computationally trivial).

In traditional navigation, the estimate of one's position and heading on the map, derived from dead-reckoning, is used to, establish the Bayesian priors on the chart features with which a currently sensed feature of the terrain may be identified. Features in grossly different directions and at grossly different distances from the direction and distance at which one takes the currently sensed feature to lie have zero prior probability of being identified with it. A navigator sailing on Long Island Sound who sees something to her east that looks like Mt. Vesuvius will never conclude that she is in the Bay of Naples, no matter how good the correspondence between what she now sees and her image of Vesuvius. Conversely, most navigators, from humans to insects, will accept a poor match between what they see and their search image of a landmark if what they now see is close to the location where they expect to see the landmark. Thus, dead-reckoning plays an important but subtle role in the recognition of the landmarks, the sightings of which are then used to correct the reckoning. This is true even for the estimation of position from star sightings (stellar fixes). The first data entry on most published worksheets for working one's sightings is the dead-reckoning estimate of one's position at the time the sighting was taken. What is computed from the sighting is the adjustment that should be made to that estimate. In some methods for working a sight, the star sighted is never even explicitly identified. It is implicitly identified from the dead reckoning estimate together with the compass bearing and elevation of the star sighted.

The scheme for representing environmental shape is also an important consideration in landmark recognition. Some schemes, for example, schemes based on axial skeletons, decompose the shape into a part hierarchy, facilitating recognition of the parts, whereas others, for example, grid-occupancy schemes, do not. Thus, some schemes help to solve the parsing problem discussed by Hirtle, while others do not.

Using landmarks. Landmarks are used to establish heading, establish position, and punctuate progress along routes (see Chap. "Cue and Goal Encoding in Rodents: A source of Inspiration for Robotics?", p. 163 and Chap. "Landmarks for Navigation in Humans and Robots", p.203). For establishing heading, distant landmarks (distal landmarks in Save et al's terminology) are best because they minimize parallax. Ideally, points of directional reference (points used to establish heading) should function independently of one's position; changes in one's position as one moves and errors in the estimate of one's static position should have no effect on the estimation of one's heading. The farther away a directional reference point is, the less parallax there will be, that is, the less the effect of variations in position on the estimate of heading. That is why distant landmarks are preferred for establishing heading. Celestial landmarks are, for practical purposes, points at infinity; they show no parallax. That is no doubt why both animals and humans prefer them as points of directional reference. Exactly the opposite considerations apply when landmarks are used to estimate position. In that case, one wants the direction of the landmark to be maximally sensitive to variation in one's position. That makes close landmarks (proximal landmarks) preferable for that purpose.

In traditional navigational language, when a landmark is used to punctuate progress along one or more routes (see Chap. "These Maps are Made for Walking – Task Hierarchy of Spatial Cognition", p. 181 and Chap. "Landmarks for Navigation in Humans and Robots", p. 203), it is called a waypoint. Waypoints parse routes into segments. They also serve to verify and correct one's estimate of how far one has progressed along the chosen route, that is, to correct the reckoning. Waypoints generally lie close beside a route. The logic of this is the same as the logic for choosing landmarks used to estimate position to be as close to the estimated position as possible. When a waypoint is close abeam (nearby on a line perpendicular to the route one is following), its angular velocity is maximal, so the information it supplies per unit time about one's position is maximal. Thus, waypoints are chosen for ease of recognition and for proximity to the route. Ease of recognition may arise either from highly distinctive features of the waypoint itself or because it has an easily and precisely estimable location relative to the larger structure of the surrounding environment, emphasizing once again the importance of understanding the principles by which this larger structure is encoded.

References

1. Blum, H.: Biological shape and visual science (Part I). Journal of Theoretical Biology 38, 205–287 (1973)
2. Blum, H., Nagel, R.N.: Shape description using weighted symmetric axis features. Pattern Recognition 10, 167–180 (1978)
3. Cheng, K., Gallistel, C.R.: Shape Parameters Explain Data from Spatial Transformations: Comment on Pearce et al (2004) and Tomassi and Polli (2004). Journal of Experimental Psychology: Animal Behavior Processes 31(2), 254–259 (2005)
4. Feldman, J., Singh, M.: Bayesian estimation of shape. In: Proceedings of the National Academy of Sciences, USA (2006)
5. Gallistel, C.R.: The organization of learning. MA: Bradford Books/MIT Press, Cambridge (1990)
6. Katz, R.A., Pizer, S.M.: Untangling the Blum medial axis transform. International Journal of Computer Vision 55, 139–153 (1996)

7. Kimia, B.B., Tannenbaum, A.R., Zucker, S.W.: Shapes, shocks, and deformation. 1. The components of two-dimenisonal shape and the reaction-diffusion space. International Journal of Computer Vision 15, 189–224 (1995)
8. Kovács, I., Féher, A., Julesz, B.: Medial-Point Description of Shape: a Representation for Action Coding and its Psychophysical Correlates. Vision Research 38, 2323–2333 (1998)
9. Leyton, M.: Symmetry, Causality, Mind. MA: MIT Press, Cambridge (1992)
10. Leyton, M.: A Generative Theory of Shape. Springer-Verlag, New York (2001)
11. Loomis, J.M., Klatzky, R.L., Golledge, R.G., Philbeck, J.W.: Human navigation by path integration. In: Golledge, R.G. (ed.) Wayfinding: Cognitive mapping and spatial behavior, Baltimore: Johns Hopkins Press (1998)
12. Menzel, R., Greggers, U., Smith, A., Berger, S., Brandt, R., Brunke, S.: Honey bees navigate according to a map-like spatial memory. In: Proceedings of the National Academy of Sciences, vol. 102, pp. 3040–3045 (2005)
13. PastergueRuiz, L., Beugnon, G., Lachaud, J.P.: Can the Ant Cataglyphis cursor (Hymenoptera: Formicidae) Encode Global Landmark-Landmark relationships in addition to isolated landmark-goal relationships? Journal of Insect Behavior 8, 115–132 (1995)
14. Pizer, S.M., Fritsch, D.S., Yushkevich, P., Johnson, V., Chaney, E., Gerig, G.: Segmentation, Registration, and Measurement of shape Variation via Image Object Shape. IEEE Transactions on Medical Imaging 18(10), 851–865 (1996)
15. Siddiqi, K., Shokoufandeh, A., Disckinson, S., Zucker, S.: Shock graphs and shape matching. International Journal of Computer Vision 30, 1–24 (1999)
16. Thrun, S.: Robotic mapping: A survey. In: Lakemeyer, C., Nebel, B. (eds.) Exploring artificial intelligence in the new millenium, p. 29. Mogan Kaufman, San Francisco (2002)
17. Thrun, S., Bücken, A., Burgard, W., Fox, D., Fröhlinhaus, T., Hennig, D., et al.: Map learning and high-speed navigation in rhino. In: KortenKamp, D., Bonasso, R.P., Murphy, R.R. (eds.) AI-based mobile robots: Case studies of successful robot systems, MA: MIT Press, Cambridge (1998)
18. Thrun, S., Burgard, W., Fox, D.: Probabilistic robots. MA: MIT Press, Cambridge (2005)

6

Geometry and Navigation

Ken Cheng

Summary. In vertebrate animals, the geometric arrangement of surfaces in an environment has been shown to play an important role in relocating a desired place. In such relocation tasks, an animal is typically first shown a target location in a rectangular enclosure. After being disoriented, it then has the task of relocating the target. Aside from the geometric shape of the enclosure, other nongeometric or featural cues are typically available. These include colours of walls, objects serving as landmarks, or smells. An often reported pattern of results is preferred reliance on the geometric cues, sometimes to the exclusion of nongeometric cues. Various axes of space, shape parameters that include principal axes and axes of symmetry, may play a role in how animals use geometric information to determine which direction is which. Some work on robotics related to the geometry literature is presented and the issue of modularity is discussed.

6.1 Introduction

To navigate to a desired place, an animal typically needs to figure out which direction is which, a problem known as determining the heading [31]. If the animal is keeping track of the (straight-line) distance and direction from the starting point of the journey, a process known as path integration [31], it needs to keep track of the heading continuously. Many animals path integrate [1, 26, 31, 42, 58]. In outdoor environments, many animals use both information in the sky and large-scale Earth-based landmarks to establish which direction is which. In insects, directional information is extracted out of the pattern of polarised light in the sky, and the neurobiological mechanisms have been worked out in some detail [55, 56]. Continuous large-scale landmarks, such as a shoreline, a line of trees, or a road are also used by honeybees to establish compass direction [30]. In fact, bees refer to follow continuous landmarks over the sky compass when the two are put in conflict. A large body of research on vertebrate navigation, however, has been conducted in restricted indoor laboratory environments, where both sky and large-scale landmarks are absent. If inertial cues based on path integration are disrupted, then the overall geometric shape of the environment seems to play an important role in reorientation. This geometry literature is now burgeoning. This chapter gives a brief overview summary of the research findings, and then discusses the theory and its relations to robotic navigation. The review of literature will be brief because a full review is available elsewhere [12].

M.E. Jefferies et al. (Eds.): Robot. & Cogn. Approach. to Spat. Map., STAR 38, pp. 145–161, 2008.
springerlink.com

6.2 Review of Empirical Literature

6.2.1 The Rotational Error: Basic Phenomenon

In most studies in the geometry literature, disoriented subjects had the task of relocating a place of reward within a rectangular enclosure. Conceptually, the cues for orientation in the arena may be divided into two kinds, the geometric and nongeometric or featural information (Fig. 6.1). Much of the interest in rotational errors hinges on this distinction. Gallistel [31] defines it as follows:

A geometric property of a surface, line, or point is a property it possesses by virtue of its position relative to other surfaces, lines, and points within the same space. A non-geometric property is any property that cannot be described by relative position alone (p. 212).

I can illustrate this using the example of the first geometry study [3] (Fig. 6.1). The rats were exposed to some food at a randomly chosen location in the rectangular arena, they got to eat some, were removed, and then had the task of digging the rest of the food out of the same location after a 90s delay, the food now buried under uniform wood chips. The rectangular shape of the arena provides the geometric cues (Fig. 6.1B). In Fig. 6.1B, points of surfaces are shown in relation to one another, with finer details omitted, such as panels in the corners. Basically, it is a rectangle, with the target location drawn on it. Also shown is one axis of symmetry through the middle of the length of the rectangle. Various axes that can be defined on the basis of geometric information, parameters based on shape, may well be of theoretical significance (see Sect. 6.3). What is not shown in Fig. 6.1B are other characteristics of the surfaces, such as having a white or black colour, or a smell of anise or peppermint. These featural cues of different modalities are shown in Fig. 6.1C. Other than features on surfaces, discrete objects also serve as featural cues in some studies. These objects do possess geometric properties, but it is the broad coarse-grained geometric properties that are thought to be most important, not fine geometric details.

The interesting error that led to the geometric/nongeometric distinction is called the rotational error (Fig. 6.1). In the artificial rectangular task environment, but almost never in the real world, defining a location solely with respect to geometric cues leads to an ambiguity in specifying the target location, whereas defining a location with respect to featural cues delivers one single unambiguous location as the target. In the rectangular space, both the correct location and the rotational error fit the geometric definition of the place. The rotational error is located at 180 ° rotation through the centre from the correct location. It is produced when the 'map' is matched to the world in the 'wrong' way, rotated 180 ° from what it should be. Making the rotational error systematically, that is, above chance levels, indicates that the animal is sometimes relying solely on geometric information for relocation. And if correct responses do not outnumber rotational errors statistically, then an animal is relying solely on geometric information in doing the task.

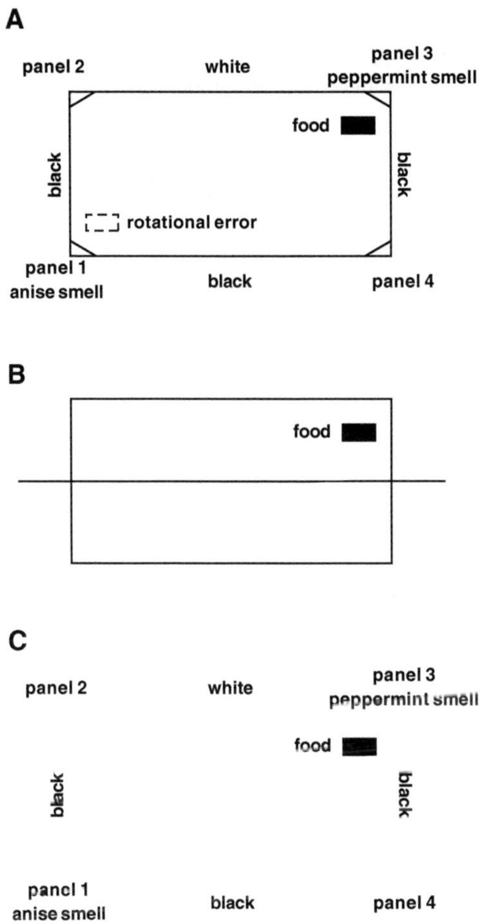

Fig. 6.1. Schematic illustration of the relocation task rats in Cheng's [3] study. A. Rats were shown some food at the target location, and were later required to go back and find the food, now buried under uniform wood chips. The arena had plenty of features other than its overall shape; these included different brightnesses of walls, different visual and tactile patterns in the corners, a nd smells emanating from some corners. The rats sometimes searched at the target location, and sometimes made the rotational error, searching at the location rotated 180 ° from the goal through the centre. B. To commit the rotational error, they must have a representation that only encoded the location of the food with respect to the overall shape of the arena. In this case, the animal has two ways of matching its record to the world. The correct way of matching (top wall of B matched to top wall of A) leads to a correct choice, while the wrong match (top wall of B matched to the bottom wall of A, with the 'map' in B rotated 180 °) leads to the rotational error. C. If the record contains featural information, it can be matched to the world in only one way. No systematic errors would result from such a record.

6.2.2 Rats

In Cheng's [3] experiments, rats made systematic rotational errors in both working-memory and reference-memory paradigms. In the working-memory paradigm, the target location could be anywhere in the arena, and a different target location was shown to the rat on each trial. Despite the multimodal featural cues, including smells in some corners, different textures and visual patterns in the corner panels, and with one long wall white and the three other walls black, the rats made many rotational errors, with correct choices not exceeding rotational errors statistically. No other systematic errors were found. Margules and Gallistel [40] replicated these results.

In the reference-memory paradigm [3], possible target locations were restricted to the corners, and the target location remained the same throughout training. Rats chose the target location far more than the rotational error, but still made systematic rotational errors. These results indicate that the featural cues were discriminable to the rats.

6.2.3 Other Mammals

The pattern of results found for rats turned out to be characteristic of other mammals tested, including human children. Subjects were typically tested in reference-memory paradigms, although in the case of children, few trials of training were given. The pattern is that at least under the right circumstances, correct locations were chosen significantly more than rotational errors, indicating that the subjects did use featural information for relocation. But the amount of rotational errors remained above chance, indicating the use of only geometric information on some trials.

For rhesus monkeys, the pattern of results depended on the size of the featural cues provided in the arena, whose size remained constant throughout [34]. Small featural cues led to the monkeys' making as many rotational errors as correct choices. Large featural cues, however, led to significantly more choices of the correct corner than the rotational error, while the amount of rotational errors continued to be systematic.

Children resemble monkeys in performing the relocation task: give them big enough featural cues, and they will use them. Early results on 18-24 month old children tested in a small (1.2 m by 1.8 m) rectangular enclosure revealed that children used the geometric cues readily, choosing mostly either the correct corner or the rotational error [35, 36]. But they made no use of featural information over four trials with the target at the same corner, making as many rotational errors as correct choices. Featural cues were the distinctive colour of a wall or distinct objects. Other experiments ruled out motivational and attentional factors, and showed that the featural cues were clearly discriminable. But later experiments with a larger enclosure (2.4 m by 3.6 m) showed that even 18-24 month olds can use featural cues, choosing the correct location significantly more often than the rotational error [38]. Rotational errors, however, remained systematic, as with rats and monkeys. Explicit manipulations of the size of the testing arena showed that it was crucial in predicting the amount of rotational errors [37]. In the small 1.2 m by 1.8 m space, children up to 5 years of age made as many rotational errors as correct choices. In the larger 2.4 m by 3.6 m space, they chose the correct location significantly more than the rotational error, with rotational errors remaining systematic, however, for 3-4 year olds, the youngest age tested.

6.2.4 Birds and Fish Use Geometry and Features

Birds and fish do not make systematic rotational errors (unless the featural cues that they encounter in training are degraded on tests), but they also can and do use geometric cues. If trained with both geometric and featural cues, chicks [49, 50], pigeons [21, 19], and two species of fish, the redtail splitfin [45] and goldfish [51], all chose mostly the correct corner, making no systematic errors of any kind. On the whole, birds performed better than fish. Well trained chicks and pigeons were almost perfect. This pattern shows that these species use featural cues well.

That birds and fish learn to use geometric cues as well is demonstrated in tests in rectangular arenas stripped of all featural cues, reduced to four walls of the same colour. Each species then solved the problem up to geometric ambiguity, choosing mostly either the target location or the rotational error. Interestingly, these species learned to use the geometric cues even when salient featural cues were available during training. The featural cues predicted the target location perfectly, while the geometric cues left a rotational ambiguity. But the availability of featural cues did not interfere with the learning of geometric cues. Basically, geometric cues seem to be learned obligatorily [12].

Both species of birds have also been tested with size tranformations of the arena. In pigeons, when the arena was made slightly smaller, the already trained birds went with the shape of the arena [20]. That is, if the long wall was to the left of the short wall at the target corner in the training space, they chose a corner in the test space where the long wall was to the left of the short wall, even though these walls were now of different absolute lengths from before. They preserved relative distances or relative lengths of walls. In two studies, chicks were trained to search in the middle of a square arena, and then tested in a larger arena [47, 48]. Chicks showed significant tendencies to search both at the centre (thus preserving relative distances or going with the shape) and at the correct absolute distance from a wall.

6.3 Theory

Gallistel [31] has presented the most thorough theoretical analysis of how geometry might be used in the relocation task. Conceiving of the relocation problem as one of using an internal record to locate a place in the external world, the problem can be divided into two parts. First, one has to align the record with the world, which is the determination of heading. Then one has to use the record to pinpoint the target location. The theory states that geometric properties play a major role in determining heading. The geometric information used is not the entire shape in all its details, but some global abstract properties, the principal axes.

6.3.1 Principal Axes

For purposes here, we need to consider only a two-dimensional space, consisting of the horizontal search surface bounded by walls. Intuitively, the first principal axis is the long axis that cuts through the middle of the length of the space. Figure 6.1B shows that in a rectangle, it is the long axis of symmetry. In mechanics, this is the axis that minimises the angular momentum when the space is rotated around the axis. In statistics, it is the principal component, the line that minimises the sums of squares of perpendicular distances from points in the space to the line. The second or minor principal axis is orthogonal to the first principal axis. Mechanically, it is the axis that maximises the angular momentum. The principal axes are clearly much reduced data compared with the metric properties contained in the shape. Nevertheless, they are metric properties based on shape, or shape parameters.

Matching the principal axes of internal record and perceived space is a global matching strategy, matching based on selected global parameters. Computing the principal axes is a determinate process, not a trial-and-error iterative process typical of feature-matching schemes based on matching many local features. Only a limited number of parameters (two axes) are to be matched, no matter how complex or large the space. The matching process is thus not subject to computational explosions. The computation can be based on coarse-scale geometry; fine geometric details are not needed. Missing out on finer details means some errors in computing a principal axis, but such errors remain small. And if the purpose of the alignment is only to point to the correct region of space to do a finer-scaled search, then small errors matter little. Finer scaled spatial information, both geometric and nongeometric, will be necessary for precise pinpointing of a target, but they are not needed for determining heading. This resilience against errors, or robustness, is a desideratum in any matching scheme.

Of course, using principal axes to determine directions can lead to a 180 ° misalignment between internal record and external space. In fact, matching on the basis of principal axes alone leaves a rotational ambiguity in all spaces, not just rectangular ones. This is because the two principal axes are orthogonal, and their poles are not marked. There are always two ways (180 ° apart) of matching up internal records of principal axes to those in the world. This is one reason why axes of symmetry, which can be curved, might make better global shape parameters for heading (Sect. 6.3.3, [6, 11]). In defense of principal axes, however, some other metric properties of the axes may serve to eliminate this ambiguity in most spaces. For instance, within the confines of a space, the two principal axes typically do not intersect at their respective centres. There is typically a long end and a short end. Hence, matching in addition the relative lengths from the intersection of the axes to the ends of the space does the job. Note that we should not take rotational errors in rectangular spaces to be evidence for the use of principal axes rather than other axes, because in a rectangular space, the same ambiguity applies to various axes of symmetry as well.

We can contrast this and similar global matching schemes (based on other types of axes) with local matching algorithms in which matches between corresponding points or features (sets of points) are sought. Local matching strategies face some problems, such as computational complexity and perceptual aliasing. As the space gets bigger, so do the number of features, leading to an explosion in the number of possible ways to match features in the world to features in the record. This makes for a large correspondence problem. Perceptual aliasing refers to ambiguity resulting from limited descriptions of what to match. For example, the limited description "door" has many possible matches in many indoor settings, as does"tree" in outdoor settings. False correspondences may be worse than useless. Perceptual aliasing may be reduced by increasing both the details in the descriptions of features, and the system's perceptual discriminative capabilities. Of course, both of these necessary tactics come at a cost.

Both the large correspondence problem and perceptual aliasing mean that small details can really matter. Small mismatches of details may lead to large errors, what amounts to a lack of robustness. For example, suppose that one door of a particular height and shade of brown on the north side of a room is to be used as a directional cue. If one errs on the colour and height, and ends up matching to the east door rather than the north door, one ends up with an enormous (90 ° error. To prevent such errors of fine details from wreaking havoc, one must match many features. This explosion is computationally expensive.

6.3.2 Transformational Experiments on the Use of Geometry

Despite the proliferation of geometry studies, no experiment addressed how geometry is used for relocation until 2004, when two papers rejecting global matching appeared, on rats [44] and on chicks [46]. Both studies used the transformational strategy [13]: Animals were trained in one shape of space and transferred on tests to different shapes. The rats were trained in a rectangular shape (without featural cues) and transferred to a kite shape [44]. Conceptually, this is formed by halving the space at a diagonal, reflecting one half, and joining the two halves at the diagonal again. The chicks were trained to search in a parallelogram arena, also without featural cues [46]. The parallelogram had 2:1 ratios in both corner angles and wall lengths. On tests, the shape was transformed to a rectangle (preserving ratio of wall lengths), a rhombus (preserving corner angles), or else reflected (preserving all metric properties but reversing left-right relations or sense). Both the rats and the chicks in all the transformed spaces still had strong biases to search at particular corners. Because the animals still searched nonrandomly in drastically changed shapes of spaces, they were clearly not matching on the basis of shape congruence.

Both Pearce et al. [44] and Tommasi and Polli [46] rejected all forms of global matching in favour of various local matching schemes (such as matching the angle of walls at a corner) and sensorimotor routines (such as finding a long wall and moving to its left end). Cheng and Gallistel [11], however, found that a strategy of using the major principal axis accounted for the data just as well. The minor principal axis was irrelevant in that it would not serve to disambiguate the rotational ambiguity in any of these training spaces. Cheng and Gallistel drew the first principal axis through each of training and transformed test spaces, and superimposed the training space on each test space, lined up at the middle of the axis. A strategy of picking out a corner on this basis accounted for all the data, without any further assumptions. In contrast, the local explanations proposed in [44] and [46] come with ad hoc qualifications.

Gallistel and I [6, 11] argued that some form of global matching is highly likely, although we did not favour principal axes as a basis for matching (discussed in Sect. 6.3.3). The reason that global matching is necessary has already been outlined. In simple experimentally constructed spaces, matching on the basis of a few local features may be possible. In the real world, even in artificially constructed worlds, this is rarely so. Finding the right tree or the right door out of an entire search space may tax the perceptual capabilities of an animal or a robot, unless the region in which to search is restricted. That is what some kind of global matching scheme does. It is not for pinpointing a location, but for picking out a rough region to search in. More narrowly targeted local processes can then effectively take over the job of pinpointing the target. A judicious combination of global and local processes can effectively cut down the explosion of computational complexity in both.

6.3.3 What Axes?

If axes are to be used, I suspect that principal axes are not the best. Although they are physically and computationally simple, as shape paramaters they miss a lot about the shape they are supposed to capture. Principal axes are necessarily straight lines at right angles to one another, and these facts may be disadvantageous. As already discussed, without additional properties used in matching, there are always two ways to match the principal axes in a record to the principal axes in the world, two matches 180 ° apart. Furthermore, having to be straight is a limitation; it means that the principal axes do not typically describe the essence of a shape. The principal axes do not capture the the S of a S-shaped or the C of a C-chaped space. Having principal axes over the entire space might also miss what we might consider as the structure of a space. Thus, a Y-shaped should have three radiating spines rather than two orthogonal axes.

A set of alternatives to principal axes are various axes of symmetry. Leyton [39] provides ample theoretical discussion. For instance, one set of axes may be formed by drawing circles that are tangent to the edges of the shape. Different formal definitions may then be used to define the axis of symmetry, such as the middles of the circles, or the midpoints of the chords connecting two corresponding points on the shape. These axes may curve as they run through the middle part of a section of space. This is an advantage, as most natural spaces, such as ponds or openings in a forest, are not rectilinear spaces, and a curved axis typically allows only one unambiguous match between the record and the world. Axes of symmetry provide a 'stick-figure' outline of the shape. Such axes can give an I-shaped space a single fairly straight axis, an S-shaped space a curved S-shaped axis, and the Y-shaped space three axes converging at the centre of the "Y". Whether such axes of symmetry (and which ones) prove to be important in biological navigation, and whether they may be useful for robotic navigation both remain to be seen.

Whatever axes or global representations of a space are used, they do not stand alone. Crucially, such global representations need to be linked to finer-scaled representations that can pinpoint a target. The finer scaled representations are likely to include both geometric and featural properties. This is evidenced well by a corpus of work on pigeons searching for a small target containing food in artificial arenas (e.g., [4, 5, 7],reviews: [8, 14]). The axes point to a region; then navigation requires appropriate look-up instructions, akin to arrows pointing to various inset maps with more details. Such an idea was already in Cheng [3]. A geometric frame encoding the broad shape, also called a geometric module, contains look-up instructions that point to various featural modules. This issue of instructional linkage goes beyond the use of geometry or scales of maps. I have biased the discussion in this chapter to map-like representations. But instructional linkages are essential even when the navigational strategies are not map-like. Insect strategies, which have been useful for robotics [29], Franz this volume], are described as a series of navigational servomechanims [10]. Navigational servomechanisms are linked together to accomplish the task. At appropriate points in the chain, one mechanism must give way to another. Contextual cues are suggested to play a major role in instructional linkage [10, 15].

6.4 Geometry and Robotic Navigation

In attempting to construct a metric representation of space in artificial intelligence, Davis [23] emphasised the need for representation of space at different scales. More global representations are necessarily coarser, a fact arising from imprecisions in representations. The same proportion of error ($\pm x\%$) translates to larger absolute errors with a more global representation. Representations of smaller regions, with smaller absolute errors, are needed for fine-scaled navigation. The need to deal with different scales of representations is also discussed elsewhere in this book (see Chaps. by Thrun, Mallot et al., Yeap, Jefferies). To my knowledge, however, the use of principal axes or other axes has not been proposed for robotic navigation.

Personal communication and a check on citations have revealed three lines of robotic work that draw some inspiration from Cheng [3]. Egerton and colleagues [24, 25] took the original notion of a geometric module and geometric frame, and based a program of robotic navigation on it. Nolfi [43] tested minimal robots in the Cheng [3] task in evolutionary simulations. Yeap and Jefferies [59], Jefferies this volume, Yeap this volume] suggested that the construction of a frame of boundaries with exits is important for a robot in building a map. Such a frame shares much with the notion of a geometric frame.

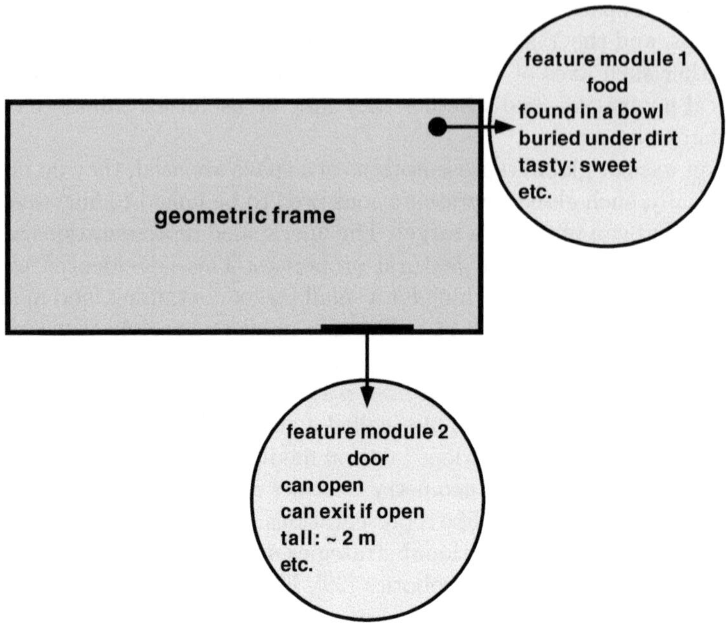

Fig. 6.2. Schematic adaptation of Egerton et al.'s [25] geometric frame plus feature modules for robotic navigation, ideas borrowed from [3]. The robot is navigating in a rectangular arena such as the one used in Cheng's [3] experiments. The geometric frame encodes only the overall shape of the space, plus labels for feature modules, two of which are shown. The robot needs to look up the contents of feature modules for further descriptions of a particular location.

6.4.1 "Biologically Inspired" Robotic Navigation

Egerton and colleagues [24, 25] took inspiration from the original geometric module [3] in designing and building navigating robots for indoor and outdoor environments. They titled their work "biologically inspired" in both publications. A sketch of their key idea (Fig. 6.2) reveals a division of the mapping system into a geometric frame and feature modules, an idea resembling the theoretical explications in Cheng [3]. The idea is that an overall frame based on the geometry of surfaces serves to point out approximately where the robot is. As with Cheng [3], the frame encodes the broad shape rather than summary axes. Feature modules to be accessed for particular locales then serve to specify nongeometric cues useful for navigation in the area.

An important reason for this division of representation is the perceptual aliasing problem already discussed. Landmarks are so plentiful in typical outdoor and indoor environments that they become an embarrassment of riches that creates problems. Too many views of different objects or scenes are not discriminable from one another. Too many landmarks may fill particular descriptions such as a door or a tree. A geometric frame has the function of reducing perceptual aliasing problems. It gets a robot to the ball park, a region at which a target may be found. The number of perceptual matches should then be greatly reduced. The robot can get away with limited perceptual abilities and limited descriptions of what it is looking for or attempting to match.

6.4.2 Representationless Solutions to Navigation

Many roboticists would want a robot to have some internal states. These states may be consulted and used in part to make decisions. They are internal representations. Notions such as axes of symmetry constitute internal representations. Agents without such representations are called reactive agents. They are basically S-R machines. Reactive agents always react to any particular external stimulus in the same way; thus, no internal representation is consulted. Typically, they have limited perceptual capabilities as well, and bear the full brunt of the perceptual aliasing problem. Nolfi [43] described a number of complex tasks that reactive agents can solve. Included among them is the Cheng [3] task. The reactive agents, however, only solve the task up to rotational errors.

Nolfi [43] engineered his reactive agents with genetic algorithms that simulate natural selection in biological evolution. A number programs for a specified robot with limited perceptual and motor capabilities were set to work on a problem of locating a target corner in a rectangular arena; this problem stayed the same throughout evolution. Those doing the best got to breed. They reproduced copies of themselves, with some variations in their codes. The rest, which failed the selection process, were thrown out. The new generation was set to work on the problem, and the selection and reproduction processes were repeated.

The reactive agents did not possess senses that can discriminate a long wall from a short wall. Those that evolved to solve the task developed various behavioural strategies for always getting to the correct kind of corner (the correct corner or the rotational error). For example, in one problem, the robot was always released at one of eight positions within the rectangular arena (chosen at random): at the centre, facing the middle of one of the four sides, or at the middle of one of the four sides, facing the centre. (These were the training conditions for the rats in [3].) One line of robots evolved a strategy of moving ahead over open space with a drift to the left. The drift ensured that it always ran into a long wall first no matter where it was started from. And then, to solve this particular problem, it was a matter of turning left when it ran into a wall, and moving along the wall until it got to a corner. Each of these routines are sensorimotor in nature. Each can be specified as a programmed reaction to a particular perceived situation, without recourse to any internal representation.

Reactive agents can do many things, including solving the Cheng [3] relocation task up to rotational errors. But of course, as acknowledged by Nolfi [43], that does not mean that rats or other animals are solving the task in the same way. In the studies on relocation in an indoor arena, only the location of search has been reported. The paths taken by the animals have not appeared in print. Nolfi's work suggests that such paths can be relevant to interpreting what strategy an animal is using.

Another important tool in [43] is evolutionary simulation. This too can be useful for understanding animal behaviour. It has been used to investigate insects' navigational strategies [22]. Neural networks were given a problem presented to insects, which was to find a target location near a salient landmark (e.g., [2, 57]). The neural networks evolved by selection and recombination in the computer simulations. (Thus, the work was all virtual, with no robots.) Successful networks were selected to reproduce by mixing their algorithms. Interestingly, the evolved solutions of bee-like networks (which were those that could move sideways) resembled bee and wasp search strategies [9, 16, 17]. They first zoomed in towards the landmark (a strategy called beaconing), and then veered off towards the goal when they were near the landmarks.

It would be interesting to run evolutionary simulations on the Cheng [3] relocation problem, but going beyond reactive agents [43] and using agents with representations. Would a strategy of using global shape parameters evolve, and if so, what parameters?

6.4.3 Frames with Exits

Yeap and Jefferies [59] were concerned with early cognitive mapping, which has the goal of getting a robot to build up a map of the environment it is navigating through. They suggest the strategy of having the robot build an early map of a local environment specifying where boundaries (such as walls) and exits are. The idea is to ignore or look past objects within a space (for the time being), and map out where the barriers are that bound a space, along with openings through the barriers, which are exits. Such a frame, called an Absolute Space Representation, allows the robot to navigate through its environment while it builds up more knowledge. I am being deliberately brief, as the chapters by Yeap and by Jefferies in this volume contain far more detailed explications of the work.

Robotics is useful for studies in animal behaviour as well as vice versa [54]. In the case of Yeap and Jefferies' Absolute Space Representation, it points out the importance of encoding exits. In the animal literature on geometry, the importance of exits has not yet been tested. If exits are important, then they might serve to disambiguate rotational symmetry, whereas other objects of equivalent size might not. Testing this hypothesis is straight forward, and deserves to be done.

6.5 Geometry and Modularity

Systematic rotational errors, found in mammals, have led to views of modularity, in which the encoding of geometry is done by a dedicated module that does not encode other kinds of information [12]. The title of the Cheng [3] paper included the phrase "a purely geometric module". It reflected the theoretical view that the labour of encoding spatial properties is divided in the brain. One unit or module has the job of encoding just the geometric properties. The module has also been called the geometric frame. It is supposed to be the main map for a space, serving the role of coordinating all spatial information. This map would contain labels for looking up other information at key locations. Important featural information is in principle accessible within the same system, via these look-up instructions. Egerton et al.'s robotic program [24, 25] was based on this idea.

The idea of modularity in encoding geometric information can be pushed further, or it can be abandoned [12]. In a strong view of modularity [52, 53], one entire navigational system is based solely on geometric information, without access to featural information via look-up rules. It was not clear to Cheng and Newcombe [12] what the entirely modular system is supposed to do. Wang and Spelke [52] write of the re-orientation process, the process of establishing which direction is which. Wang and Spelke [53] on the other hand, contradict this claim, and write that it is the relocation process that is modular, the step in which the surrounding landmarks are used to pinpoint the target location. In either case, success at using featural information, which is also clearly demonstrated in mammals, comes about via the use of other modules than the geometric module. One such module is view-based matching [52].

Modularity is not needed to account for rotational errors. Various performance factors may be invoked [12, 41]. The fact that both featural and geometric information are used in all animals tested, at least under the right circumstances, makes it unnatural to positseparate stores for geometric and featural information. Furthermore, one would want to encode the geometric relations between features as well as the geometric relations of points as points; surely the distance and direction (metric relations) among key features matter. Rotational errors might arise because features are not encoded (for instance, because they are harder to learn than the broad geometric shape), or because the animal does not put much weight on them, even if they are encoded.

My current view is a mixture of all these views [6]. Geometric and featural information are stored together in one system. But one (or more) process of extracting major axes of space, as already discussed, operates on the geometric information in the representation. Axes are shape parameters defined by the geometry of shape, not dependent in any way on colours or smells. The use of such axes can sometimes lead to rotational confusions akin to cognitive illusions. This process alone can account for rotational errors, without the need for further modularity.

Modularity is a key notion in cognitive science [18, 27, 28, 31, 32, 33, 41]. It might be some time before the issue of modularity and geometry is settled.

6.6 Concluding Remarks

In summary, the importance of geometric information in navigation arose from a systematic error that rats made in a relocation task in a rectangular arena [3]. The rat confused locations that stood in the same geometric relation with respect to the shape of the arena. These locations are clearly distinguishable if nongeometric information such as the grey levels on walls or smells in the corners is used to define a location. This work has led to research on the relocation task in a number of other vertebrate animals, including fish, chicks, pigeons, monkeys, and human children and adults. Mammals make systematic rotational errors under some circumstances. The other vertebrates do not make systematic rotational errors, but prove to be adept at learning geometric information nevertheless. The systematic rotational errors have led to various and differing views on modularity and geometric information, an issue that is still unsettled [12]. Gallistel [31] has suggested that the direction determining process is based not on the entire shape, but on the shape parameters of the principal axes of space. Gallistel and I have suggested recently that other axes, various axes of symmetry, may be better candidates [6, 11]. The ideas in Cheng [3] have a played a role in various programs of robotic navigation ([25, 43, 59], Jefferies this volume, Yeap this volume).

References

1. Biegler, R.: Possible uses of path integration in animal navigation. Anim. Learn. Behav. 28, 257–277 (2000)
2. Cartwright, B.A., Collett, T.S.: Landmark Learning in Bees. Journal of Comparative Physiology A 151, 521–543 (1983)
3. Cheng, K.: A purely geometric module in the rat's spatial representation. Cognition 23, 149–178 (1986)
4. Cheng, K.: Some psychophysics of the pigeon's use of landmarks. Journal of Comparative Physiology A 162, 815–826 (1988)
5. Cheng, K.: The vector sum model of pigeon landmark use. Journal of Experimental Psychology: Animal Behavior Processes 15, 366–375 (1989)
6. Cheng, K.: Reflections on geometry and navigation. Connection Science 17, 5–21 (1990)
7. Cheng, K.: The determination of direction in landmark-based spatial search in pigeons: A further test of the vector sum model. Animal Learning and Behavior 22, 291–301 (1994)
8. Cheng, K.: Landmark based spatial memory in the pigeon. In: Medin, D.L. (ed.) The psychology of learning and motivation, vol. 22, pp. 1–21. Academic Press, New York (1995)
9. Cheng, K.: How honeybees find a place: lessons from a simple mind. Animal Learning and Behavior 28, 1–15 (2000)
10. Cheng, K.: Arthropod navigation: Ants, bees, crabs, spiders finding their way. In: Wasserman, E.A., Zentall, T.R. (eds.) Comparative cognition: Experimental explorations of animal intelligence, pp. 189–209. Oxford University Press, Oxford (2006)
11. Cheng, K., Gallistel, C.R.: Shape parameters explain data from spatial transformations: Comment on Pearce et al (2004) and Tommasi and Polli (2004). Journal of Experimental Psychology: Animal Behavior Processes 31, 254–259 (2005)
12. Cheng, K., Newcombe, N.S.: Is there a geometric module for spatial orientation? Squaring theory and evidence. Psychonomic Bulletin and Review 12, 1–23 (2005)
13. Cheng, K., Spetch, M.L.: Mechanisms of landmark use in mammals and birds. In: Healy, S. (ed.) Spatial representation in animals, pp. 1–17. Oxford University Press, Oxford (1998)
14. Cheng, K., Spetch, M.L., Kelly, D.M., Bingman, V.P.: Small-scale spatial cognition in pigeons. Behav. Proc. 72, 115–127 (2006)

15. Collett, T.S., Fauria, K., Dale, K.: Contextual cues and insect navigation. In: Jeffery, K.J. (ed.) The neurobiology of spatial behaviour, pp. 31–47. Oxford University Press, Oxford (2003)
16. Collett, T.S., Rees, J.A.: View-based navigation in Hymenoptera: multiple strategies of landmark guidance in approach to a feeder. Journal of Comparative Physiology A 181, 47–58 (1997)
17. Collett, T.S., Zeil, J.: Places and landmarks: an arthropod perspective. In: Healy, S. (ed.) Spatial representation in animals, pp. 18–53. Oxford University Press, Oxford (1998)
18. Coltheart, M.: Modularity and cognition. Trends Cognition Science 3, 115–120 (1999)
19. Kelly, D.D, Spetch, M.L.: Pigeon's encoding of geometric and featural properties of a spatial environment. Journal of Comparative Psychology 112, 259–269 (1998)
20. Kelly, D.D, Spetch, M.L.: Pigeons encode relative geometry. Journal of Experimental Psychology: Animal Behavior Processes 27, 417–422 (2001)
21. Kelly, D.D, Spetch, M.L.: Reorientation in a two-dimensional environment: II. Do pigeons (Columba livia) encode the featural and geometric properties of a two-dimensional schematic of a room? Journal of Comparative Psychology 118, 384–395 (2004)
22. Dale, K., Collett, T.S: Using artificial evolution and selection to model insect navigation. Current Biology 11, 1305–1316 (2002)
23. Davis, E.: Representing and acquiring geographical knowledge, Pitman, London (1986)
24. Egerton, S., Callaghan, V.: From mammals to machines: towards a biologically inspired mapping model for autonomous mobile robots. In: Proc. International Conference on Intelligent Autonomous Systems, Venice, Italy (2000)
25. Egerton, S., Callaghan, V., Chernett, P.: A biologically inspired mapping model for autonomous mobile robots. In: Mohammadin, M. (ed.) New frontiers in computational intelligence and its applications, pp. 20–29. IOS Press, Amsterdam (2000)
26. Etienne, A.S., Jeffery, K.J.: Path integration in mammals. Hippocampus 14, 180–192 (2004)
27. Fodor, J.A: The modularity of mind. MIT Press, Cambridge, MA (1983)
28. Fodor, J.A: The mind doesn't work that way. MIT Press, Cambridge, MA (2001)
29. Franz, M.O., Mallot, H.A.: Biomimetic robot navigation. Robotics Autonomous System 30, 133–153 (2006)
30. Von Frisch, K., Lindauer, M.M: Himmel und Erde in Konkurrenz bei der Orientierung der Bienen [Sky and Earth in competition in the orientation of bees]. Naturwissenschaften 41, 245–253 (1954)
31. Gallistel, C.R.: The organization of learning. MIT Press, Cambridge, MA (1990)
32. Gallistel, C.R.: The replacement of general-purpose learning models with adaptively specialized learning modules. In: Gazzaniga, M.S. (ed.) The new cognitive neurosciences, MIT Press, Cambridge, MA (2000)
33. Gallistel, C.R.: The principle of adaptive specialization as it applies to learning and memory. In: Kluwe, R.H., Lüer, G., Rösler, F. (eds.) Principles of human learning and memory, Birkhäuser Verlag, Berlin (2002)
34. Gouteux, S., Thinus-Blanc, C., Vauclair, J.: Rhesus monkeys use geometric and nongeometric information during a reorientation task. Journal of Experimental Psychology: General 130, 505–519 (2001)
35. Hermer, L., Spelke, E.: A geometric process for spatial representation in young children. Nature 370, 57–59 (1994)

36. Hermer, L., Spelke, E.: Modularity and development: the case of spatial reorientation. Cognition 61, 195–232 (1996)
37. Learmonth, A.E., Nadel, L., Newcombe, N.S.: Children's use of landmarks: implications for modularity theory. Psychology Science 13, 337–341 (2002)
38. Learmonth, A.E., Newcombe, N.S., Huttenlocher, J.: Toddlers' use of metric information and landmarks to reorient. Journal of Experimental Child Psychology 80, 225–244 (2001)
39. Leyton, M.: Symmetry, Causality, Mind. MIT Press, Cambridge, MA (1992)
40. Margules, J., Gallistel, C.R.: Heading in the rat: determination by environmental shape. Animal Learning and Behavior 16, 404–410 (1988)
41. Newcombe, N.S.: The nativist-empiricist controversy in the context of recent research on spatial and quantitative development. Psychology Science 13, 395–401 (2002)
42. Newcombe, N.S., Huttenlocher, J.: Making space: the development of spatial representation and reasoning. MIT Press, Cambridge, MA (2000)
43. Nolfi, S.: Power and limits of reactive agents. Robotics Autonomous System 42, 119–145 (2002)
44. Pearce, J.M., Good, M.A., Jones, P.M., McGregor, A.: Transfer of spatial behavior between different environments: Implications for theories of spatial learning and for the role of the hippocampus in spatial learning. Journal of Experimental Psychology: Animal Behavior Processes 30, 135–147 (2004)
45. Sovrano, V.A., Bisazza, A., Vallortigara, G.: Modularity as a fish (Xenotoca eiseni) views it: Conjoining geometric and nongeometric information for spatial reorientation. Journal of Experimental Psychology: Animal Behavior Processes 29, 199–210 (2003)
46. Tommasi, L., Polli, C.: Representation of two geometric features of the environment in the domestic chick (Gallus gallus). Animal Cognition 7, 53–59 (2004)
47. Tommasi, L., Vallortigara, G.: Searching for the center: spatial cognition in the domestic chick (Gallus gallus). Journal of Experimental Psychology: Animal Behavior Processes 26, 477–486 (2000)
48. Tommasi, L., Vallortigara, G., Zanforlin, M.: Young chicks learn to localize the center of a spatial environment. Journal of Comparative Physiology A 180, 567–572 (1997)
49. Vallortigara, G., Pagni, P., Sovrano, V.A.: Separate geometric and non-geometric modules for spatial reorientation: Evidence from a lopsided animal brain. Journal of Neuroscience 16, 390–400 (2004)
50. Vallortigara, G., Zanforlin, M., Pasti, G.: Geometric modules in animals' spatial representations: a test with chicks (Gallus gallus domesticus). Journal of Comparative Psychology 104, 248–254 (1990)
51. Vargas, J.P., López, J.C., Salas, C., Thinus-Blanc, C.: Encoding of geometric and featural information by goldfish (Carassius auratus). Journal of Comparative Psychology 118, 206–216 (2004)
52. Wang, R.F., Spelke, E.S.: Human spatial representation: insights from animals. Trends Cognition Science 6, 376–382 (2002)
53. Wang, R.F., Spelke, E.S.: Comparative approaches to human navigation. In: Jeffery, K.J. (ed.) The neurobiology of spatial behaviour, Oxford University Press, Oxford (2003)
54. Webb, B.: What does robotics offer animal behaviour? Animal Behaviour 60, 545–558 (2000)

55. Wehner, R.: The polarization-vision project: championing organismic biology. In: Schildberger, K., Elsner, N. (eds.) Neural basis of behavioural adaptation, G Fischer, Stuttgart (1994)
56. Wehner, R.: Desert Ant Navigation: GHw Miniature Brains Solve Complex Tasks. Journal of Comparative Physiology A 189, 579–588 (2003)
57. Wehner, R., Räber, F.: Visual Spatial Memory in Desert Ants, Cataglyphis Bicolor (Hymenoptera: Formicidae). Experientia 35, 1569–1571 (1979)
58. Wehner, R., Srinivasan, M.V.: Path integration in insects. In: Jeffery, K.J. (ed.) The neurobiology of spatial behaviour, Oxford University Press, Oxford (2003)
59. Yeap, W.K., Jefferies, M.E.: Computing a Representation of the Local Environment. Artificial Intelligence 107(2), 265–301 (1999)

7

Cue and Goal Encoding in Rodents: A Source of Inspiration for Robotics?

Etienne Save, Vincent Hok, Sophie Renaudineau, Carole Parron, and Bruno Poucet

Summary. To navigate in their environment, rodents are able to rely on a variety of behavioral strategies. The most flexible strategies result from their ability to form spatial representations that encode information about spatial cues and about important places (nest, goals, etc). In the present chapter, we address the issues of cue and place encoding in the brain and suggest that they are crucial processes for behavioral flexibility and adaptation to environmental changes. First, it is suggested that, due to a different spatial distribution (distant *vs.* nearby) or nature (allothetic *vs.* idiothetic) of spatial cues, animals use and have to coordinate the use of multiple spatial reference frames. This involves activation of various brain regions including the hippocampus and neocortical areas. In particular, location-specifc activity of hippocampal neurons (place cells) has been shown to be controlled by different reference frames. Second, we present new data suggesting that activity of prefrontal cortex neurons reflects goal encoding. It is concluded that the knowledge of these mechanisms in animals may be a source of inspiration to improve the adaptive capacities of navigating robots.

7.1 Space and Navigation

All our everyday actions take place in time and space. The nature of space, a philosophical issue that has stirred up the minds since ancient times [40, chapter 1], remains enigmatic. Nevertheless, space is, above all, a substantial property of the world with which we have to deal permanently. Virtually all animal species are equipped with mechanisms that allow perception and processing of space. Spatial behaviors, e.g. exploration, orientation, navigation, etc., result from activation of these mechanisms. Moving in the environment is crucial for survival (to fulfill fundamental needs such as hunger, thirst, etc.). Thus, spatial behaviors have a very strong adaptive value for all species. They have evolved so as to allow animals to gain independence relative to environmental changes.

Current studies of spatial behaviors in animals owe much of their conceptual background to the work of O'Keefe and Nadel [40]. Following Tolman's early suggestion [63], these authors basically proposed that animals are able to form a representation of their environment in the brain, based on the encoding of multiple spatial relationships between landmarks. Using an allocentric representation (or spatial map), animals are able to exhibit flexible behavior, i.e. to adapt their behavior to environmental changes. This endows them with the capacity of using shortcuts, detours, or navigate to places from new starting locations, etc.

M.E. Jefferies et al. (Eds.): Robot. & Cogn. Approach. to Spat. Map., STAR 38, pp. 163–180, 2008.
springerlink.com

The issue of patial mapping is also relevant for roboticists whose purpose is to build autonomous mobile robots. Indeed, to be autonomous, a robot must exhibit a number of abilities that characterize animal navigation based on an map-like representation. Thus, in the present chapter, we address the issue of how space is represented in the brain by examining two basic aspects of spatial mapping, namely cue encoding in different reference frames and goal encoding. Lesion and electrophysiological studies in rodents reveal that these functions correspond to activation of several brain regions that may be part of a large functional network devoted to spatial cognition.

7.1.1 Spatial Perception and Strategies

The strategy an animal uses to navigate is determined in part by the sensory information available. There are two sorts of information: idiothetic information originating from the animal's own movement (proprioceptive, vestibular information), and allothetic information originating from the environment (visual, auditory, somatosensory, or olfactory information). Using only idiothetic cues, an animal is able to continuously calculate its position relative to a reference location (e.g. the nest) and eventually return in a straight path to this location. Signals generated by linear and angular acceleration are integrated during its journey, therefore allowing the animal to maintain and update a vector oriented toward the reference location. This strategy is called path integration and has been described in many different species from ants to humans [14, for a review]. Path integration does not allow flexible navigation since it offers the animal only one possible path to return to the nest. However, this is undoubtedly a strategy which is a fundamental part of the spatial behaviour of rodents. Indeed, it allows the maintenance of minimal navigational ability in absence of allothetic cues, i.e. in darkness. Functionally, it is assumed that rats use path integration when they have to explore a part of their environment they have not encountered before. Because "spatial mapping takes time" [66], initial excursions in an unfamiliar area have to be performed independently of allothetic cues. Rats therefore rely on path integration to generate a direct trajectory back to the nest. Due to accumulation of errors resulting from successive estimation of linear and angular movements however, path integration allows accurate navigation only across short distances [15].

Exploration allows the animal to progressively acquire some spatial knowledge of its whole environment [62, 44]. This supports the early hypothesis that attributed a role to exploratory activity in building and updating spatial maps [40]. For example, during exploration rats are able to memorize the spatial arrangement of a group of objects located in their environment. Experimentally, formation of a spatial representation is probed by examining the effects of changing the spatial relationships between objects. Modification of the object configuration induces an increase of exploration directed toward the displaced objects [62, 56]. Such renewal of exploration indicates that the spatial change has been detected and identified. This suggests that there exists in the brain a mechanism that performs a comparison between the familiar configuration and the new configuration, allowing detection of a mismatch. The reactivation of exploration is assumed to reflect an updating process of the representation.

Through exploration, an animal also encodes multiple spatial relationships between distant cues. The spatial map thus allows the animal to use a global frame of reference to encode important places and derive their spatial relationships. Because it is independent on the animal's position, the spatial map places few restrictions on the trajectories across that space, thus enabling flexible navigation. For example, a rat could be able to reach a familiar place from virtually any novel starting location. Navigation based on a spatial map is classically tested in the Morris water maze task [34]. The rat is released at the periphery of a circular tank containing opaque water and has to swim until it comes across a submerged, i.e. not visible, platform. In a few trials from different starting places, the rat learns the position of the platform. That the animals exclusively rely on distal room cues to reach the goal is tested during a probe trial with the platform removed. Rats usually spend more time in the area that contained the platform during training relative to other equivalent areas in the pool, thus exhibiting clear place learning ability. In addition, a rat is still able to reach the goal in a straight path when released from a starting place it has never experienced before.

Although a spatial map appears to be the most powerful mechanism to allow efficient navigation in any situation, it is likely that, when appropriate, animals use alternative strategies that require less cognitive demand than spatial mapping. In the most favorable situation, the goal is visible from the animal's starting place. Navigation thus implies that the animal decreases the distance between itself and the goal. Similarly, would the goal be an olfactory or auditory source, the rat would reach it by following the concentration gradient. In large environments however, the goal may be located beyond direct perception. A possible navigational strategy is then to learn a route. To do so, the animal has to memorize a sequence of associations between stimuli (landmarks) and movements (turn right, turn left, go ahead, etc). Navigation based on a route can be very fast as long as the animal can recall correctly the whole sequence. However, if a landmark is missing, the sequence is interrupted and the animal fails to reach the goal. This strategy does not allow adaptation to environmental changes. Note that path integration, visually-guided and route strategies are based on the processing of egocentric spatial relationships, i.e. relationships that are encoded in a reference frame centered on the animal whereas spatial mapping is based on the processing of allocentric spatial relationships.

7.2 Cue Encoding and Spatial Reference Frames

7.2.1 Why Use Different Reference Frames?

Space appears as a continuous, coherent dimension within which we perform our body or navigation movements. In fact, our sensation of a single spatial continuum results from the integration of multiple spaces each with its own reference frame. The reference frames are generated from a collection of stable environmental or internal cues that may be used as coordinate systems to compute locations and movements. A place can be memorized in different systems of reference which are usually coherent, thus allowing construction of an overall integrated spatial representation. Two main classes of reference frames can be considered, one is based on the processing of idiothetic information and the other on the processing of allothetic information. Although these two frames may be used independently to guide spatial behavior and allow specific strategies (see Sect. 7.1.1), it is assumed that an animal's navigational strategy results from a complex interaction between allothetic and idiothetic cues. Allothetic information has been recently shown to be encoded in different frames of reference, based on the location, either proximal or distal, of environmental landmarks. Proximal landmarks are usually three dimensional objects located in the animal's locomotor space, which the animal can approach and closely investigate. In contrast, distal cues are stimuli that are located in the remote environment. They thus have a two dimension appearance and cannot be explored by the animal. For example, in the water maze task, distal cues are typically cues provided by the experimental room (posters, cabinets, windows, etc.) whereas proximal cues would be objects placed in the pool. Distal cues have been shown to be readily used by animals to perform various spatial tasks [13, 60]. In contrast, although "spatially separated intramaze cues can also serve as place cues" [38], it seems that using proximal cues for place learning requires more extensive training [20, 22, 59]. This suggests that encoding spatial information using proximal and distal cues activates distinct processes.

7.2.2 Hippocampal Place Cells

How the brain encodes the different frames of reference and manages their interaction is a major issue that has been investigated in the last ten years. Most studies have examined the contribution of the hippocampus in the use of reference frames using system and unit recording level approaches. The assertion that the hippocampus plays a crucial role in spatial cognition dates back to O'Keefe and Nadel's work [40]. These authors proposed that spatial maps are implemented in the hippocampus, based on data showing that hippocampal lesions disrupt spatial learning and above all, based on the existence of place cells [39]. Place cells are pyramidal cells located in the CA1 and CA3 sub-fields of the rat hippocampus. They are characterized by location-specific firing, that is they tend to fire rapidly when the animal enters a restricted area in the environment, called either the place field or the firing field [35, 46, 3, for reviews]. Place cells are virtually silent when the animal is outside the place field. Simultaneous recording of a large population of cells shows that an environment is entirely mapped at a neural level and can be described as a unique spatial pattern of place fields. Because there is some degree of overlapping between place fields, each location corresponds to activation of a large amount of place cells. Thus, functional dynamics of the place cell system likely results from a interaction between spatial and temporal firing properties of cell populations.

7.2.3 Place Cell Activity is Controlled by Allothetic and Idiothetic Cues

Place cell activity has been demonstrated to be controlled by environmental cues. In a classical experiment, a rat has to continuously explore a circular arena to retrieve small food pellets that drop from a feeder fixed to the ceiling. Continuous exploration allows correct sampling of unit activity in all locations of the arena [37]. The arena is located in a cue-controlled environment, i.e. surrounded by opaque curtains, and the only available cue is a large white cardboard (the "cue card") covering 100° of arc of the arena wall. A first recording session is conducted to identify the location of the place field relative to the cue-card. Ninety degree rotation of the cue-card in absence of the animal produces an equivalent rotation of the place field in a second recording session, thus showing that the cue-card exerts a control over place field location. Place cell activity is also strongly influenced by environmental features such as the geometric shape of the arena. Transporting a rat from a circular-shaped to a square-shaped apparatus usually induces a dramatic change in the spatial pattern of place fields. Some place cells begin to fire in a different location (e.g. south instead of north) while others stop firing. New fields, developed by previously silent cells, also appear. This phenomenon, known as remapping, reflects the capacity of the place cell system to encode separate representations of distinct environments. Each specific representation is reactivated when the animal recognizes the corresponding environment [36, 42]. Recent studies revealed that remapping may not occur on the first exposure to a new environment but may result from learning [31].

Most popular spatial tasks (e.g. water maze, etc.) are based on the use of visual cues while other sensory information, such as olfactory, auditory, tactile, movement-related information are made irrelevant or eliminated. Thus, one often assumes that rats mainly rely on visual information to form spatial maps. However, studies have accumulated indicating that various sensory modalities collectively contribute to place cell firing. For example, we have shown that blind rats that never had visual experience still exhibited normal place cell activity when adults [51]. Indeed, place fields were found to be controlled by objects located at the periphery of a circular arena. It was concluded that vision is not crucial for normal development of the place cell system and that other modalities may have compensated for the lack of vision. In particular, rats may have used conjointly tactile, olfactory, and idiothetic information to allow for place field stability. In another study, we compared the stability of place fields in darkness when olfactory cues deposited by the animal were either eliminated or available [52]. The results revealed that elimination of olfactory cues yielded unstable place fields. In particular, most cells stopped firing. Overall, these result are consistent with the idea that different sensory modalities can sustain place field stability. When they provide coherent information, these modalities act in combination and are able to compensate each other. Thus, the place cell system appears as an opportunist system whose function is to maintain spatial mapping in spite of environmental changes.

Among the sensory information that plays an important role in maintaining the stability of place fields, idiothetic information has been neglected for a long time. The idea that place fields are controlled by such information was initially supported by two observations. First, place cell activity was found to be influenced by vestibular inputs. Rotation of the whole apparatus at a speed that is detected by the rat vestibular system, resulted in fields remaining stable relative to the experimental room reference frame [58, 67]. In contrast, for rotation speed below threshold for vestibular detection, place fields were found to rotate with the apparatus as the vestibular system was unable to detect the rotation of the apparatus. Second, place fields were found to persist after the room light had been turned off, i.e. in darkness, suggesting that the lack of visual inputs was compensated for by other sensory information including idiothetic cues [47]. In Save et al.'s study (2000), we also investigated the persistence of place fields in darkness to determine how robust this persistence would be during a prolonged dark period. Manipulating sensory information indicated that the fields remained stable provided olfactory cues (self-deposited odors) were available. However, olfactory cues by themselves are not sufficient to support spatial mapping [28]. Thus, combining olfactory and idiothetic cues may be necessary to keep stable spatial mapping.

A number of studies have suggested that allothetic and idiothetic reference frames interact to control place field location. Gothard and co-workers [20] recorded place cells as the animal was trained to shuttle on a linear track between a starting box and a fixed reward location. During the journey to the goal, the starting box could be moved to any of five possible locations (including the initial location) along the track so that after visiting the goal, the rat returned to the box which was then in a new location. When the box was moved toward the goal, a mismatch occurred in outbound journeys between the rat's location as estimated by path integration and its actual location relative to environmental cues. Gothard et al. found that in all journeys, place fields located on the initial part of the journey were controlled by path integration (relative to the starting box) whereas place fields located on the final part of the journey were controlled by external cues (Figure 7.1). For mismatch situations, place fields located at a distance from the box were found to shift from an internal (path integration) to an external (environmental) reference frame. This correction process was interpreted as revealing a competing interaction between the two reference frames.

One way to disentangle the respective influence of idiothetic and allothetic cues on place cells is to conduct "conflict" experiments in which the two kinds of information are in discrepancy. For example, Jeffery and O'Keefe [25] examined which kind of information, allothetic or idiothetic predominated in controlling place cell activity when a distal stimulus (a card) and the rat were rotated by different angles (90 and 180°). Prior to rotation trials, a number of training trials were made during which the rats experienced manipulation of the card (90° rotation). "Uncovered" rats could see the card moving from trial to trial and "covered" rats could not see the card moving. In conflict trials, place cells in uncovered rats displayed a switch from a control by idiothetic cues to a control

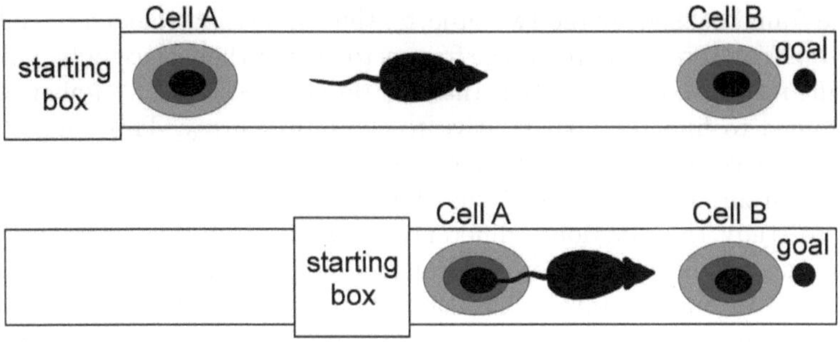

Fig. 7.1. A schematic of the experiment by Gothard et al. [20]. Rats were trained to shuttle on a linear track between a starting box and a food cup (goal). During outward journeys, the starting box could be moved to a different location (bottom). On the initial portion of the track, cells fired at a constant distance from the starting box (place field of cell A) whereas on the final portion, they fired relative to fixed cues (place field of cell B), suggesting a control by path integration and environmental cues, respectively.

by the card whereas in covered rats, the card predominated in controlling the place fields. This suggests that for uncovered rats the card was unreliable for anchoring place fields, thus leading the animal to switch to idiothetic cues.

In studies using conflicts, the environmental cue is generally rotated while the animal remains in a stationary arena. It was reported that for moderate mismatch (45°), place fields remain under the control of the visual cue [27, 50]. In contrast larger mismatch (e.g. 180°) results in the fields remaining stable suggesting a control by idiothetic cues [50] or in the formation of a new representation (remapping), suggesting a re-organization of the spatial map with respect to the environmental reference frame [20, 6, 27].

Overall, it is now acknowledged that place field stability basically results from a dynamic interaction between idiothetic cues and allothetic cues. Some authors have proposed that idiothetic cues are a primary drive for place cells resulting from the intrinsic wiring of the hippocampus whereas, allothetic cues are progressively embodied in the representation through experience-dependent modifiable connections [32].

7.2.4 Far and Near: Two Allocentric Reference Frames Based on Distal Cues and Proximal Objects

Although most spatial tasks are based on utilization of distal cues, a number of studies have suggested that rats may also rely on proximal cues to form an allocentric spatial representation. In a study by Gothard and colleagues [20], rats were trained to locate a goal relative to two objects placed directly in the circular arena that also contained distal cues, i.e. cues attached to the arena wall. Most place cells were found to fire as a function of a specific reference frame. Some of them were bound to the arena frame and others were bound to the reference frame defined by the two objects, thus indicating that place cells may use different environmental reference frames to anchor their place fields. Another interesting result from this study is that the rats needed a great number of trials to be trained to find the goal relative to the object array. This suggests that using a reference frame based on proximal objects is more difficult than using a reference frame based on distal cues for place learning. Thus, an important issue is to study the functional difference between these two reference frames so as to get some insight on how environmental spatial information is encoded in the brain. In our laboratory, Cressant et al. examined whether the control that objects directly placed in the arena may exert on place fields is similar to that exerted by distal cues. The recording arena contained three distinct objects that were placed in a central position and formed an equilateral triangle [10]. A procedure similar to that previously used to determine the control exerted by a two dimension cue card on the angular position of place fields was used [36]. Thus, following an initial recording session, the effect on place fields of rotating the object array 90° was tested. The main finding is that the objects did not control the angular position of place fields, i.e. rotation of the object array did not induce equivalent rotation of the fields. In contrast, control was obtained when *a)* a wall-fixed cue card was added to the set of object, *b)* the three objects were clustered and located near the periphery of the arena and, *c)* the three objects were placed at the periphery, against the wall, and formed either an equilateral [10] or an isosceles [11] triangle configuration. Thus, the results demonstrate that place fields can be controlled by the object array. However, only when located near or at the periphery of the apparatus do objects appear to be used as landmarks, thus suggesting that their location is critical to their landmark status. Objects may be taken as landmarks when they delimit a large enough locomotor space. Because they induce more important parallax effects than distal cues, they are less reliable and may not be spontaneously used. Alternately, slender objects may not enable the animal to disambiguate the geometrically equivalent places in a circular arena, although the array is made of distinct objects forming an isosceles triangle configuration. This hypothesis is supported by the observation that the animal may neglect the identity of the objects [2] and that asymmetry of the array may be less easily detected when the objects are at a close distance rather than far away [10, 11].

7.2.5 Effects of Conflicts Between Distal and Proximal Cues

In a study by Shapiro and co-workers [57], rats were trained to explore an elevated cross maze surrounded by several distal cues and covered by inserts (proximal cues) of different texture and visual aspect on each arm. They reported that rotation of the proximal and distal cues in opposite direction produced remapping in the majority of cells, thus reflecting that these cells encoded the relationships between the two kinds of cues. A smaller proportion of cells were found to be bound to distal cues and still fewer were bound to proximal cues. This was interpreted as reflecting a hierarchy of influence of available cues on place cell activity. Furthermore, the responses of small ensembles of simultaneously recorded neurons were discordant: some cells were seen to remap whereas other cells were seen to be controlled by either distal or proximal cues [61, 26, 5]. That only a few cells were controlled by proximal cues in Shapiro et al.'s study may be accounted for by the fact that the animals paid little attention to these 2D features. Thus, the hierarchy of influences proposed by these authors may be specific to their study.

Recently, we examined the possibility that the hierarchy may be different if the proximal cues available to the animal were more relevant for its behavior [48, unpublished data]. To address this issue, we recorded place cells while the rats explored an elevated circular arena containing three different objects (proximal cues) and surrounded by curtains holding three large distinct stimuli (distal cues). After two initial standard recording sessions, a double rotation (made while the rat was in its homecage) of the proximal and distal cues was made, with the two sets of cues rotated 90° in opposite directions. This resulted in a conflict between the distal and proximal reference frames (Figure 7.2). It was hypothesized that, if place fields are under control of the overall configuration of proximal and distal cues, then they should remap. Similarly, if place fields are under control of either proximal or distal cues, then they should rotate along with the set of cues they are anchored to. Lastly, if place fields are under control of background cues (e.g. folding or shades in curtains, etc.), then they should remain stable relative to the room. Of the 111 recorded place cells, 66 % were found to remap. This included cells whose fields shifted to an unpredictable location, cells that stopped firing, and cells that were formerly silent and that developed new fields. Eighteen percent of cells were found to be bound to proximal objects whereas a smaller proportion (10 %) was controlled by distal cues. Note that some cells remained stable relative to the room (7 %). Our results suggest that most cells encode the spatial relationships between proximal and distal cues and are thus consistent with Shapiro et al.'s findings [57]. Nevertheless, among the cells that responded to the double rotation by "following" one kind of cue, a majority were controlled by proximal objects. In contrast, Shapiro et al. [57] had obtained control mostly by the distal cues. This suggests that the hierarchical representation of cues within reference frames and therefore the relative strength of these frames is not pre-established but is flexible, depending on the nature of the cues available and, likely, on the task.

A nice demonstration that the animal is able to manage simultaneously multiple reference frames comes from Fenton and colleagues who developed a place avoidance task [15]. Rats were trained to avoid footshocks that were delivered when entering a portion of a circular arena that could be stable or rotated. Rotation of the arena allowed dissociation of two reference frames, a reference frame defined relative to the room and a reference frame defined relative to the arena. In the rotating condition, the rats were thus trained to avoid two target areas, one being in register with the room reference frame (non rotating target area) and the other being in register with the arena reference frame (rotating target area). Place cell recordings in this situation revealed that some fields were bound to the room reference frame and other to the arena reference frame.

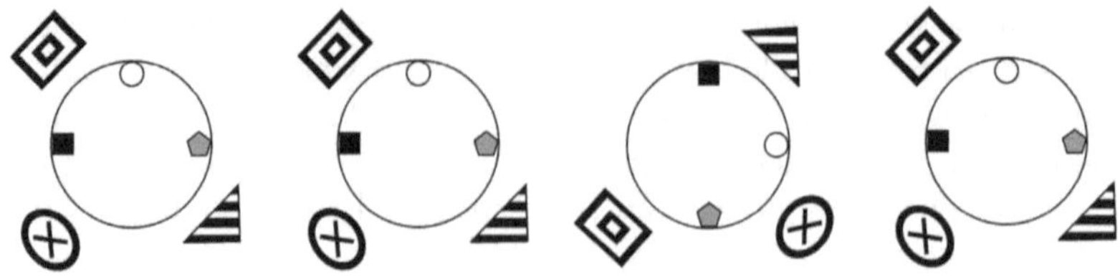

Fig. 7.2. Experimental design of Renaudineau et al.'s study. The circular arena containing 3 objects (shown as black square, white circle and gray pentagone) was surrounded by curtains where 3 distal cues were attached (shown as 3 different patterns). Upon isolation of place cells, four successive 16-minute recording sessions were conducted. Between each session, the rat was returned to its home cage. The double rotation resulted in a 180° mismatch between the objects and the distal cues.

Overall, these results support the hypothesis of distinct processes for coding different reference frames [68].

7.2.6 Encoding of Proximal and Distal Cues Depends on Multiple Brain Areas

Although hippocampal place cells play a pivotal role in spatial information processing in rodents, spatial information is not generated *ex nihilo* in the hippocampus. The hippocampus receives a large amount of inputs from cortical areas, via the entorhinal cortex. In particular, the associative parietal cortex has long been identified as an important area for the processing of spatial information [54, for a review]. To investigate the nature of the interactions between the parietal cortex and the hippocampus, we examined the effects of parietal cortical lesions on place cell firing as the rat performed pellet-chasing in a circular arena containing three objects placed at the periphery, and forming a isosceles triangle configuration [53]. Manipulation of environmental cues included 90° rotation of the objects (in the absence of the animal) to examine cue control and removal of the objects (in the presence of the animal). In the latter test, it is usually found that place fields remain stable relative to their position before removal, thus indicating that the animals are able to maintain a spatial representation by using other cues such as surface (e.g. olfactory) and idiothetic cues [52]. This aspect was strongly altered in parietal rats. Thus, when the objects were removed, place fields in parietal rats shifted from an arena to a room reference frame whereas in control rats they remained aligned with the arena reference frame, therefore suggesting that the associative parietal cortex plays a role in the establishement of the reference frame provided by proximal cues. Such results are consistent with a previous lesion study showing that rats with parietal lesions were unable to locate a submerged platform in the water maze when they had to rely on objects directly placed into the pool, whereas they exhibited control-like place learning abilities when they had to use distal room cues [55]. This is also coherent with the fact that encoding of a configuration of objects located in the arena was disrupted by parietal lesions [56]. Interestingly, rats with hippocampal lesions are impaired in using both proximal and distal cues [55] whereas rats with entorhinal lesions were impaired in using distal cues while the use of proximal cues was unaffected [41]. These results are consistent with the hypothesis of two processing systems, one devoted to the encoding of proximal cues and the other to distal cues. These systems are mediated by distinct functional networks that may converge to the hippocampus. What other structures may be part of each system has to be determined. Another issue is to investigate the respective contribution of these structures and their interaction in cue processing.

7.3 Goal Encoding

Any representational system that guides the animal's behavior [30] must incorporate not only the animal's position but also the goal location, thus allowing the animal to select efficient paths toward the goal. So far, how the goal is represented in the brain is poorly understood. For example, we do not know whether goals are encoded in the hippocampus (by place cells) along with other locations or elsewhere in the brain. Hollup et al. [24] have recorded place cells in rats trained to find a submerged platform in an annular water maze. Activity was monitored as the animals performed a probe test with the platform removed. They found that firing fields accumulated in the segment of the annulus corresponding to the previous platform location. This suggests that the goal is encoded in the hippocampus in the form of an over-representation of place fields that differentiates it from other places. Some studies have indicated that place fields tend to accumulate at reward locations. For example, place fields were found to shift to a new corner of a square apparatus when the reward was delivered at that location [4]. In contrast, other studies failed to find any evidence of goal encoding by place cells. For example, Lenck-Santini et al., recorded place cells as the rats performed a place preference task in a circular arena [29]. In this task, the animal is trained to locate an unmarked "trigger" zone and stay in this zone for 2 sec. Satisfying this condition triggers a food dispenser fixed to the ceiling, thus allowing delivery of a small food pellet that drops in the arena. Upon landing, the pellet rolls randomly on the arena floor and ends its course at an unpredictable location. The animal then searches for the pellet, eats it and returns to the trigger zone for another trial [49]. In Lenck-Santini et al.'s study, the only available cue was a white cue card attached to the arena wall. The trigger zone could be located at a distance or very close from the cue card thus allowing use of place navigation or cue navigation strategies, respectively. In neither situation, however, did we found an accumulation of place fields at the trigger location as compared with other locations in the arena. Overall, it is not clear whether there is a representation of the goal in the hippocampus. Thus, although it cannot be ruled out that this structure mediates some aspects of goal encoding, one has to search correlates of such processes in other brain areas.

Among these areas, the medial prefrontal cortex (mPFC) including the prelimbic and infralimbic areas is a good candidate. Indeed, this structure has been strongly involved in goal-directed behavior [12]. That the mPFC and the hippocampus are functionally related is bolstered by the existence of a direct monosynaptic, LTP-modifiable connection originating from the ventral hippocampus [17]. This suggests that some spatial signal is transmitted from the hippocampus to the mPFC. One possible effect of this connection may be to endow prefrontal neurons with the ability to exhibit location-specific firing in relation to the goal. To test this hypothesis, Poucet [43] recorded prefrontal units as the rats performed a pellet-chasing task in a circular arena but failed to observe location-specific firing. Note however, that pellet chasing is not a navigation task since the animal has to simply wander about in the environment. Because it is possible that frontal neurons are displaying location-specific activity when the rat is explicitly trained to navigate in space, we trained rats in the place preference task. We found that a substantial amount of cells in the prelimbic/infralimbic areas had clear spatial correlates [23]. Interestingly, the firing fields of prefrontal neurons were not homogeneously distributed across the arena. Two goal zones were more represented than the rest of the arena, namely, the "trigger" zone (the zone that the rat had to reach to trigger release of the reward) and the "landing" zone (the zone located under the food dispenser, where the pellet drops). The spatial discharge of prefrontal neurons was characterized by large and noisy firing fields, that were markedly different from the small and crispy fields of hippocampal place cells. This observation suggests that the two structures might have complementary roles. In addition to the spatial dimension of goal encoding likely provided by the hippocampus, the prefrontal representation of the goal may also allow integration of motivational, and emotional dimensions through its interactions with the amygdala complex and the nucleus accumbens [9]. Overall, these results support the hypothesis that prefrontal neurons encode goals and suggest that the mPFC is part of a functional network that allows animals to select an appropriate strategy and to generate efficient paths toward the goal [19].

7.4 What Might Be Useful for Robots?

To be autonomous a mobile robot needs to be implemented with mechanisms allowing adaptation to changes. Current mobile robots are capable to some extent of learning and interacting with their environment to exhibit flexible behavior. However, they are generally tested in specific controlled laboratory environments while performing specific tasks, conditions that require relatively little behavioral flexibility. Thus, versatile robots have to be endowed with larger adaptive capacities.

In the present chapter, we have shown that cue and goal encoding are fundamental processes that contribute to behavioral flexibility in animals. They allow the animal to continue navigating effectively in spite of environmental changes. The functional properties that result from these processes may enable robots to exhibit greater adaptation and autonomy. We thus suggest a number of properties that may be relevant to robot adaptive navigation.

7.4.1 The Conjoint Use of Allothetic and Idiothetic Cues

Perception in robots is based on multiple sensors that provide information on their environment. Thus, like in animals, two sources of information, idiothetic and allothetic, are available to the robot. Idiothetic information corresponds to odometry and allothetic information is provided by laser range finders, sonars or vision. Most biologically-inspired models and robots (for example in the *animat* approach) use a combination of allothetic and idiothetic information to exhibit spatial behavior. Based on this combination, they are endowed with the capacity of constructing a topological or metric map, localizing themselves within this map and planning paths to reach goals [65, 18, 33]. One consequence of integrating these two kinds of information is that they can compensate for each other to some extent. Allothetic cues may compensate for cumulative errors resulting from the use of idiothetic information. On the contrary, idiothetic cues may serve to disambiguate distinct locations that appear the same to the robot's sensors. Such mechanisms however, do not necessarily reflect an actual adaptive response to environmental changes but rather apply to planned situations occuring in a stable environment. Adaptation to unexpected changes requires more sophisticated interactions.

Animal studies have suggested that allothetic and idiothetic information are processed in such an interactive way that animals can readily rely on one or the other or both kinds of information to maintain navigational ability, *depending on which is more appropriate at the time.* In other words, permanent and flexible interactions between different sources of information are key processes for adaptation to environmental changes. Among changes, those that yield conflict between idiothetic and allothetic information are not frequent, however. In animals and robots, the several reference frames are usually congruent. In contrast, situations in which a category of sensory information becomes suddenly unavailable are more likely to occur. For example, when visual information comes to be lacking or irrelevant, olfactory and idiothetic information may be used to maintain a stable spatial representation. Another critical situation that requires complex interaction between idiothetic and allothetic information is when the animal faces an unfamiliar environment (see Sect. 7.1.1). Thus, elaboration and use of a spatial representation in unstable environments requires integration of multimodal information that is coordinated, dynamically-adjusted and experience dependent.

7.4.2 The Distinction Between Proximal and Distal Cues

The distinction between proximal and distal cues is particularly relevant for navigation in large, i.e. natural, environments. These two kinds of cues do not provide redundant information but rather complementary spatial information. Due to motion parallax effects, distal cues such as distant mountains, provide the animal with more stable directional information but poorer positional information than proximal cues. In contrast, proximal cues such as nearby rocks or trees, provide more precise positional information but poorer directional information than distal cues [41]. Thus, maintaining a goal-directed trajectory is aided more by distal cues whereas accurate place-learning is aided more by proximal cues. As far as we know, the functional distinction between proximal and distal cues is not usually implemented in mobile robots. Detection and categorization processes that are needed to perform such discrimination are likely not trivial. This is probably not critical for navigation in laboratory-sized environments but may improve accurate long distance travels in larger environments.

7.4.3 Distributed Organization of Space Representation in the Brain

Spatial navigation results from the interaction of multiple systems which are mediated by multiple brain regions [64]. Among these regions, the hippocampus plays a central role but recent work has emphasized the importance of a cooperation between the hippocampus and the neocortical areas. Thus, functions such as cue and goal encoding require permanent cortico-hippocampal interactions. For instance, cue encoding involves contribution of numerous cortical areas including, primary, parietal and entorhinal cortices and goal encoding involves the contribution of prefrontal areas. Such a distributed organization reflects not only the intricacy of the processes subserving spatial navigation but may be also crucial for flexibility.

Do robots need to be implemented with a functional architecture resembling as close as possible to the biological organization to exhibit autonomy? The increasing number of models of the hippocampus and its place cells, allowing robotic implementation of spatial representations, are consistent with this view [65, 8, 19]. These models constitute a basis for more complex neural architectures involving cortical and sub cortical modules such as that proposed by Banquet and colleagues in which hippocampo-prefronto-basal interactions were incorporated [1, 45]. Overall, implementing mechanisms inspired by biological systems may endow robots with sufficient autonomy to deal with unexpected environmental changes and achieve their mission. It may also help to understand the nature of the processes that enable animals to be autonomous.

References

1. Banquet, J.-P., Gaussier, P., Quoy, M., Revel, A., Burnod, Y.: Cortico-hippocampal maps and navigation strategies in robots and rodents. In: Hallan, B., Floreano, D., Hallam, J., Hallam, B., Hages, G., Meyer, J.-A. (eds.) From animals to animats, vol. 7, pp. 141–150. MIT press, Boston (2002)
2. Benhamou, S., Poucet, B.: Landmark use by navigating rats (rattus norvegicus) contrasting geometric and featural information. J. Comp. Psychol. 112, 317–322 (1998)
3. Best, P.J., White, A.M., Minai, A.: Spatial processing in the brain: the activity of hippocampal place cells. Ann. Rev. Neurosci. 24, 459–486 (2001)

4. Breese, C.R., Hampson, R.E., Deadwyler, S.A.: Hippocampal place cells: stereotypy and plasticity. J. Neurosci. 9, 1097–1111 (1989)
5. Brown, J.E., Skaggs, W.E.: Concordant and discordant coding of spatial location in populations of hippocampal CA1 pyramidal cells. J. Neurophysiol. 88, 1605–1613 (2002)
6. Bureš, J, Fenton, A.A., Yu, K., Zinyuk, L: Place cells and place navigation. Proc. Natl. Acad. Sci. 94, 343–350 (1997)
7. Burgess, N., O'Keefe, J.: Spatial Models of the Hippocampus. In: Arbib, M.A. (ed.) The Handbook of Brain Theory and Neural Networks, 2nd edn., pp. 468–472. MIT press, Cambridge (2002)
8. Burgess, N., O'Keefe, J.: Neural representations in human spatial memory. Trends Cogn. Sci. 7, 517–519 (2003)
9. Cardinal, R.N., Parkinson, J.A., Hall, J., Everitt, B.J.: Emotion and motivation: the role of the amygdala, ventral striatum, and prefrontal cortex. Neurosci. Biobehav. Rev. 26, 321–352 (2002)
10. Cressant, A., Muller, R.U., Poucet, B.: Failure of centrally placed objects to control the firing fields of hippocampal place cells. J. Neurosci. 17, 2532–2542 (1997)
11. Cressant, A., Muller, R.U., Poucet, B.: Further study of the control of place cell firing by intra-apparatus objects. Hippocampus 9, 423–431 (1999)
12. Dalley, J.W., Cardinal, R.N., Robbins, T.W.: Prefrontal executive and cognitive functions in rodents: neural and neurochemical substrates. Neurosci. Biobehav. Rev. 28, 771–784 (2004)
13. Eichenbaum, H., Stewart, C., Morris, R.G.: Hippocampal representation in place learning. J. Neurosci. 10, 3531–3542 (1990)
14. Etienne, A.S., Jeffery, K.J.: Path integration in mammals. Hippocampus 14, 180–192 (2004)
15. Fenton, A.A., Stuchlik, A, Kaminsky, Y., Zahálka, A, Bureš, J: Place avoidance by path integration is only accurate for about 7 m without useful visual and substratal cues. Soc. Neurosci. Abst. 24, 1685 (1998)
16. Fenton, A.A., Wesierska, M., Kaminsky, Y., Bureš, J: Both here and there: simultaneous expression of autonomous spatial memories in rats. Proc. Natl. Acad. Sci. 95, 11493–11498 (1998)
17. Ferino, F., Thierry, A.M., Glowinski, J.: Anatomical and electrophysiological evidence for a direct projection from Ammon's horn to the medial prefrontal cortex in the rat. Exp. Brain Res. 65, 421–426 (1987)
18. Filliat, D., Meyer, J.-A.: Map-based navigation in mobile robots. I. A review of localization strategies. J. Cog. Syst. Res. 4, 243–282 (2003)
19. Gaussier, P., Revel, A., Banquet, J.-P., Babeau, V.: From view cells and place cells to cognitive map learning: processing stages of the hippocampal system. Biol. Cybern. 86, 16–28 (2002)
20. Gothard, K.M., Skaggs, W.E., McNaughton, B.L.: Dynamics of mismatch correction in the hippocampal ensemble code for space: interaction between path integration and environmental cues. J. Neurosci. 16, 8027–8040 (1996)
21. Gothard, K.M., Skaggs, W.E., Moore, K.M., McNaughton, B.L.: Binding of hippocampal CA1 neural activity to multiple reference frames in a landmark-based navigation task. J. Neurosci. 16, 823–835 (1996)
22. Gould-Beierle, K.L., Kamil, A.C.: The use of local and global cues by Clarck nutcrackers, Nucifraga Columbiana. Anim. Behav. 52, 519–528 (1996)
23. Hok, V., Save, E., Lenck-Santini, P.-P., Poucet, B.: Coding for spatial goals in the prelimbic/infralimbic area of the rat frontal cortex. Proc. Natl. Acad. Sci. 102, 4602–4607 (2005)

24. Hollup, S.A., Molden, S., Donnet, J.G., Moser, M.-B., Moser, E.I.: Accumulation of hippocampal place fields at the goal location in an annular watermaze task. J. Neurosci. 21, 1635–1644 (2001)
25. Jeffery, K.J., O'Keefe, J.: Learned interaction of visual and idiothetic cues in the control of place field orientation. Exp. Brain Res. 127, 151–161 (1999)
26. Knierim, J.J.: Dynamic interactions between local surface cues, distal landmarks, and intrinsic circuitry in hippocampal place cells. J. Neurosci. 22, 6254–6264 (2002)
27. Knierim, J.J., Kudrimoti, H.S., McNaughton, B.L.: Interactions between idiothetic and external landmarks in the control of place cells and head direction cells. J. Neurophysiol. 80, 425–446 (1998)
28. Lavenex, P., Schenk, F.: Olfactory cues potentiate learning of distant visuospatial information. Neurobiol. Learn Mem. 68, 140–153 (1997)
29. Lenck-Santini, P.-P., Muller, R.U., Save, E., Poucet, B.: Relationships between place cell firing fields and navigational decisions by rats. J. Neurosci. 22, 9035–9047 (2002)
30. Lenck-Santini, P.-P., Save, E., Poucet, B.: Evidence for a relationship between place-cell spatial firing and spatial memory performance. Hippocampus 11, 377–390 (2001)
31. Lever, C., Wills, T., Cacucci, F., Burgess, N., O'Keefe, J.: Long-term plasticity in hippocampal place-cell representation of environmental geometry. Nature 416, 90–94 (2002)
32. McNaughton, B.L., Barnes, C.A., Gerrard, J.L., Gothard, K., Jung, M.W., Knierim, J.J., Kudrimoti, H., Qin, Y., Skaggs, W.E., Suster, M., Weaver, K.L.: Deciphering the hippocampal polyglot: the hippocampus as a path integration system. J. Exp. Biol. 199, 173–185 (1996)
33. Meyer, J.-A., Filliat, D.: Map-based navigation in mobile robots. II. A review of map-learning and path planning strategies. J. Cog. Syst. Res. 4, 283–317 (2003)
34. Morris, R.G. M.: Spatial localization does not require the presence of local cues. Learn Motiv. 12, 239–260 (1981)
35. Muller, R.U.: A quarter of a century of place cells. Neuron 17, 813–822 (1996)
36. Muller, R.U., Kubie, J.L.: The effects of changes in the environment on the spatial firing of hippocampal complex-spike cells. J. Neurosci. 7, 1951–1968 (1987)
37. Muller, R.U., Kubie, J.L., Ranck, J.B.: Spatial firing pattern of hippocampal complex-spike cells in a fixed environment. J. Neurosci. 7, 1935–1950 (1987)
38. O'Keefe, J: Hippocampus function: does the working memory hypothesis work? Should we retire the cognitive map theory? Behav. Brain Sci. 2, 339–343 (1979)
39. O'Keefe, J., Dostrovsky, J.: The hippocampus as a spatial map. Preliminary evidence from unit activity in the freely moving rat. Brain Res. 34, 171–175 (1971)
40. O'Keefe, J., Nadel, L.: The hippocampus as a cognitive map. Oxford University Press, London (1978)
41. Parron, C., Poucet, B., Save, E.: Entorhinal cortex lesions impairs the use of distal but not proximal landmarks during navigation in the rat. Behav. Brain Res. 154, 345–352 (2004)
42. Paz-Villagràn, V., Save, E., Poucet, B.: Coding of separate and connected spatial contexts by place cells. Eur. J. Neurosci. 20, 1379–1390 (2004)
43. Poucet, B.: Searching for spatial unit firing in the prelimbic area of the rat medial prefrontal cortex. Behav. Brain Res. 84, 151–159 (1997)
44. Poucet, B., Chapuis, N., Durup, M., Thinus-Blanc, C.: A study of exploratory behavior as an index of spatial knowledge in hamsters. Anim. Learn. Behav. 14, 93–100 (1986)

45. Poucet, B., Lenck-Santini, P.-P., Hok, V., Save, E., Banquet, J.-P., Gaussier, P., Muller, R.U.: Spatial navigation and hippocampal place cell firing: the problem of goal encoding. Rev. Neurosci. 15, 87–109 (2004)
46. Poucet, B., Lenck-Santini, P.-P., Save, E.: Drawing parallels between the behavioural and neural properties of navigation. In: Jeffery, K.J. (ed.) The neurobiology of spatial behaviour, pp. 187–198. Oxford University Press, Oxford (2003)
47. Quirk, G.J., Muller, R.U., Kubie, J.L.: The firing of hippocampal place cells in the dark depends on the rat's recent experience. J. Neurosci. 10, 2008–2017 (1990)
48. Renaudineau, S., Poucet, B., Save, E.: Effects of a conflict between proximal objects and distal cues on place cell activity (unpublished data)
49. Rossier, J., Schenk, F., Kaminsky, Y., Bureš, J: The place preference task: a new tool for studying the relation between behavior and place cell activity in rats. Behav. Neurosci. 114, 273–284 (2000)
50. Rotenberg, A., Muller, R.U.: Variable coupling to a continuous viewed stimulus: evidence that the hippocampus acts as a perceptual system. Phil. Trans. R Soc. Lond. B 352, 1505–1513 (1997)
51. Save, E., Cressant, A., Thinus-Blanc, C., Poucet, B.: Spatial firing of hippocampal place cells in blind rats. J. Neurosci. 18, 1818–1826 (1998)
52. Save, E., Nerad, L., Poucet, B.: Contribution of multiple sensory information to place field stability in hippocampal place cells. Hippocampus 10, 64–76 (2000)
53. Save, E., Paz-Villagràn, V., Alexinsky, T., Poucet, B.: Functional interaction between the parietal cortex and hippocampal place cell firing in the rat. Eur. J. Neurosci. 21, 522–530 (2005)
54. Save, E., Poucet, B.: Hippocampal-parietal cortical interactions in spatial cognition. Hippocampus 10, 491–499 (2000a)
55. Save, E., Poucet, B.: Involvement of the hippocampus and associative parietal cortex in the use of proximal and distal landmarks for navigation. Behav. Brain Res. 109, 195–206 (2000b)
56. Save, E., Poucet, B., Foreman, N., Buhot, M.C.: Object exploration and reactions to spatial and non-spatial changes in hooded rats following damage to parietal cortex or hippocampal formation. Behav. Neurosci. 106, 447–456 (1992)
57. Shapiro, M.L., Tanila, H., Eichenbaum, H.: Cues that hippocampal place cells encode: dynamic and hierarchical representation of local and distal stimuli. Hippocampus 7, 624–642 (1997)
58. Sharp, P.E., Blair, H.T., Etkin, D., Tzanetos, D.B.: influences of vestibular and visual motion information on the spatial firing pattern of hippocampal place cells. J. Neurosci. 15, 173–189 (1995)
59. Sutherland, R.J., Dyck, R.H.: Place navigation by rats. Can J. psychol. 38, 322–347 (1984)
60. Suzuki, S., Augerinos, G., Black, A.H.: Stimulus control of spatial behavior in the eight-arm maze in rats. Learn Motiv. 11, 1–18 (1980)
61. Tanila, H., Shapiro, M.L., Eichenbaum, H.: Discordance of spatial representation in ensembles of hippocampal place cells. Hippocampus 7, 613–623 (1997)
62. Thinus-Blanc, C., Bouzouba, L., Chaix, K., Chapuis, N., Durup, M., Poucet, B.: A study of spatial parameters encoded during exploration in hamsters. J. Exp. Psychol: Anim. Behav. Proc. 13, 418–427 (1987)
63. Tolman, E.C.: Cognitive maps in rats and men. Psychol. Rev. 55, 189–208 (1948)

64. Touretzky, D.S., Redish, A.D.: Theory of rodent navigation based on interacting representations of space. Hippocampus 6, 247–270 (1996)
65. Trullier, O., Wiener, S.I., Berthoz, A., Meyer, J.-A.: Biologically artificial navigation systems: review and prospects. Prog. Neurobiol. 51, 483–544 (1997)
66. Whishaw, I.Q.: Spatial mapping takes time. Hippocampus 8, 122–130 (1998)
67. Wiener, S.I., Korshunov, V.A., Garcia, R., Berthoz, A.: Inertial substratal and landmark cue control of hippocampal CA1 place cell activity. Eur. J. Neurosci. 7, 2206–2219 (1995)
68. Zinyuk, L., Kubik, S., Kaminsky, Y., Fenton, A.A., Bureš, J: Understanding hippocampal activity by using purposeful behavior: place navigation induce vs. place cell discharge in both task-relevant and task-irrelevant spatial reference frames. Proc. Natl. Acad. Sci. 97, 3771–3776 (2000)

8

These Maps Are Made for Walking – Task Hierarchy of Spatial Cognition

Sabine Gillner and Hanspeter A. Mallot

Summary. Spatial behaviours and abilities do not form a monolithic module of cognition but can be subdivided into a hierarchy of behaviours, mechanisms, and representations. This hierarchical structure is a result of cognitive evolution. Therefore, the ordering of the individual modules will follow the general rules of phylogeny. In particular, the complexity of spatial tasks to be solved by an organism and the behaviours evolved as adaptation to these tasks is of great relevance. In this paper, we present an approach to spatial hierarchy based on the complexity of the tasks, rather than on the complexity of the underlying mechanisms. Individual levels of the task hierachy are discussed from a theoretical point of view and specific experimental examples are given. In conclusion, hierachies based on tasks seem to differ from representational hierarchies in three respects, the treatment of landmarks, the role of metric information, and the relation of language and space.

8.1 Introduction

Among the cognitive abilities found in the animal kingdom, spatial cognition is probably the most common one. Animals with rather limited capacities in domains such as object recognition, communication, or problem solving may still be able to perform surprisingly well in spatial tasks. As an example, consider the desert ant *Cataglyphis* [50] which is able to solve complex navigational tasks with a minimum of processing power. The key to this efficiency seems to be task specificity, i.e. the reduction of the information processing machinery to the very essentials required to perform the behaviour. Other information processing abilities, which are not required in navigation behaviour, are not implemented in the ant's brain. Task specificity is a general theme of systems generated by evolution and will therefore not be limited to ants.

In this paper, we discuss human spatial cognition with an eye on evolution. Of course, this is not a new idea, other authors already pointed out that navigational mechanisms found in animals may also be operating in humans [49, 48]. For spatial cognition in birds and mammals, a "parallel map theory" has been presented, suggesting the evolution of modern cognitive maps from a large scale bearing map and a local scene map [28]. The goal of such enterprises is to eventually come up with a phylogenetic tree of spatial behaviours showing different capacities as traits arising at some point in evolution and developing from there on. Since existing cognitive systems are the result of this evolution, understanding

M.E. Jefferies et al. (Eds.): Robot. & Cogn. Approach. to Spat. Map., STAR 38, pp. 181–201, 2008.
springerlink.com

cognitive evolution will also enhance the understanding of human spatial cognition as it stands today.

Investigations into the evolution of behavioural traits generally suffer from two problems. First, we have virtually no fossil records of behaviour. Second, the two Darwinian driving forces of evolution, i.e., adaptative value of traits on the one hand and variation in the genes controlling those traits on the other hand, are only loosely coupled via a complex chain of steps including the sensory, neural, and motor system of the animal. In this situation, evidence for the evolutionary relatedness of traits found in two animal species has to come both from the selection side of evolution, i.e. tasks and behaviour, and from the genetic variation side of evolution, as it shows in the underlying neural mechanisms. Both sides can be linked by the idea of *evolutionary scaling*: evolutionary sequences of cognitive (or other) traits must involve small steps that can be realized by genetic variation, but at the same time lead to some gain in adaptivity. Therefore, we suggest that evolutionary hierarchies of cognitive abilities must take into account both a hierarchy of mechanisms and a hierarchy of tasks.

A similar conclusion can be drawn from the synthesis of behaving systems, i.e. from robotics. In an influential paper, Brooks [5] discussed the question of how to build a robot with complex behavioural abilities. Rather than adding up information processing modules to form a general problem solving machine, Brooks suggested to structure the problem from specific "task-achieving behaviours". Once the robot is able to perform some basic behaviour, more complex ones can be added in what Brooks termed the subsumption architecture. We think that this is also a useful approach to investigate animal and human spatial cognition. If the correct phylogenetic sequence of the evolution if cognitive tasks is identified for a given species, the underlying neural mechanisms can be expected to follow some sort of subsumption architecture.

Fig. 8.1. The Hexatown virtual environment. Street segments are of equal length and meet at angles 120 degrees. Three landmarks (buildings) are located around each junction. Subjects can move through the environment by selecting "ballistic" movement sequences (60 degree turns or translations of one street segment) initiated by clicking a button [20].

This goal, however, seems still far down the line. In this paper, we will review a series of behavioural experiments on human cognition that have been designed in the general logic sketched out above. Most of the experiments have been carried out in various versions of the "Hexatown" virtual environment depicted in Fig. 8.1. We will extend on the general logic in the individual sections. As a result, we will suggest a task hierarchy for spatial behaviour starting from recognition of special places ("home") and leading on to excursions from and homing to this place, adding more places, chaining of excursions into routes, recombining segments of known routes to novel routes, planning of alternative routes and usage of large scale spaces.

8.2 Recognizing Places

The most basic task in spatial behaviour is probably recognizing places. Even in simple search behaviour, the goal has to be recognized once it has been reached. In its simplest form, place recognition may be restricted to one or a few special places such as nest entries or feeding sites, but more complex spatial memories will contain larger numbers of known places. Place recognition has to rely on some sort of landmark information, i.e. sensor data characteristic of each place. Thus, the problem of place recognition is largely identical to the problem of landmark recognition. Let us define a landmark as a piece of sensory information, characteristic of a place, that is stored in memory and used in place recognition. Based on this definition, three questions about landmarks can be asked:

1. Depth of processing: What codes are generated from the sensory input to be stored in memory?
2. Landmark selection: Which parts of the input information are used to form landmark codes?
3. Landmark usage: How are landmarks used in spatial behaviour?

For the last question, we follow the distinction between guidance (piloting) and direction (recognition-triggered response) [40, 48]. In place recognition, landmarks are used as guidance to pinpoint the location of a place, not as pointers, directing the navigator elsewhere. Landmarks used as pointers to other places ("direction") will be considered in Sect. 8.4. We will now turn to the discussion of landmark processing and landmark selection.

8.2.1 Landmarks and Depth of Processing

The most general account of landmark information is the notion of "local position information", defined as the sum of all sensory inputs perceivable at a certain location [48]. Although this definition is not restricted to the visual modality, local views or snapshots will often be the most important type of local position information. In an extended environment, the local position information is a vector-valued function of position and pose (orientation of body and sensors); the components of the vector are the sensor readings currently obtained. For the visual modality, the local position information is the image or view locally obtained. These views, parameterized with the agent's position and pose, thus form a manifold containing all visual landmark information available in the environment ([16] and Franz et al., this volume). More specific kinds of landmarks can be derived by applying various amounts of image processing (see Table 8.1). For robots, low-frequency Fourier components od the panoramic image can be used to bring to snapshots into register for subsequent image comparison [46]. Honey-bees have been shown to use raw image information ("snapshots") and find places by matching currently visible and remembered snapshots [7]. The usage of raw snapshot information in humans has not been clearly demonstrated. However, Christou and Bülthoff [11] have shown the landmark objects are recognized faster and more accurately if they are presented in a familiar rather than in a novel orientation, indicating that views rather than 3D objects are remembered in the navigation task.

In a mechanism called the geometric module ([9] and Cheng, this volume), rats determine their position from local depth maps, i.e. the distances to the surrounding walls. This information is also derived from visual input but requires more elaborate processing. In rats, possible cues to determine the distance of walls include motion parallax, the elevation of the lower edge of the wall in the rat's visual image, and maybe some stereopsis. In humans, it was shown that young children [25] as well as non-attentive adults [26] also use geometric information to find a hidden object in a room.

In everyday language, a landmark is an identified, nameable object or sight, whose recognition in an image requires a fully-fledged object recognition system. This type of information indeed increases navigation performance [24]. The relation between landmark recognition and object recognition in general is not entirely clear. While many computational problems are identical, neural processing of the two recognition systems in humans shows an interesting dissociation [29]: The recognition of attentively perceived objects is associated with activation in the right fusiform cortex, whereas objects at decision points are processed mainly in the parahippocampal region, indicating a difference in the neural processing underlying landmark- and object recognition.

The level with the largest depth of processing in the sketched landmark hierarchy consists in the assignment of names to the recognized landmarks. Nameability of landmarks may be more relevant in spatial language and direction giving, than in actual navigation.

8.2.2 Landmark Selection

As compared to the complete view manifold, landmark information must be limited in two ways. First, only parts of each image should be stored as a memory. Second, landmarks should be recorded only at selected observer locations. Both selections will be guided by the following criteria:

1. **Salience:** Landmarks should be well recognizable. In a verbal recall task, named objects are usually those standing out from the environment in size, shape, or functionality [2].

Table 8.1. Depth of processing in landmark recognition

input	image processing	output
view manifold	local recording	raw snapshot at discrete points
	depth estimate	depth signature (geometric module)
	object recognition	identified landmarks
	associating a text	named landmark

2. **Relevance:** Landmarks should be remembered at places where navigational decisions are required. If adult subjects are asked to choose objects with potential landmark value they tend to name objects in the vicinity of road crossings [1] where a decision has to be made. Note that the distinction of landmark and object recognition in [29] was based on this criterion.
3. **Permanence:** The landmark and its position should be constant over time. In children, navigational errors have been shown to result from choosing as landmarks salient but non-permanent objects such as fancy cars, which may be gone when visiting their original location again [12]. Rat head direction cells follow the more distant of two independently moving landmarks even if they cover the same visual angle. A possible interpretation of this prefrence is that the more distant object will be larger and therefore less likely to move [55].

8.2.3 Eye-Movements in a Navigation Task

A behavioural approach to landmark selection and landmark usage is the study of eye-movements in navigation tasks. As compared to landmark naming tasks, behavioural measures should be more general, allowing to access also non-named landmarks in the sense of Table 8.1. Furthermore, even if a particular landmark is verbally reported, this does not proove its actual usage in the navigation task. Behavioural measures of landmark usage therefore seem desirable.

The relevance of eye-movement studies for analysing sequential tasks has been demonstrated in every-day activities such as preparing tea or sandwiches [32, 23]. Here, eye-movements have been shown to precede an ongoing behaviour by several milliseconds such that features relevant for the next subtask are fixated. Furthermore, it has been known for decades that eye-movements are task-dependent, also when directed at static images or paintings [54]. When asked different questions about the content of a painting, subjects' eye-movements vary widely in ways that seem to be indicative of the visual features used in answering these questions. Turning to navigation tasks, we therefore hypothesize that objects which are used as landmarks are fixated more often than non-landmarks.

In an experiment using a modified version of Hexatown, Jin and Gillner used a route learning paradigm while measuring eye movements [30]. At each junction, one unique building was placed together with two identical instances of a so-called distractor. The distractor is just one object used repetitively throughout

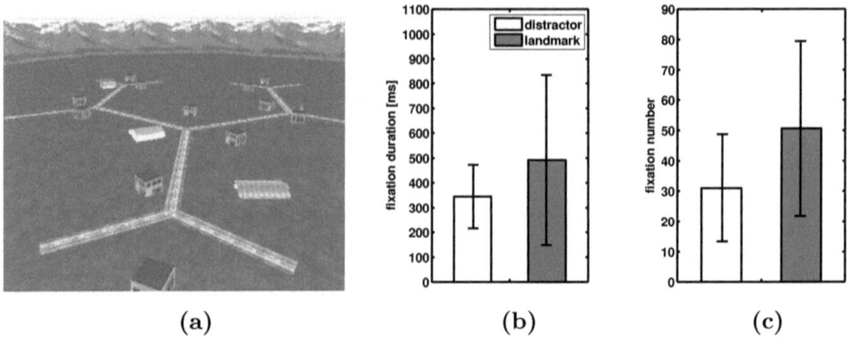

Fig. 8.2. (a) Screenshot of the experiment of Jin and Gillner [30]. In a modified version of Hexatown, one unique building (landmark) and two instances of the distractor building are placed at each junction. In a navigation task, subjects directed their gaze longer **(b)** and more often **(c)** towards landmarks than to distractors. Figures are averages over all landmark objects (filled columns) and all instances of the distractor object (open columns).

the whole environment and therefore does not provide landmark information. In this experiment, subjects were driven passively along a certain route which had to be replicated in a later testing phase. It turns out that subjects directed their gaze more often and longer to landmarks than to distractors (cf. Fig. 8.2). This result is consistent with the idea that eye-movements are indicative of landmark usage. In an ongoing experiment, we address the question whether the removal of frequently fixated objects from the environment leads to reduced navigation performance.

8.3 Homing: Returning to Places

8.3.1 Landmark-Based Homing

Places in an open space, lacking proximal cues, can be remembered by the configuration of distal cues. Jacobs et al. [27] have suggested that this performance is based on a map-like representation storing the positions of the individual landmark cues together with the goal location. In a virtual environment representing a courtyard with four different side walls, human subjects were asked to locate a hidden goal location on the floor. In the test condition, one or more of the surrounding walls were removed. Interestingly, no significant performance drop was found as long as at least one wall remains.

An alternative mechanism for landmark-based homing is the snapshot guidance discussed in Sect. 8.2. In order to exclude object-based landmark information, we designed a virtual environment representing a circular room with grey floor and ceiling and a smooth, featureless cycle of colors covering the wall [21]. Subjects were placed in the virtual room and asked to look around and remember their location. They were then virtually replaced ("teleported") to another location and asked to home to the previously inspected place. Results show that homing in this featureless environment is possible. In addition, if the color contrast of the wall pattern is reduced, homing performance drops, as is predicted by quantitative models of snapshot-based homing (e.g., [16, 46]). On the other hand, if object-based landmarks are introduced, homing performance is improved [6]. We conclude that visual homing can be based on landmark information obtained by different amounts of visual processing, including featureless snapshot information as well as recognized landmark objects.

8.3.2 Homing by Path Integration

Path integration is a concept well investigated in the animal kingdom. It is based on the perception of egomotion, which, besides landmarks, is the second major source of spatial information. It is also the simplest mechanism allowing short-cut or pointing behaviour. The occurrence of these behaviours is often taken as evidence for a metric cognitive map, which is considered a much more sophisticated level of representation in our hierarchy (see Sect. 8.6).

The discussion of metric information in spatial memory can be clarified by the distinction between a working and a reference memory for space. In simple path integration, instantaneous egomotion estimates are vectorially added to an egocentric representation of the start position such that the current distance and direction of the start point are always available. The home vector thus consitutes a working memory, it does not provide information on places visited along the route. This sort of path integration is best studied in ants but has been shown to occur also in many other animals and in humans (see [50, 37, 39] for review). Some evidence for metric knowledge in spatial long term memory, i.e. a metric cognitive map, will be dicussed later.

There seems to be no direct demonstration of the usage of plain home vectors by humans. Still, humans are able to navigate to a goal, relying on egomotion perception. The predominant cue for perceiving egomotion seems to be vision, as subjects relying on proprioreception and vestibular input, i.e. blind-folded or blind subjects make substantial errors [34]. If human subjects base their homing on optical flow information, the performance is much better, as long as the visual input is presented on a display with a large (e.g., 180°) horizontal field of view [42].

As an alternative to continuous updating of the home vector, Fujita et al. [17] have suggested that humans remember the walked path in working memory, while home vectors are calculated only when required (encoding error model). This idea was originally based on an analysis of homing errors in the Loomis et al. study [34]. However, the decrease in performance for more complex paths, which is prediced by the encoding error model, could not be demonstrated experimentally [52].

While this last result argues for continuous updating and is thus consistent with the insect-like home-vector idea, there is further evidence that humans remember more than the vector to the home position. Blind-folded subjects were seated on a robot platform and moved passively for short distances. Later, they were asked to reproduce the trip by controlling the robot with a joystick. Besides

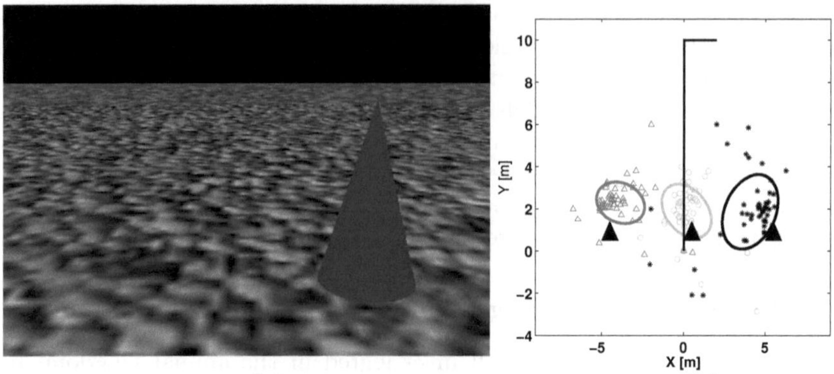

(a) (b)

Fig. 8.3. **(a)** Screenshot of the homing-experiment [19]: a cone (originally colored in red) served as a landmark in an otherwise empty environment, the floor was textured to obtain optic flow information. This view has been taken from the starting point. After the passage of the landmark it was not visible to the subjects again until they started homing. In part of the trials the landmark has been translated to the right or to the left **(b)** Homing positions and corresponding scatter ellipses. Light grey (∘): no translation of the landmark, middle grey (△): translation to the left, black (∗): translation to the right. The big black triangles indicate the three possible positions of the landmark, only one position was occupied during each trial.

the distances, subjects also reproduced the velocity profile, indicating that the distance together with temporal information had been stored, presumably in some sort of longterm memory [4]. To what extent this metric knowledge enters long-term memory, however, is an open question.

8.3.3 Integration of Egomotion and Landmark Knowledge

In experiments where subjects had to solve a simple triangle-completion task, Gillner and Jin [19] showed that landmark information dominates path integration. In these experiments a prominent landmark was placed in the vicinity of the starting point (cf. Fig. 8.3). After the subject passed the landmark, the landmark was moved several meters away from its original position, without informing the subject. The final homing positions indicated that the subjects rely more on the information of the landmark than on the optic flow information. Landmark dominance is not complete, however, as is indicated by the scatter ellipses in Fig. 8.3.

8.4 Travelling a Known Route

The next level in the task hierarchy is the storage and reproduction of multistep routes, where each step can be considered a homing to a place functioning as an

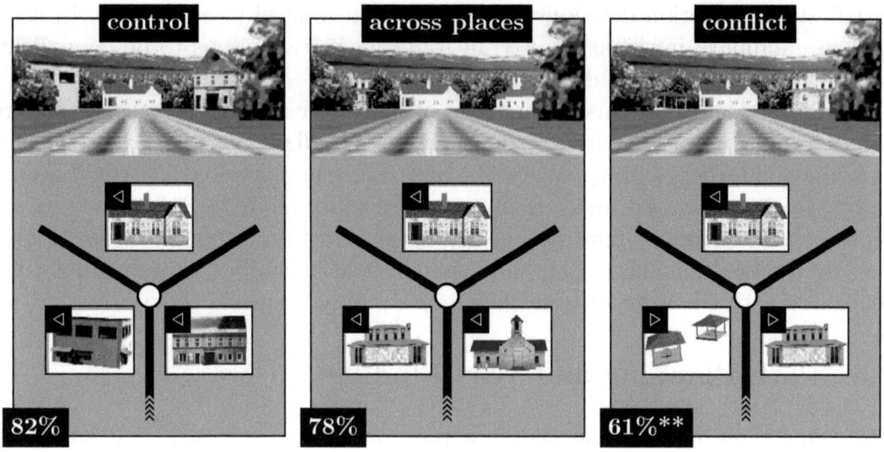

Fig. 8.4. Route leraning experiment by Mallot and Gillner [36]. After learning a route in Hexatown, subjects are tested by releasing them at some point on the route and translating them towards an adjacent place. Here the subjects are asked to decide whether the route continues left or right. In the control condition (no landmark replacements) 82 % of 160 decisions (40 subjects, 4 decisions at different places) were correct. Landmark replacements had no effect as long as all landmarks had been associated with the same movement decision during the training phase (middle panel). If landmarks are combined that "point into different directions", a significant reduction in performance is found. We conclude that place recognition is not required in route behaviour. Rather, recognition of individual landmarks, or views, suffices.

intermediate gial. Route knowledge can be modelled as a sequence of stimulus-response (S-R) associations or recognition-triggered responses, which results in a stereotyped behaviour. Alternatively, S-R-S associations have been suggested, predicting that a navigator not only generates a motor response at a certain location but also forms some expectations about the ongoing location [48].

What is the landmark information used in route navigation? In experiments with landmark transpositions after route learning, we showed that the behavioural response is triggered by the recognition of individual landmark objects or views, not of the configurations of objects making up a place [36]. When learning a route, each object together with its retinal position when viewed from the decision point (left peripheral, central, right peripheral) is associated with a movement triggered by the recognition of this object. When objects from different places are recombined in a way that their associated movements are consistent (i.e. all objects are associated with the movement decision "go right"), no effect in subjects' way finding performance was found. A decision was evaluated arbitrarily as "correct", if a subject chose the movement decision associated with the central view. If, however, objects are combined in inconsistent ways (i.e. one object is associated with the movement decision "go left", the other two objects with the movement decision "go right"), subjects get confused and the distribution of motion decisions approaches chance level (see Fig. 8.4). In conclusion, landmark information used in route navigation seems to be dependent on the viewing and travel direction.

Further evidence for this dependence of place recognition on travel direction comes from experiments with a spatial priming paradigm [38, 44]. If the sequence of prime and targets corresponds to the occurrence in a previously learned route, the target is recognized faster than control stimuli, indicating that the spatial representation contains information about the direction of travel. Interestingly, this "route direction effect" [44] has a specific spatial component and can not be replicated with the same objects shown in a purely temporal order.

8.5 Path-Planning and Way-Finding

The behaviours described so far allow the approach of a known goal following a formerly learned route. But surviving in a changing environment requires an adaptive and flexible behaviour, by developing new solutions based on present knowledge. In the domain of spatial cognition these flexible behaviours are (i) the planning of novel routes by recombining sections of known routes and (ii) the inference of metric information for finding shortcuts and detours. In this section, we will discuss the first idea, which is also known as topological navigation. Metric navigation will be discussed in Sect. 8.6.

8.5.1 Topological Shortcuts: Recombining Route Sections to Novel Routes

The planning of pathes depends on the goals of the current excursion and on explicit, or declarative, knowlegde of space. The memory structure supporting this sort of flexible route-planning, rather than stereotyped route-following, is called a cognitive map in the definition of O'Keefe and Nadel [40]. Direct evidence for the disctinction between route-following and finding of novel routes was recently presented by [22]. If subjects are requested to replicate a certain route several times, neural activation is found predominatly in the caudate nucleus. In contrast, hippocampal activation is found in a way-finding task, where novel routes had to be infered from memory.

In an experiment using the Hexatown environment (Fig. 8.1), we demonstrated that human navigators indeed have the ability of recombining sections of known routes into novel ones [20]. In a way-finding paradigm, subjects were released at some location in Hexatown and asked to find certain landmarks shown them as a print-out on a sheet of paper. Each individual search corresponded to a route learning task. For each of twelfe different route tasks, we measured the number of trials need to reach a criterion. The results show that routes performed later in the task sequence are learned faster, indicating that some goal-independent knowledge was transferred from the known routes to the novel tasks. At the same time, the persistence of stereotyped recognition-triggered response behaviour could also be demonstrated. That is to say, if errors were made, they often resulted from taking the stereotyped route decision at a given place.

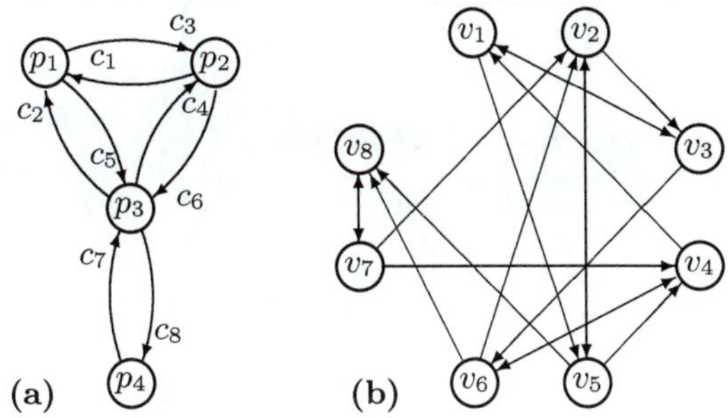

Fig. 8.5. Place- and view-graph after [43]. **(a)** Simple maze shown as a directed graph with places p_i and corridors c_j. **(b)** Associated view-graph where each node v_i corresponds to one view, i.e. one directed connection in the place graph. Edges indicate view sequences occurring in locomotion.

8.5.2 Graph Memory

Topological shortcut behaviour can be modelled by a graph structure integrating the knowledge from a number of routes [31, 43]. Two alternative graph structures (place graph and view graph) are presented in Fig. 8.5. Although the landmark replacement experiment [36] (Fig. 8.4) presents some evidence for the view-graph approach, the distinction between the two graph types is experimentally brittle. However, another implication of both graph models is much more accessible by empirical research. As compared to true maps (such as street maps printed on a sheet of paper), graph maps can much more easily accommodate uncomplete or inconsistent knowledge.

An experiment addressing the issue of inconsistent knowledge was performed by Steck and Mallot [45]. In order to study the integration of different types of landmark information, distal landmarks, placed on a mountain ridge surrounding the village, were added to the Hexatown environment. In this configuration, various strategies can be used to find a goal: subjects could ignore the distant landmarks altogether, they could rely on the distant ones exclusively, or they could use both types in combination. Steck and Mallot tried to identify these strategies by replacing the distant landmarks after learning (cue-conflict experiment), so that different patterns of movement decisions could be expected for each of the above strategies. Results indicate that different strategies were used by different subjects and by the same subject at different decision points. In a second experiment, one landmark type was removed from the maze after learning. In this cue-reduction experiment, subjects who had relied on the now removed landmark type when doing the cue-colict experiment, were still able to use the previously neglected landmark type. This indicates that both types of information were present in memory but one was ignored in the cue-conflict situation (see Fig. 8.6). In spite of both informations being present, landmark transpositions had not been reported in the first experiment.

Fig. 8.6. Panoramic views of Hexatown used in the interaction experiments [45]. *Top:* Training condition: both distant (mountain peak, distant tower in the left panel) and local landmarks (building) are visible. *Middle:* "Dawn" condition. Only the silhouette of the landscape and tower (distant landmarks) are visible. *Bottom:* "Night" condition. Only local landmarks are visible. Subjects who ignored one landmark type (distant or local) in a landmark transposition experiment with cue conflict, were still able to use the previously ignored landmark type in these environments.

8.5.3 Path Planning

On a behavioural level, path planning is the most important competence relying on declarative memory. Surprisingly, the number of investigations in the field of path planning is rather low. There are some results which show that humans try to minimize the effort during path planning—on a physical as well as on a cognitive level. Gärling and Gärling [18] investigated the shopping behaviour of pedestrians and showed that most shoppers prefer to first choose the location farthest away, probably due to the fact that by doing this they minimize the effort of carrying their shopping goods. Christenfeld [10] showed that subjects prefer routes with the longest initial segment in the right direction. Subjects delayed a turning decision as long as possible in order to keep the cognitive load low. That is, they do not turn until they have to.

Wiener and co-workers [51, 53] carried out a series of experiments indicating that path planning is not based on place knowledge alone but also relies on knowledge of regions in the environment. These regions could be defined by

Fig. 8.7. Experimental setup of the path planning experiment. *left:* Schematic map, places are marked by numbers, regions by the grey rectangles. *middle:* bird's eye view of the environment *right:* Subjects perspective within an experiment, looking on a landmark.

natural borders like streets or rivers as well as by their specific function, such as topical sections of an exhibition park. In the experiments, Wiener et al. used a regular, rectangular environment (Fig. 8.7) constructed in virtual reality, consisting of places and connecting streets. Each place is marked by one unique landmark, which was invisible until the subject enters a small neighborhood of the landmark. Regions are defined as two islands, each containing semantically grouped landmarks. The landmarks on one island were all cars, whereas the landmarks on the second island were of the category "animals".

After an exploration phase, subjects had to plan and execute the shortest routes from a given starting place to a goal. For each route there exist at least two alternative solutions which did not differ by means of the overall length. The difference between the two routes is the distance they covered in the target region. The variable of interest was the subject's preference to approach the target region as fast as possible. The tendency to reach the target region as soon as possible is significantly above chance level, indicating that subjects base their planning decisions not only on place knowledge but also on knowledge about regions. Wiener et al. suggested a *fine-to-coarse* planning strategy which postulates a focal representation in spatial working memory. In this focal map, places from the currently visited region are represented individually while more distant regions are only represented by an overall "region node". Planning toward these region nodes predicts the effects found in the experiments.

8.6 Finding Cross-Country Shortcuts

8.6.1 Metric Place Knowledge

Leaving known routes and finding cross-country shortcuts or using detours caused by a blocked path requires the knowledge of metric relations in an environment. Metric information can be obtained via path integration which was discussed as in Sect. 8.3.2 as a kind of working memory. In this section we will focus on behavioural data showing that humans are able of gathering metric information which is encoded in a rather more enduring manner, i.e. in longterm memory. Ecologically valid behaviours, which cannot be explained without the assumption of metric information are rare, the most common paradigms in psychological experiments are pointing to an invisible goal or judging the metric distance between two locations in space.

The accuracy of pointing to unseen targets is typically between 20° to 49° [41, 8] which is rather imprecise. Imagine trying to reach a goal 100 m away, this direction accuracy would result in deviations up to 119 m to the right or to the left of the goal location. Therfore, metric information is only useful in combination with other spatial cues. For example, in channelled environments, metric information on an ordinal scale may support movement decisions like "go right" or "go left".

In a long-term memory task similar to triangle completion, Foo et al. [15] trained subjects on two legs of a triangle. The corner places of the triangle could be recognized from independent landmark cues. After learning the individual legs, subjects were released at the end of one leg and asked to walk to the end of the other leg, thus shortcutting over the unknown third leg of the triangle. Results show that this task can be performed, but again with fairly high error rate. Still, some metric information seems to be included in spatial long-term memory.

8.6.2 Using Compasses

The performance of shortcutting behaviour can be improved by the usage of compass information. By compass information, we mean a cue to heading or body orientation, independent of path integration, and available everywhere in the environment. Except from nautical compasses, such information is at least approximately provided by distant landmarks, geographical slant, or the sun azimuth in connection with the time of day. Using compass information will improve path integration since the heading direction need not be inferred from rotation increments but can be read from the compass. Indeed, desert ant path integration relies heavily on the well-known polarization compass of insects [50]. Storing compass information in long-term memory should improve performance in pointing to invisible landmarks and the production of sketch-maps.

As one type of compass information, Restat et al. [41] investigated the role of geographical slant in the Hexatown virtual environment (Fig. 8.8). The whole environment was either planar or slanted by an angle of 4° in one of two directions. Subjects could interact with the virtual environment by pedalling with force-feedback on a bicycle simulator (translation) or by hitting buttons (discrete rotations in 60° steps). After memory acquisition, spatial knowledge was accessed by three tasks: (i) finding routes between certain landmarks, either the return path of formerly learned routes (return-path in Fig. 8.9) or novel-paths; (ii) pointing from various positions to the learned goals; (iii) choosing the more elevated of two presented landmarks. The number of navigation errors (wrong motion decisions with respect to the goal) was significantly reduced in the slanted conditions. Furthermore, Restat et al. found that subjects were able to point more accurately to currently invisible targets in the slanted virtual environments (cf. Fig. 8.9). The number of correct answers for judging the relative

Fig. 8.8. Overview of the three conditions for the slant experiment [41]. **Left:** map of the environments. Landmarks indicated by numbers have been used as goals in the exploration phase and as targets in the pointing phase. **Right:** subjects perspective. Each row shows the three pictures projected on a 180 degree screen. The images are projected with a small overlap; therefore the discontinuities visible here are not present in the actual experiment. The picture shows the view from the place with object 5 in the direction of the street towards the only adjacent place. **Top row** shows the *Flat* slant condition. **Middle row** shows the *Northeast* slant condition. **Bottom row** shows *Northwest*.

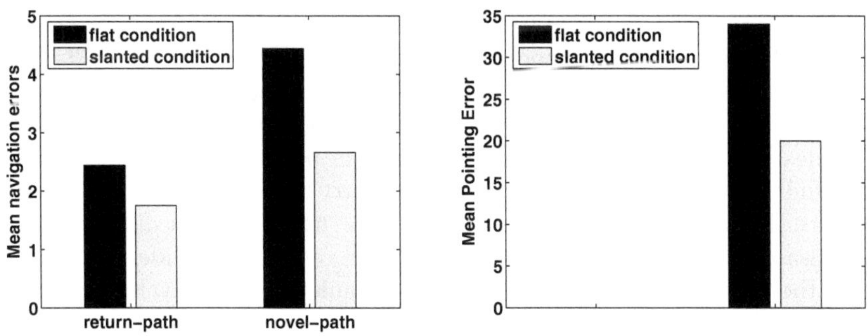

Fig. 8.9. Results of the slant-experiment: Both, the navigation performance (left) and the ability to point to objects (right) are enhanced, if subjects explored a slanted environment compared to a flat condition

height of two landmarks in the slanted conditions was between 80 - 90 %, i.e. it was highly above chance level. There was a correlation between the height difference and the reaction time for this task: with increasing height difference the reaction time was decreased.

In summary, we conclude that metric information is present in spatial long-term memory, probably for all three dimensions of space. However, this information may be rather noisy. Still, it helps improving navigation performance.

8.7 Communication About Space

Communication about space is not in itself a navigational behavior. Still, by giving route instructions or publishing maps, we do support navigational behavior in other people and influence their ways of route planning and spatial reasoning. In building spatial memories, it is conceivable that namebale information is more readly encoded and used.

One line of evidence concerning the relation of language and spatial behavior comes from the work of Levinson and co-workers on intercultural comparisons, see [35]. For example, the Tenejapa language from India does not use egocentric spatial descriptions (like "left" or "right"), but uses allocentric verbal description ("west", "east"). Native speakers of Tenejapan and Dutch were tested in non-linguistic spatial tasks, where subjects were asked to compare simple stimuli (cards) after a whole-body rotation (see Fig. 8.10). The Tenejapa speakers used an allocentric reference frame also in the non-verbal task whereas the Dutch speakers used an egocentric reference frame. Levison concludes from these experiments that language not only reflects the way humans think about spatial relation but also influences spatial behavior. It should be noted, however, that the role of the verbal instruction given to the subjects is not entirely clear. It is well conceivable that the phrase "most similar card" means different things to speakers of different language while the underlying representation of space is indeed the same.

A second line of evidence concerns the spatial memories built from texts as compared to memories built from visual inspection or active exploration. Behavioral differences between verbal and visual acquisition groups are minor, leading to the idea of an amodal spatial memory used in both cases (e.g., [3]). On the other hand, neural activities during spatial imagery tasks performed in memories acquired either verbally or visually from looking at maps, do show differences [47].

The usability of verbal directions for actual navigation provides yet another view on the interaction of space and language. Daniel et al. [13] evaluated "good" and "poor" route description by means of navigational performance of naive subjects using these descriptions for wayfinding. The effectiveness of spatial phrases depended on their ability to connect actions to landmarks, which fits very well in the above mention hierarchical structure of spatial memory.

In an evolutionary view, language-based navigational performances build on the pre-lingual wayfinding hierarchy found already in animals. These abilities seem to be preserved in what has been called the spatial core knowledge of the

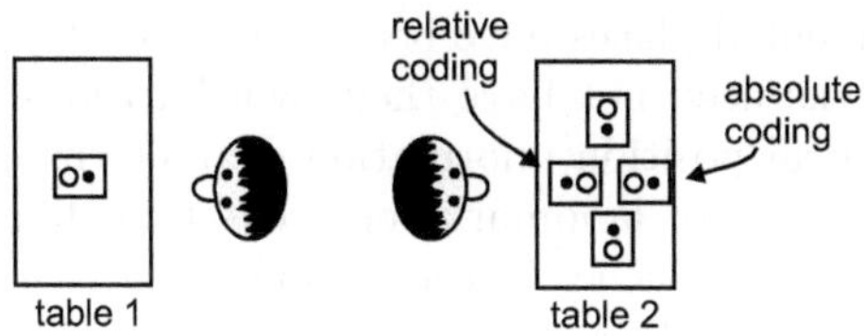

Fig. 8.10. After learning a card on table 1, subjects were asked to turn 180° to table 2. They then indicated which card on table 2 was most similar to the card on table 1. Coding in an allo– or egocentric coordinate system will lead to different choices. Redrawn from Levinson [33].

human cognitive system [14]. Simple reasoning abilities such as needed for route planning are already part of the wayfinding hierachy. However, language adds on this its advanced capacities for general reasoning as well as the ability of communication about space.

8.8 Discussion

8.8.1 Repertoires vs. Ontologies

In this paper, we have described a behavioural approach to spatial cognition, characterized by the hierarchy of spatial tasks summarized in Table 8.2. From this task hierarchy, a hierarchy of representations can be derived that is similar to the semantic hierarchy of Kuipers [31] or the way-finding hierarchy of Trullier et al. [48]. However, when compared to hierarchies based primarily on the complexity of representations (or "ontologies"), different views emerge with respect to three issues, landmarks, metric information, and language.

Landmarks. In the task hierarchy approach, a landmark is piece of data taken from the sensory input and stored as a property, or label, of a place representation. The spatial representation may thus consist only of place and action

Table 8.2. Tasks and representations in spatial cognition

Task	*Representation*
recognizing places	*snapshot, depth signature* (Sect. 8.2)
finding home after excursions	*vector; guidance* (Sect. 8.3)
following a route	*chains of recognition triggered responses* (Sect. 8.4)
recombining route segments	*graphs (networks) of S-R-S associations* (Sect. 8.5)
route selection and planning	*fine-to-coarse planning in focal map* (Sect. 8.5.3)
cross-country shortcuts	*metric embedding of places* (Sect. 8.6)
communicating about space	*naming places and actions* (Sect. 8.7)

tokens, where all prepresented places have been visited by the observer. In contrast to places, landmarks need not have their own locations in the cognitive map but are pieces of local position information characterizing the place they were sensed from. This view of landmark representation differs from the the occpancy grid and SLAM-approaches in robot navigation (see Thrun, this volume), where landmarks are localized objects with coordinate values. The local-position-information approach to landmarks has two advantages. First, it allows for relatively low depth of processing in landmark recognition. Second, views or vistas, such as the lining up of objects from a particular viewpoint or lines of sight can easily be treated as landmarks.

Metric Information. In an ontological approach, it is tempting to treat metric information such as distances and angles as one type of data that jointly enters the hierarchy at some level of complexity. In the task hierarchy, it seems that metric information is relevant at different task levels to different extent. Path integration is a simple way of using metrics for returning to a nest or "home" after an excursion. Metric embedding of topological knowledge, i.e. assigning coordinates to places is a much more advanced mechanism required in finding what we called cross-country shortcuts. Clearly, many animals with simple path-integration abilites do not have an elaborated long-term memory of space, let alone a metric embedding of known places.

Language and Space. The relation of language and space is a matter of ongoing debate. From an evolutionary point of view, it is clear that most, if not all, spatial abilities are already found in animals lacking language. Therefore, spatial behaviour cannot generally be based on language abilities. Spatial language, then, is a further task level concerned with spatial cooperation and communication about space, i.e. a task level at the intersection of spatial and social behaviour. The same is probably true for map drawing, another ability relevant mostly for communication. Clearly, this "spatio-social" task level plays an extremly important role in human spatial behaviour, for example in finding our way to a destination we never have visited before. Still, we suggest that this ability belongs to task level that should be well distinguished from way-finding behaviour as defined in the animal literature.

8.8.2 Other Hierarchies

One final point that we would like to make is that hierachies can be defined along various criteria, which, as discussed above, will not always lead to the same results. Here, we have focussed on the hierarchy of tasks and compared it to the hierarchy of mechanisms and representations derived mainly from computational arguments. Two additional axes that have been proposed for scaling navigational mechanisms are ontogenetic development and the sequence of knowledge acquisition in a new environment. These different hierarchies and their relation to each other are of course to be elucidated by future research. However, one unifying approach is to look at such hierarchies as paths through a broader evolutionary tree, whose leaves are the spatial abilities of the respective species.

References

1. Allen, G., Kirasic, K.C., Siegel, A.W., Herman, J.F.: Developmental issues in cognitive mapping: The selection and utilization of environmental landmarks. Child development 50, 1062–1070 (1979)
2. Appleyard, D.: Why buildings are known: A predictive tool for architects and planners. In: Broadbent, G., Bunt, R., Llorens, T. (eds.) Meaning and Behavior in the Built Environment, pp. 135–161. John Wiley & Sons, Chichester (1980)
3. Avraamides, M.N., Loomis, J.M., Klatzky, R., Golledge, R.C.: Function equivalence of spatial representations derived from vision and language: evidence from allocentric judgements. Journal of Experimental Psychology: Learning, Memory, and Cognition 30, 801–814 (2004)
4. Berthoz, A., Israël, I., Georges-François, P., Grasso, R., Tsuzuku, T.: Spatial memory of body linear displacement: What is being stored? Science 269, 95–98 (1995)
5. Brooks, R.A.: A robust layered control system for a mobile robot. IEEE Journal of Robotics and Automation RA-2, 14–23 (1986)
6. Burk, J., Gillner, S., Mallot, H.A.: Visual homing accuracy in humans. In: Ilg, U.J., Bülthoff, H.H., Mallot, H.A. (eds.) Dynamic Perception: Workshop of the GI section 'Computer Vision', Akademische Verlagsgesellschaft Aka, Berlin, pp. 57–62 (2004)
7. Cartwright, B.A., Collett, T.S.: How honey bees use landmarks to guide their return to a food source. Nature 295, 560–564 (1982)
8. Chance, S.S., Gaunet, F., Beall, A.C., Loomis, J.M.: Locomotion mode affects the updating of objects encountered during travel: the contribution of vestibular and proprioceptive inputs to path integration. Presence: Teleoperators and Virtual Environments 7, 168–178 (1998)
9. Cheng, K.: A purely geometric module in the rat's spatial representation. Cognition 23, 149–178 (1986)
10. Christenfeld, N.: Choices from identical options. Psychological Science 6, 50–55 (1995)
11. Christou, C., Bülthoff, H.H.: View dependence in scene recognition after active learning. Memory & Cognition 27, 996–1007 (1999)
12. Cornell, E.H., Heth, C.D., Alberts, D.M.: Place recognition and wayfinding by children and adults. Memory & Cognition 22, 633–643 (1994)
13. Daniel, M.P., Tom, A., Manghi, E., Denis, M.: Testing the value of route directions through navigational performance. Spatial Cognition and Computation 3(4), 269–289 (2003)
14. Dehaene, S., Izard, V., Pica, P., Spelke, E.: Core knowledge of geometry in an amazonian indigene group. Science 311 (2006)
15. Foo, P., Warren, W.H., Duchon, A., Tarr, M.J.: Do humans integrate routes into a cognitive map? map- versus landmark-based navigation of novel shortcuts. Journal of Experimental psychology: Learning, Memory, and Cognition 31, 195–215 (2005)
16. Franz, M.O., Schölkopf, B., Mallot, H.A., Bülthoff, H.H.: Learning view graphs for robot navigation. Autonomous Robots 5, 111–125 (1998)
17. Fujita, N., Klatzky, R.L., Loomis, J.M., Golledge, R.G.: The encoding-error model of path-way completion without vision. Geographical Analysis 25, 295–314 (1993)
18. Gärling, T., Gärling, E.: Distance minimization in downtown pedestrian shopping. Environment and Planning A 20, 547–554 (1988)
19. Gillner, S., Jin, Y.: Integration of landmark information and optic flow in humans. In: Würtz, R.P., Lappe, M. (eds.) Dynamic Perception, Ruhr–Universität Bochum, Akademische Verlagsgesellschaft Aka GmbH, Berlin. pp. 29–34 (2002)

20. Gillner, S., Mallot, H.A.: Navigation and acquisition of spatial knowledge in a virtual maze. Journal of Cognitive Neuroscience 10, 445–463 (1998)
21. Gillner, S., Weiß, A.M., Mallot, H.A.: Visual place recognition and homing in the absence of feature-based landmark information. Cognition (submitted)
22. Hartley, T., Maguire, E.A., Spiers, H.J., Burgess, N.: The well-worn route and the path less traveled: distinct neural bases of route following and wayfinding in humans. Neuron 37, 877–888 (2003)
23. Hayhoe, M., Ballard, D.: Eye movements in natural environments. Trends in Cognitive Sciences 9, 188–194 (2005)
24. Heft, H.: The role of environmental features in route–learning: Two exploratory studies of way–finding. Environmental Psychology and Nonverbal Behavior 3, 172–185 (1981)
25. Hermer, L., Spelke, E.S.: A geometric process for spatial reorientation in young children. Nature 370, 57–59 (1994)
26. Hermer-Vazquez, L., Spelke, E.S., Katsnelson, A.S.: Sources of flexibility in human cognition: dual-task studies of space and language. Cognitive psychology 39, 3–36 (1999)
27. Jacobs, W.J., Thomas, K.G.F., Laurance, H.E., Nadel, L.: Place learning in virtual space. II. Topographical relations as one dimension of stimulus control. Learning and Motivation 29(4), 288–308 (1998)
28. Jacobs, L.F., Schenk, F.: Unpacking the cognitive map: The parallel map theory of hippocampal function. Psychological Review 110(2), 285–315 (2003)
29. Janzen, G., van Turennout, M.: Selective neural representation of objects relevant for navigation. Nature Neuroscience 7(4), 673–677 (2004)
30. Jin, Y.: An Eye-Gaze Tracking Approach for Landmark Recognition in Human Navigation. PhD Dissertation, Eberhard-Karls-Universität Tübingen, `http://www.uni-tuebingen.de/cog`
31. Kuipers, B.: The spatial semantic hierarchy. Artificial Intelligence 119, 191–233 (2000)
32. Land, M.F., Hayhoe, A.: In what ways do eye movements contribute to everyday activities? Vision Research 41, 3559–3565 (2001)
33. Levinson, S.C.: Frames of reference and Molyneux's question: Crosslinguistic studies. In: Bloom, P., Peterson, M.A., Nadel, L., Garrett, M.F. (eds.) Language and Space, pp. 109–169. The MIT Press, Cambridge, MA (1996)
34. Loomis, J.M., Klatzky, R.L., Golledge, R.G., Cicinelli, J.G., Pellegrino, J.W., Fry, P.A.: Nonvisual navigation by blind and sighted: Assessment of path integration ability. Journal of Experimental Psychology: General 122, 73–91 (1993)
35. Majid, A., Bowerman, M., Haun, D.B.M., Levinson, S.C.: Can language restructure cognition? the case for space. Trends in Cognitive Sciences 8, 108–114 (2004)
36. Mallot, H.A., Gillner, S.: Route navigation without place recognition: what is recognized in recognition–triggered responses? Perception 29, 43–55 (2000)
37. Maurer, R., Séguinot, V.: What is modelling for? A critical review of the models of path integration. Journal of theoretical Biology 175, 457–475 (1995)
38. McNamara, T.P., Halpin, J.A., Hardy, J.K.: Spatial and temporal contributions to the structure of spatial memory. Journal of Experimental Psychology: Learning, Memory, and Cognition 18, 555–564 (1992)
39. McNaughton, B.L., Battaglia, F.P., Jensen, O., Moser, E.I., Moser, M.-B.: Path integration and the neural basis of the 'cognitive map'. Nature reviews Neuroscience 7, 663–678 (2006)
40. O'Keefe, J., Nadel, L.: The hippocampus as a cognitive map. Clarendon, Oxford, England (1978)

41. Restat, J., Steck, S.D., Mochnatzki, H.F., Mallot, H.A.: Geographical slant facilitates navigation and orientation in virtual environments. Perception 33, 667–687 (2004)
42. Riecke, B.E., van Veen, H.A.H.C., Bülthoff, H.H.: Visual homing is possible without landmarks - a path integration study in virtual reality. Presence: Teleoperators and Virtual Environments 11, 443–473 (2002)
43. Schölkopf, B., Mallot, H.A.: View–based cognitive mapping and path planning. Adaptive Behavior 3, 311–348 (1995)
44. Schweizer, K., Herrmann, T., Janzen, G., Katz, S.: The route direction effect and its constraints. In: Freksa, C., Habel, C., Wender, K.F. (eds.) Spatial Cognition. LNCS (LNAI), vol. 1404, pp. 19–38. Springer, Heidelberg (1998)
45. Steck, S.D., Mallot, H.A.: The role of global and local landmarks in virtual environment navigation. Presence. Teleoperators and Virtual Environments 9, 69–83 (2000)
46. Stürzl, W., Mallot, H.A.: Efficient visual homing based on Fourier transformed panoramic images. Robotics and Autonomous Systems 54, 300–313 (2006)
47. Tlauka, M., Keage, H., Clark, C.R.: Viewing a map versus reading the description of a map. Cognitive Science 29, 807–818 (2005)
48. Trullier, O., Wiener, S.I., Berthoz, A., Meyer, J.-A.: Biologically based artificial navigation systems: Review and prospects. Progress in Neurobiology 51, 483–544 (1997)
49. Wang, R.F., Spelke, E.S.: Human spatial representation: insights from animals. Trends in Cognitive Sciences 6, 376–382 (2002)
50. Wehner, R.: Desert ant navigation: how miniature brains solve complex tasks. JCPA 189, 579–588 (2002)
51. Wiener, J.M., Mallot, H.A.: 'Fine-to-coarse' route planning and navigation in regionalized environments. Spatial Cognition and Computation 3, 331–358 (2003)
52. Wiener, J.M., Mallot, H.A.: Path complexity does not impair visual path integration. Spatial Cognition and Computation 6(4), 333–346 (2006)
53. Wiener, J.M., Schnee, A., Mallot, H.A.: Use and interaction of navigation strategies in regionalized environments. Journal of Environmental Psychology 24, 475–493 (2004)
54. Yarbus, A.L.: Eye Movements and Vision. Plenum Press, New York (1967)
55. Zugaro, M.B., Berthoz, A., Wiener, S.I.: Background, but not foreground, spatial cues are taken as references for head direction responses by rat anterodorsal thalamus neurons. The Journal of Neuroscience 21(RC154), 1–5 (2001)

9

Landmarks for Navigation in Human and Robots

Stephen C. Hirtle

Summary. One determinant of navigation concerns the ability to use landmarks. However, despite wide acceptance of the concept of landmarks, there is considerable debate as to what is meant by the term 'landmark' and how landmarks are used to assist in navigation. Sorrows and Hirtle [30] introduced a tripartite theory of landmarks that can be applied to navigation by humans in real and electronic spaces. Their approach was to classify landmarks along three dimensions: visual, semantic, and structural. These dimensions can be defined independently for navigation in physical space and for navigation in electronic spaces, such as the World Wide Web. It is argued in this paper, that the same framework can be extended to robot navigation, but with the realization of the dimensions appearing quite different in robotics world. The term landmark remains a fundamental concept of navigation and can provide a theoretical bridge between scientific camps of researchers.

9.1 Introduction

In this paper, the role of landmarks is examined for two distinct populations, humans and robots. Through this analysis, we will delineate the varied definitions and uses of landmarks. However, rather than produce just a list of differences, it is argued that such a synthesis can add to the theory of landmarks and identify important areas of cross-fertilization for research.

Landmarks have proven to be one of the most fundamental concepts in building models of navigation and spatial representation [1, 7, 16, 25]. For this chapter, it is assumed that a landmark is an object or location external to the observer, which serves to define the location of other objects or regions. This definition, while general in scope, will ignore uses in the literature where a landmark is defined to represent any point or location in space. Landmarks, as defined here, need to be extracted from a rich environment and will be stored for use in later navigation or identification. What makes a landmark useful to an agent is dependent on a large number of factors including the nature of the space, the goal of the agent, the representational database, and computational complexity.

In this chapter, we begin with a look at landmarks in human spaces and present a general theory of landmarks from Sorrows and Hirtle [30]. The examination of landmarks for humans covers both landmarks in real-space and in cyberspace. Research on landmarks for robotic spaces is examined in Sect. 3. The final section concludes by revisiting the general theory of landmarks and discussing how each field can learn from advances in the other.

M.E. Jefferies et al. (Eds.): Robot. & Cogn. Approach. to Spat. Map., STAR 38, pp. 203–214, 2008.
springerlink.com

Fig. 9.1. Example of a strong visual landmark from the University of Pittsburgh campus

9.2 Landmarks in Human Spaces

The ability to navigate in an environment is dependent upon one's ability to form a spatial representation of that environment, and landmarks play a key role in the creation of such a cognitive map [16]. A landmark is an object or location external to the observer, which serves to define the location of other objects or regions. Determining an exact definition of a landmark is difficult [23], but there have been several recent survey papers that have developed useful frameworks for understanding landmarks [7, 30]. Heth et al. [10] describe two ways landmarks are fundamental to navigation. First, landmarks are the memorable cues, which are selected along a path, particularly in learning and recalling turning points along the path. Second, landmarks enable one to encode spatial relations between objects and paths, enabling the development of a cognitive map of a region. This distinction can also be described as landmark-goal relationships, where landmarks are cues along a path to a goal, and landmark-landmark relationships, which provide a global understanding of the environment for navigation, even in lower species [21].

9.2.1 A Theory of Landmark Use by Humans

Sorrows and Hirtle [30] have extended the typologies of landmarks to include three distinct categories: visual, structural and semantic. These overlapping categories are described in detail below.

Visual landmarks are those that are visually distinctive from the surrounding environment, as shown in Fig. 9.1. These landmarks are easily recognized by individuals who are unfamiliar and familiar with the environment. Visual landmarks are noted by the contrast they provide with the surrounding environment and most likely are noted for their prominent location.

Semantic landmarks are those in which the meaning stands out, as shown in Fig. 9.2. The photograph in Fig. 9.2 shows a doorway to a building on the University of Pittsburgh campus that is visually non-distinct. However, the doorway

Fig. 9.2. Example of a semantic landmark as the door is visually and structurally unimportant, but leads to the primary registration office for students

becomes important to students on campus as it leads to the main registration office on campus. Such landmarks had been called cognitive landmarks in the original article by Sorrows and Hirtle [30], but the equivalent term semantic landmarks, coined by Raubal and Winter [24], is now preferred. They have also been referred to as symbolic landmarks by Portugali [22]. A semantic landmark may have a well-defined role in the environment, such as an information kiosk. They also may be quite personal with meaning to only a few individuals, who are very familiar with the space, such as the copy room or coffee lounge in an office complex. Portugali [22] has extended his definition of symbolic landmarks to include legendary landmarks, such as the Verona balcony where the famed dialogue between Romeo and Juliet was said to have taken place. It is a landmark to locals and tourists alike, even if the actual conversation never occurred at this spot, but is only mentioned in the legend.

Structural landmarks are those in which the location is central in terms of the topology of the space. Structural landmarks, such as a downtown plaza or a major transit stop, are highly accessible and often used as point of departure or arrival. Structural landmarks can be seen in terms of a network graph indicating the connectivity of the nodes. Dupont Circle in Washington, DC is a well-known structural landmark where ten streets converge around a traffic circle. It is interesting to note that an organization of art galleries in the Dupont Circle area chose to represent the circle by the road network, as shown in Fig. 9.3, and not the visually distinctive fountain in the center of the circle. Thus, Dupont Circle has become known as a landmark for its structural significance.

These three dimensions of landmarks, visual, semantic, and structural, can be present in a single landmark and together boost the "landmarkness" of a location. At the same time, many well-known landmarks are landmarks because of a strong value on only one of the dimensions.

(a) (b)

Fig. 9.3. A logo, shown on the left, chosen by the Galleries of Dupont Circle (http://artgalleriesdc.com/), based on the structural relationships of 10 roads converging on the circle, rather than on the visual image of fountain in the center of the circle as shown on the right. Reprinted with the permission of (a) the Galleries of Dupont Circle, Washington, DC and (b) Stapleton-Gray Associates, Inc.

9.2.2 Landmarks in Cyberspace

Sorrows [29] goes on to describe how each of these three classes of landmarks can apply not only to physical environments, but also to cyberspace and the World Wide Web (WWW). In an electronic space, visual landmarks are visually distinctive pages, such as a home page of a University, which has a distinctive look and feel from all other pages on a website. A semantic landmark may look similar to other pages, but for an individual has critical significance, which makes it stand out. Finally, a structural landmark might provide a large number of in and/or out links, which makes it particularly useful in navigating through a site. The structural dimension has been used by Mukherjea and Hara [18] to define a landmark as a node which is important to the user because it helps to provide an understanding of both the organization and the content of that part of the information space. Glenn and Chignell [6] describe landmarks as part of a symbol system which is both visual and semantic, and in which the visual and semantic functions are intricately tied. Although these and other definitions of landmarks in the WWW seem compatible, a key problem exists in how to determine specifically what nodes are landmark nodes. Algorithms have been proposed which use the connectivity of a node, the frequency of use of a node, and the depth of the node in the local WWW directory structure.

Sorrows [29] has used memory experiments for cyberspace in a novel way to demonstrate the use of landmarks. Among other tasks, she asked subjects

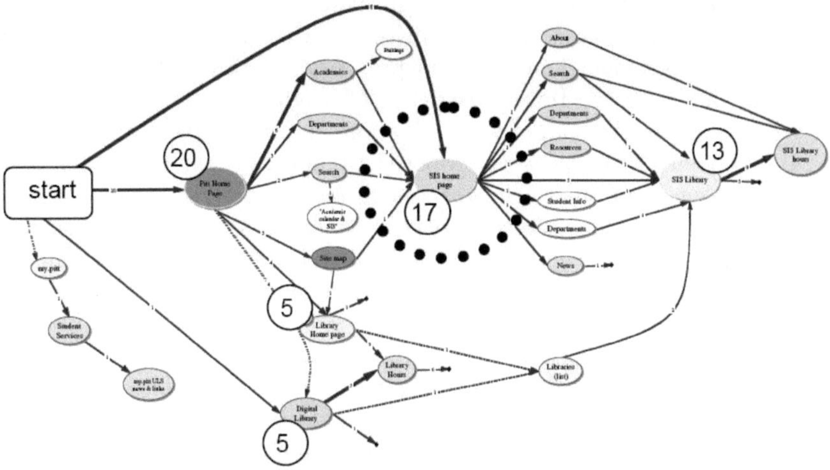

Fig. 9.4. A path graph showing a dominant landmark page, circled in small, black dots. Modified from [29]. Used with permission.

to recall well-known paths in cyberspace, just as one might ask someone to give directions from memory to a well-known building on campus. Figure 9.4 shows a path graph for the answer to one of the questions: "How would find the holiday hours for the SIS library at the University of Pittsburgh." The circles indicate web pages that were recalled. Each solid link indicates a sequence of a correct link recalled by at least one subject. Dotted links indicate recall of imaginary links; that is, links that don't exist, but subjects assume that they must. Of importance for this discussion are the paths and the small numbers, which indicate the number of subjects recalling a specific page. Note that no matter which path the subjects describe, most end up passing through a central node for the School of Information Sciences homepage, circled in small, black dots. Thus, this page forms a central landmark for these subjects, from which additional, varied paths are taken. Berendt and Brenstein [2] show a comparable graphical technique from a tool called STRATDYN, which can also highlight the search strategy used by participants while browsing websites. Finally, Sorrows [29] showed how the landmark pages from the path graphs are also ranked high in terms of the three dimensions of visual, semantic and structural distinctiveness.

9.2.3 An Application to Direction Giving

In this section, an application of how landmarks might be used in direction giving is described in detail. Of course, this is just one use of landmarks by humans. Golledge [7] makes the distinction between landmarks as a navigational aid and landmarks as an organizing concept of space. The second meaning has been called an anchor point [7] or reference point [25]. In the sense of an organizing concept, landmarks act to facilitate environmental understanding. Neighborhoods or regions will often emerge around one or more landmarks, with the landmark then serving as a superordinate feature in hierarchical representation of space [7]. This sense, which is not discussed further in this chapter, leads to districts named for the prominent landmark, such as Capitol Hill or Dupont Circle in Washington or the Water Tower District in Chicago.

In terms of using landmarks for navigation, Raubal, Winter and Nothegger [20, 24] extended the framework proposed by Sorrows and Hirtle [30] to the practical problem of automatically generating local landmarks to enrich wayfinding instructions. The dimensions of visual attraction, semantic attraction, and structural attraction were considered the additive components to determine the overall "landmarkness" of a building. Each of these ideas was further quantified as described below.

Visual attraction was measured by the visual saliency of an object in regard to its surroundings, through the measurement of four variables: facade area, shape, color and visibility. Shape was measured in two different ways. The shape factor was the ratio of height/width, so tall buildings would have strong visual attraction. Shape deviation was measured by percentage not covered by the bounding rectangle. A perfectly rectangular facade would result in a shape deviation of zero, while an irregular shaped building would result in a shape deviation greater than zero. Color was measured by the three RGB values. The final variable included in visual attraction was visibility, which is an area from which the building is visible to pedestrians. Raubal and Winter [24] note that it is possible to differentiate between day landmarks and night landmarks, by including illumination. However, this extension has not been incorporated into their model. The visual saliency is derived from the geo-referenced images provided by TeleInfo at http://www.teleinfo.de.

Semantic attraction was measured in two ways: cultural and historical importance, and explicit signage on the building indicating its name or purpose. The current implementation codes both of these variables with the Boolean value of true or false, but could be expanded to a predefined scale of significance.

Finally, the structural attraction was measured by Lynch's elements of nodes and boundaries [14]. Nodes were measured by the sum of the weighted incoming and outgoing paths from the landmark, where weights increase with the prominence of the path. Thus, a landmark on a corner is going to be more notable than landmarks in the middle of a block. Boundaries are measured by the effort to cross the boundary by multiplying the size of the cell by the ratio of the long to short side.

An overall value of landmark saliency is then calculated by the weighted sum of these three measures. Given a shortest path, which is generated using standard algorithms, this method can extract the most salient landmarks to integrate into the route description.

9.3 Landmarks for Robots

Within the robotics literature, the term "landmark" is also widely used. For some authors, the term is used in a very general way to refer to almost any

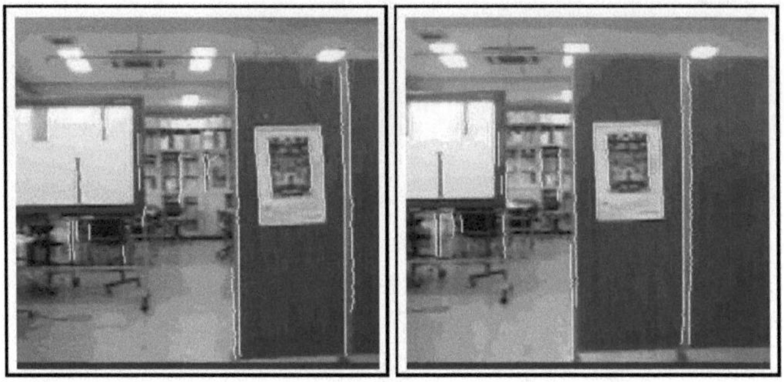

Fig. 9.5. Five vertical line segments identified as potential landmarks based on stereo matching. Reprinted from [17] with permission from IEEE.

point in the space. However, of more interest here, are those that use it more selectively, referring to critical decision points or points used to help orient the robot in space. Landmarks can also provide a kind of ground truth, similar to how a surveyor's mark provides alignment between the actual environment and a map of the environment.

In terms of selective landmarks, Thrun [31] makes a distinction between artificial landmarks (often bar codes) that have been placed in the environment and natural landmarks, which need to be discovered (e.g., [5, 26]). Natural landmarks can consist of a variety of possible cues in the environment. Moon, Miura, and Shirai [17] propose a method of selecting natural landmarks consisting of vertical line segments on a planar surface, as these tend to be more stable than points. An example is shown in Fig. 9.4, where five vertical line segments are identified through stereo matching of the images. Other researchers do not restrict the type of visual landmark to any one class, but instead examine the environment for distinctive visual features. For example, Sim and Dudek [27] use principal components analysis to generate a low-dimensional description of possible landmarks, which place no restrictions on the landmark position in the world. Greiner and Isukapalli [8] model the environment using real-world objects that are also useful in navigation, such as doors and corners within a building. In addition, pictures on the wall can serve as landmarks, just as a sign in a street environment might serve as the significant landmark for an intersection.

While much of the research on natural landmarks is using cameras and visual pattern recognition of a scene to determine novel or important characteristics, several authors argue that machine-learning techniques, such as Bayesian approaches or neural networks, will do a superior job of uncovering landmarks in sensor data [31]. Finally, some authors have mixed artificial and natural landmarks by identifying unique planar textures found in door signs, posters, and light switches (e.g., [15]). Figure 9.5 shows the stored landmarks in the top row and a remapping based on camera angle in the bottom row.

Together, these studies can be described by the same typology that was outlined above, consisting of visual, structural and semantic landmarks. Visual

Fig. 9.6. Stored landmarks in the top row and a remapping based on camera angle in the bottom row. Reprinted from [15] with permission from Elsevier.

Table 9.1. Typology of landmark type by navigational goal given by [30]. Note that each landmark type is excluded from one navigational goal.

	Type of Landmark		
Navigation	Visual	Semantic	Structural
To a specific known new goal		×	×
To a familiar goal	×	×	
In an unfamiliar environment	×		×

landmarks might be extended to the category of sensory landmarks, where sensors can be any of a variety of different modalities, including vision and sonar. Structural landmarks are easiest to describe in the topology of the space. These landmarks are key decision points for navigation, including doorways and intersecting hallways. The final category of semantic landmarks may appear to be more difficult to translate, but still has a strong correspondence. There are often objects in the environment, which need to be known for their content, such as a recharging station or a mail drop location, where a particular kind of event is to occur. Such locations would serve as semantic landmarks for the robot. Just as semantic landmarks for humans must be learned, the semantic landmarks for robots may need additional information, such as a bar code, or an initial acquisition phase to be properly learned.

The relationship between navigational task and types of landmarks used are shown in Table 9.1. As noted in the table, semantic landmarks for robots are useful only in well-learned spaces, whereas visual and structural landmarks are important in novel spaces. This becomes clear if you equate semantic landmarks with barcodes, structural landmarks with topography, and visual landmarks with visual features. Robots mapping a new underground mine or involved in a search and rescue mission, such as those exploring the devastation after the World Trade Center attacks on Sept 11, 2001, would be navigating through an unfamiliar environment. Barcoding would be of little help in determining where the robot has been and or where the robot should go, as there is no pre-established knowledge or ground truth. In contrast, robots performing repetitive navigation tasks, such as an automated mail delivery robot, would be most effective following semantic and visual cues placed in the environment.

9.4 Discussion

The tripartite theory of landmarks first introduced by Sorrows and Hirtle [30] classifies landmarks along three dimensions: visual, cognitive or semantic, and structural. These dimensions can be defined independently for navigation in physical space and for navigation in electronic spaces, such as the World Wide Web. It is argued in this paper, that the same framework can be extended to robot navigation. While there are some differences, the relation between landmark type and landmark function remains fixed across each of the target domains.

So what can each community learn from the other? First, robotics researchers have a much stronger track record of generating optimal sets of landmarks, as the ability to pick an optimal set of robotic landmarks is a particularly difficult problem. Much of the robotics literature on landmarks has focused on methods for the extraction of landmarks from a set of potential candidate landmarks (e.g., [8, 15, 17, 31]). Surprisingly, there are relatively few studies in the human navigation literature that have focused on automated techniques for the extraction of useful landmarks. One notable exception is the recent work, using ID-3 to generate the best set of predictors given a large set of attributes that characterize potential landmarks in a neighborhood [4]. The use of data mining and machine learning techniques with human data is still relatively new. More research in this area could prove to be beneficial, especially if complemented with user studies measuring the usefulness of the automatic generation of landmarks. In contrast, human researchers are more likely to take a multifaceted view of landmarks and their role in navigation [7, 24, 30]. The work of Madsen and Anderson [15], which mixes natural and artificial landmarks, is reflective of the possible extensions that should be considered by robotics researchers.

In this chapter the discussion has focused on landmarks, but clearly landmarks are just one representation object to be modeled in the navigation process. Extensions of landmarks are worth considering from both camps. For example, the notion of examining the environment for local landmarks has been extended by Simhon and Dudek [28] to include local islands of reliability. These are based on identifying distinctive regions, where in a sense the region becomes a landmark, just as a human traveler might use a town square as a important landmark region. Simhon and Dudek's implementation builds on more general theories of cognitive mapping of Yeap [32] and of Kuipers and Byun [13].

Landmarks can also result in difficulties for navigation. Unreliable or moveable landmarks are not useful for repeated navigation in the same space. Part of the human developmental process is to recognize that useful landmarks are those that are permanent, stable and visible from multiple viewpoints [10]. In fact, one reason that online navigation systems, such as www.mapquest.com, do not use landmarks is the problem of keeping the database up-to-date. The road network is also dynamic, but the changes to the road network in terms of both topology and identifiers are far less frequent then the changes to the buildings, signs and other potential landmarks within the space. Thus, the old tale of turning left where the red barn use to be, would become common place without careful consideration to the difficult problem of updating the data tables and reparameterizing the landmark space on a regular basis.

Finally, as specific applications are pursued in both the human and the robot domains (e.g., [3, 9, 19]), additional principles of navigation and of landmarks will emerge. For example, Klippel et al. [12] has considered the use of schematic information for mobile-based navigation systems, which typically use very small screens with limited readability. In such environments, landmarks must be used sparingly and detailed instructions will be replaced with sparse instructions, such as head toward the church, or the equivalent visual command known as a map gesture [11]. Such impoverished instructions can only make sense in the context of some local intelligence to make intermediate navigational decisions, which both camps of researchers need to explore further.

References

1. Allen, G.L., Kirasic, K.C., Siegel, A.W., Herman, J.F.: Developmental issues in cognitive mapping: The selection and utilization of environmental landmarks. Child Development 56, 1062–1070 (1979)
2. Berendt, B., Brenstein, E.: Visualizing Individual Differences in Web Navigation: STRATDYN, a Tool for Analyzing Navigation Patterns. Behavior Research Methods. Instruments, Computers 33, 243–257 (2001)
3. Cheverst, K., Davies, N., Mitchell, K., Friday, A., Efstratiou, C.: Developing a context-aware electronic tourist guide: Some issues and experiences. In: Proceedings of CHI 2000, Netherlands (2000)
4. Elias, B.: Extracting landmarks with data mining methods. In: Kuhn, W., Worboys, M.F., Timpf, S. (eds.) COSIT 2003. LNCS, vol. 2825, pp. 375–389. Springer, Heidelberg (2003)
5. Franz, M.O., Schölkopf, B., Georg, P., Mallot, H.A., Bülthoff, H.H.: Learning view graphs for robot navigation. In: Johnson, W.L. (ed.) Proc. 1. Intl. Conf. on Autonomous Agents, pp. 138–147. ACM Press, New York (1997)
6. Glenn, B.T., Chignell, M.H.: Hypermedia: Design for browsing. In: Hartson, H.R., Hix, D. (eds.) Advances in Human-Computer Interaction, Ablex, Norwood,NJ (1992)
7. Golledge, R.G.: Human wayfinding and cognitive maps. In: Golledge, R.G. (ed.) Wayfinding behavior: Cognitive mapping and other spatial processes, pp. 5–45. Johns Hopkins Press, Baltimore, MD (1999)
8. Greiner, R., Isukapalli, R.: Learning to select useful landmarks. In: Proceedings of 1994 AAAI Conference, Menlo Park, CA, pp. 1251–1256. AAAI Press / MIT Press (1994)
9. Hertzberg, J., Kirchner, F.: Landmark-based autonomous navigation in sewerage pipes. In: Proceedings of the First Euromicro Workshop on Advanced Mobile Robots, pp. 68–73. IEEE Computer Society Press, Los Alamitos (1996)

10. Heth, C.D., Cornell, E.H., Alberts, D.M.: Differential use of landmarks by 8- and 12- year-old children during route reversal navigation. Journal of Environmental Psychology 17, 199–213 (1997)
11. Hirtle, S.C.: The use of maps, images and "gestures" for navigation. In: Freksa, C., Brauer, W., Habel, C., Wender, K.F. (eds.) Spatial Cognition II: Integrating Abstract Theories, Empirical Studies, Formal Methods, and Practical Applications, pp. 31–40. Springer, Berlin (2000)
12. Klippel, A., Richter, K.F., Barkowsky, T., Freksa, C.: The cognitive reality of schematic maps. In: Meng, L., Zipf, A., Reichenbacher, T. (eds.) Map-based Mobile Services - Theories, Methods and Implementations, pp. 57–74. Springer, Berlin (2005)
13. Kuipers, B., Byun, Y.T.: A robot exploration and mapping strategy based on a semantic hierarchy of spatial representations. Robotics and Autonomous Systems 8, 47–63 (1991)
14. Lynch, K.: The image of the city. MIT Press, Cambridge (1960)
15. Madsen, C.B., Andersen, C.S.: Optimal landmark selection for triangulation of robot position. Robotics and Autonomous Systems 23, 277–292 (1998)
16. Mark, D.M., Freksa, C., Hirtle, S.C., Lloyd, R., Tversky, B.: Cognitive models of geographical space. International Journal of Geographical Information Science 13, 747–774 (1999)
17. Moon, I., Miura, J., Shirai, Y.: On-line selection of stable visual landmarks under uncertainty. In: Proceedings of the 1999 IEEE/RSJ International Conference on Intelligent Robots and Systems, vol. 2, pp. 781–786. IEEE Press, Piscataway, NJ (1999)
18. Mukherjea, S., Hara, Y.: Focus+context views of world-wide web nodes. In: Hypertext 1997: The Eighth ACM Conference on Hypertext, pp. 187–196. ACM Press, New York (1997)
19. Nolfi, S.: Evolving non trivial behaviors on real robots: A garbage collecting robot. Robotics and Autonomous Systems 22, 187–198 (1997)
20. Nothegger, C., Winter, S., Raubal, M.: Selection of salient features for route directions. Spatial Cognition and Computation 4, 113–136 (2004)
21. PastergueRuiz, L., Beugnon, G., Lachaud, J.P.: Can the Ant Cataglyphis cursor (Hymenoptera: Formicidae) Encode Global Landmark-Landmark relationships in addition to isolated landmark-goal relationships? Journal of Insect Behavior 8, 115–132 (1995)
22. Portugali, J.: The i in gis. In: GIScience 2000, pp. 28–31. Savannah, Georgia, USA (2000)
23. Presson, C.C., Montello, D.R.: Points of reference in spatial cognition: Stalking the elusive landmark. British Journal of Developmental Psychology 6, 378–381 (1988)
24. Raubal, M., Winter, S.: Enriching wayfinding instructions with local landmarks. In: Egenhofer, M.J., Mark, D.M. (eds.) GIScience 2002. LNCS, vol. 2478, pp. 243–259. Springer, Heidelberg (2002)
25. Sadalla, E.K., Burroughs, W.J., Staplin, L.J.: Reference points in spatial cognition. Journal of Experimental Psychology: Human Learning and Memory 5, 516–528 (1980)
26. Scharstein, D., Briggs, A.: Real-time recognition of self-similar landmarks. Image and Vision Computing 19, 763–772 (2001)
27. Sim, R., Dudek, G.: Mobile robot localization from learned landmarks. In: IROS. Proceedings of IEEE/RSJ Conference on Intelligent Robots and Systems, Victoria, BC, vol. 2, pp. 1060–1065 (1998)

28. Simhon, S., Dudek, G.: Selecting targets for local reference frames. In: ICRA 1998. Proceedings of the IEEE International Conference on Robotics and Automation, vol. 4, pp. 2840–2845. IEEE Computer Society Press, Los Alamitos (1998)
29. Sorrows, M.E.: Recall of Landmarks in Information Space. PhD thesis, Doctoral Dissertation, University of Pittsburgh (2004)
30. Sorrows, M.E., Hirtle, S.C.: The nature of landmarks for real and electronic spaces. In: Freksa, C., Mark, D.M. (eds.) COSIT 1999. LNCS, vol. 1661, pp. 37–50. Springer, Heidelberg (1999)
31. Thrun, S.: Finding landmarks for mobile robot navigation. In: ICRA 1998. Proceedings of the IEEE International Conference on Robotics and Automation, vol. 2, pp. 958–963. IEEE Computer Society Press, Los Alamitos (1998)
32. Yeap, W.K.: Towards a Computational Theory of Cognitive Maps. Artificial Intelligence 34(3), 297–360 (1988)

10

Learning Cognitive Maps: Finding Useful Structure in an Uncertain World

Eric Chown and Byron Boots

Summary. In this chapter we will describe the central mechanisms that influence how people learn about large-scale space. We will focus particularly on how these mechanisms enable people to effectively cope with both the uncertainty inherent in a constantly changing world and also with the high information content of natural environments. The major lessons are that humans get by with a "less is more" approach to building structure, and that they are able to quickly adapt to environmental changes thanks to a range of general purpose mechanisms. By looking at abstract principles, instead of concrete implementation details, it is shown that the study of human learning can provide valuable lessons for robotics. Finally, these issues are discussed in the context of an implementation on a mobile robot.

10.1 Introduction

One of the key debates surrounding the relationship between human cognitive mapping and robotics focuses on whether there is any common ground. After all, robots and humans have very different capabilities. Human cognitive mapping relies on a highly developed visual system that can recognize objects and landmarks with ease. Computer vision, by contrast, provides inaccurate and often inefficient perception for robots. As a result, robotic systems frequently rely on other sensors including laser range finders, infrared switches, and sonar arrays. Indeed, one of the apparent advantages robots enjoy is that they can be fitted with a wide range of sensing devices that give them abilities humans lack; some prominent examples are accurate range estimation and dead-reckoning. Despite many other advantages – astonishingly fast processors and flawless information retrieval – robotic navigation abilities still pale in comparison to those found in humans. The gap between human and computer vision promises to remain large for the foreseeable future. The obvious questions remain: What can robotics gain, as a field, from the study of human cognitive mapping? How may human cognitive mapping be informed by the practical considerations found in robotics?

Despite many differences between cognitive mapping and robotics, there are success stories indicating that a cross-pollination of ideas between these two fields may produce exciting results. One influential development, the topological representation, lies at the heart of virtually every theory of cognitive mapping. A topological representation consists of elements corresponding to visual landmarks, and connections between elements indicating the order in which the

M.E. Jefferies et al. (Eds.): Robot. & Cogn. Approach. to Spat. Map., STAR 38, pp. 215–236, 2008.
springerlink.com

landmarks have been experienced. Path planning is performed by extracting a sequence of landmarks from a starting location to a goal. Inspired in part by cognitive mapping, topological representations are now commonly used in robot navigation. Given the differences in human and robotic perceptual capabilities, landmarks recognized by robots differ significantly from landmarks typically recognized by humans. In this case one of the basic ideas inherent in cognitive maps, that landmarks are connected in a network, is so powerful that it works well despite the differences in the perceptual abilities of robots and humans. This should not be surprising: cognitive maps are capable of supporting effective navigation in extremely diverse conditions including times when vision is only of marginal help, such as in fog or at night. Because humans face so many possible environments, and because effective navigation is so important to survival, humans have necessarily evolved highly adaptive spatial representations. A key aspect of this type of adaptation is reliance on multiple representational paradigms – in the absence of good landmarks, for example, humans are able to navigate extremely well using other visual information. One of the primary theses of this chapter is that potential gain for robotics in drawing ideas from cognitive mapping will come from identifying the characteristics that make human spatial representations so flexible and adapting them to robotic platforms and tasks. Thrun et al., for example, found that a hybrid topological-metric map converged faster and was more accurate than a pure metric approach [28]. Landmarks make global metric maps more useful because of two properties – 1) They occur at fixed locations, and 2) They are unique – not because of anything specific about how human object recognition works.

Another example of cross-pollination of ideas between cognitive mapping and robotics was the development of the *gateway* construct in the PLAN (Prototypes, Locations, and Associative Networks) [6] and R-PLAN (Robot-PLAN) [13] architectures. Informally, a gateway is a point where a person leaves one region of space and enters another. Gateways can be recognized by certain environmental characteristics usually involving a sensory occlusion followed by an opening. Since they mark exits and entrances, gateways are extremely useful for parsing large environments into smaller, more manageable chunks. Gateways have a variety of other cognitive uses including providing the basis for more metric representations of space than are possible with a purely topological model. In robotics gateways have proven very useful in indoor environments [16] where they can be used to anchor representations and can also serve as de facto landmarks in topological representations. As with topological representations, the basic notion of gateway was taken from a cognitive mapping theory but has been modified in accordance with the capabilities of robots (gateways were originally identified purely with sonar).

What makes the gateway example even more interesting is that it was a case where robotics helped to drive cognitive theory. In this case a roboticist (David Kortenkamp) was interested in implementing a cognitive theory on a robot but did not think that cognitive mapping theory was sufficiently well developed to serve as the basis for a working model. The questions he asked the theorists forced them to fill holes in their theory. Some of the questions touched on what has been a long debate in the spatial representation community about whether human representations of space are coded egocentrically or are viewpoint independent. The practical considerations of robot implementation cast the problem in a new light. One model at the time [3], not very different from many of the grid-style representations in use today, stored "views" of space at fixed intervals. The strategy was not adaptive in that it required an enormous amount of computation and storage, but it raised questions about which views of space a person would be most likely to store. A second influential model of the time [31] explicitly focused on exits and entrances. Bringing the two robotic strategies together helped lead to the development of the gateway idea which in turn brought support to the egocentric model of spatial representation. Since that time there has been growing evidence that humans have egocentric representations [25].

The events surrounding the conception of the gateway illustrates the potential value of robotics to cognitive theory. From the perspective of a cognitive theorist an implementation is not only a chance to test out the theories, but it is also an opportunity to push theory into unexpected directions. Cognitive mapping theory offers strategies proven capable of dealing with issues that robots still struggle with, including scaling to very large environments (e.g. outdoor environments), and dealing with highly dynamic surroundings. An important common element of the implementations discussed here is the abstraction of ideas. In both cases the cognitive strategies were implemented in ways that are most appropriate for the computational strengths of robots instead of a slavish imitation of human physiology. From this perspective, implementations should be appropriate to the physical realities of the agent and robotics should look to cognitive mapping as a source of abstract ideas about how to address some of the difficulties arising as robots move into larger and more varied environments.

A common practice in robotics is allowing a robot to explore its environment in idealized conditions (e.g. no people, good lighting, etc.) before it becomes operational. A well known example of this is Minerva, a robot that provided tours of the Smithsonian [27]. Minerva was able to thrive in the chaos of a busy museum because it had previously acquired an excellent internal map of the environment. Ideally robots will begin to drop such restrictions and have the ability to learn new environments on the fly, even in the face of less than ideal conditions. To do this robots will need to improve how they learn. One way to accomplish this might be to draw on theories of how humans learn and represent environments, and distill strategies that robots could adopt in ways suitable to their own abilities. The bulk of the rest of this chapter is devoted to discussing how humans learn about large-scale space. This is done with an eye towards lessons that could be used in robot implementations. Such work is part of the co-authors' ongoing research and an example is briefly described in Sect. 10.4. It must be noted, however, that the techniques outlined in this chapter are not offered as replacements for what is currently done in robotics. In most applications, particularly in the near term, robots will perform better using standard methodologies specifically tailored to their abilities. The strategies put forth in this chapter come from a general purpose architecture, namely the human cognitive architecture, and therefore would be best suited for extremely general purpose robots.

10.2 The Creation of Spatial Structure

The theories presented in this chapter are presented in the context of the PLAN model of cognitive mapping [6]. Complete coverage of the material is well beyond the scope of a single chapter so we will necessarily present a simplified view in the name of getting the important ideas across. This is keeping with the following theme: robotics is not likely to be informed by the low level details of how the human cognitive architecture works, but rather by some of the abstract ideas inherent within it.

In PLAN, as in several other theories of cognitive mapping, it has been proposed that humans have two fairly different modes of spatial functioning based upon two distinct underlying representations. One representation is called a *route map* and corresponds to the topological network of landmarks already discussed. The other is often called a *survey map* and is more metric in character. Chown et al. theorized that each representation roughly corresponds to the processing done in the two distinct pathways found within the human visual system [30], namely the *what* and *where* pathways (more recently these have been labeled the *ventral* and *dorsal* pathways [19]). The "what" pathway is mainly concerned with object recognition. In spatial terms the objects are landmarks, which form the basis for topological networks or route maps. The "where" pathway is more directly spatial and deals with where things are in relationship to each other and to the observer. The theory in PLAN is that the metric quality of survey maps arises from the fact that information stored is roughly equivalent to actual views of the environment. The corresponding representational structures are called *local maps*. Just as landmarks are connected together to form route maps, local maps are connected together to provide the basis for survey maps. In PLAN the networks of local maps are called *R-Nets*, while more global survey maps are called *regional maps* or *R-Maps*. Although the basic elements that comprise route and survey maps are different, the way learning occurs to build each representation is essentially the same.

This section is divided into two major parts. In the first part pure associative learning is discussed. While it is the case that both survey maps and route maps have associative components, the focus will be on route maps since they are purely associative and survey maps have additional machinery. Associative learning is particularly well suited to dealing with changing environments so an emphasis will be placed upon that in the discussion. In the second section the focus turns to survey maps, and in particular how they are adapted to help people cope with large-scale environments.

The associative learning model discussed in this chapter is based upon a learning rule originally proposed by Hebb in *The Organization of Behavior* [8]. Hebb's rule applies to learning at the level of neurons, but he also provided a mechanism

– the cell assembly – to bootstrap the rule up to the "symbolic" level. Hebb's rule, was actually a mechanistic version of an associative learning rule that goes back to William James [10]. Both proposed that when two elements (be they neurons, cell assemblies or other) are simultaneously active, then an associative link between them is strengthened. In spatial terms the cognitive elements are generally landmarks (or local maps) and the associative connections between them provide the structure that forms them into coherent representations of space.

Associative learning and topological networks are already well established tools in computer science and robotics. While associative learning rules are fairly common, they rarely realize the full potential of association on display in the human cognitive architecture. By itself an associative learning rule specifies that two landmarks near each other should be linked. The subtlety and flexibility of human learning is due to the variable strength of the associative links. The next subsection begins with the pure associative rule and then proceeds to discuss some of the critical ways that the human cognitive architecture can modify the associative process as suitable to the situation at hand. In particular, emphasis is placed on modifications that help humans deal with dynamic, and very large, environments.

10.2.1 The Synthesis of Structure Through Association

In *The Organization of Behavior* Hebb showed how his learning rule could generate neural structure called cell assemblies. The basis for this idea is that when an object in the world is seen, neurons corresponding to its component features will become active. Hebb's learning rule will then tend to bind these neurons together into an assembly. As the object is subsequently seen, these bindings will become stronger and the neurons that comprise the individual features will begin to act as a unit. The associative nature of this structure is such that even when the object is only partially seen, the tight internal connections of the cell assembly will tend to activate the entire structure.

The same principles hold at the level of cognitive maps. When two landmarks are simultaneously active (as would be the case when one landmark follows another in a journey) an associative bond will be strengthened between them (based on the learning rule at the neural level). Just as the features of an object can activate each other in a cell assembly, so too can the landmarks that comprise a cognitive map. This property is especially advantageous because it can be used both to remember and to plan journeys. For example, a recent journey will effectively be indexed by its starting point. Once that cognitive element is activated it will naturally tend to activate the subsequent landmark in the journey.

As landmarks present themselves in multiple journeys this sequential structure will begin to form a network. The critical step is that the landmark in common is recognized as being the same even though it may be viewed from an altogether different orientation. The ability to create networks of landmarks gives an organism a great deal of power in planning routes. Specifically, it may become possible to extract routes that were never explicitly experienced (see Fig. 10.1).

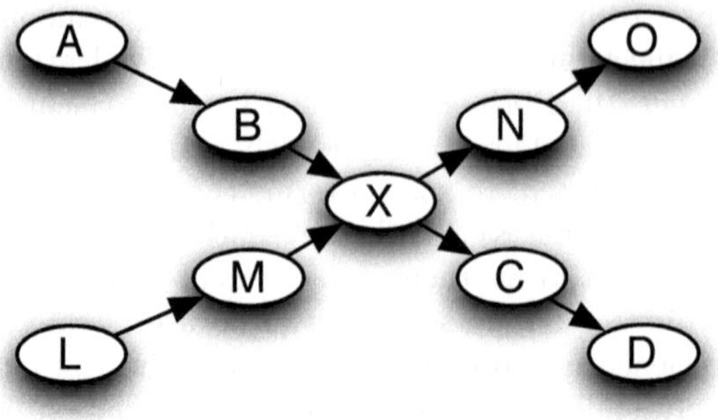

Fig. 10.1. Network creation. After experiencing sequences A-B-X-C-D and L-M-X-N-O a network can be created provided that X can be recognized as common to both sequences. This creates the possibility of extracting routes such as A-B-X-N-O that have never been experienced.

One of the biggest challenges in building a useful internal representation of a natural environment is that most environments are not static, but change over time. The human brain appears to have multiple mechanisms in place to cope with the difficulties this presents. The most important is the nature of the connections made between representations. These connection strengths are not binary, but exhibit a graded continuum. At first blush it might seem that the strength of a connection between elements should represent a kind of probability that the two elements will co-occur in the future. It is well known, however, that humans do not make decisions in a manner consistent with a simple underlying model of probability [18] and there are a number of very good reasons why this is not the case. Connection strength should reflect what is meaningful as well as what is experienced. For example, an encounter with a bear may not be probable, but it is meaningful enough that it is best to overestimate the chances of running into the bear in the same location in the future. For these reasons, changes in connection strength are not based purely upon statistical factors, but are also impacted by heuristic factors. The basic building blocks of associative learning, contiguity and repetition, are statistical – the more times two things are experienced together the stronger the linkage between them – but there are a number of modifiers in the human cognitive architecture. The rest of this subsection is devoted to some of the heuristic factors that impact associative learning.

Compensatory Learning

The most obvious problem with a pure associative learning rule is that all it does is strengthen connections. Over time even elements that co-occur infrequently will become strongly connected. This problem was found in the earliest simulation of Hebb's rule in which learning consisted solely of strengthening connections; the inevitable result was that eventually everything was strongly connected to everything else [23].

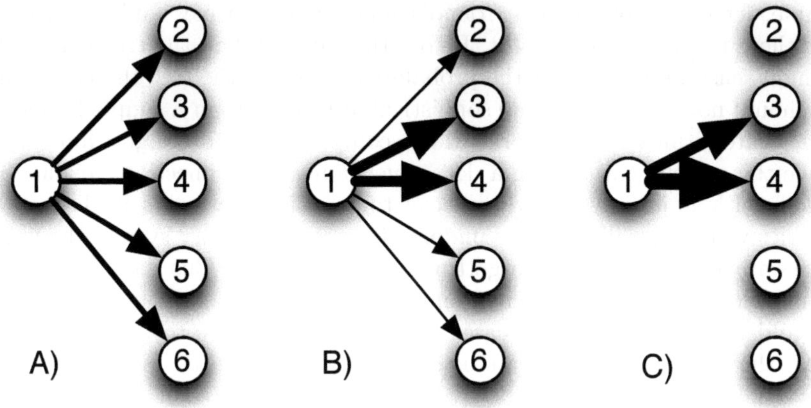

Fig. 10.2. Compensatory Learning. The thickness of the arrows denotes the strength of connection between elements. (**a**) node 1 is uniformly connected to all of the other nodes; (**b**) after 1 has co-occurred with nodes 3 and 4 the linkages between them have been increased while the link strengths to the other nodes have decreased; (**c**) it is possible that some of the connections will eventually disappear altogether because they have lost all of their strength.

A solution to the problem of ever increasing connection strength between elements, one that naturally addresses some of the challenges presented by dynamic environments, is to view connection strength as a fixed resource. Under this view, learning can be seen as a reallocation of resources based on experience: when connection strength is increased between two cognitive elements, it must also be decreased between others. A learning rule of this type is called *compensatory*. Consider a pair of associative elements that do not ever co-occur. Since the elements never co-occur the connections between them never increase. Further, any connection strength between them will occasionally decrease as other connections are strengthened. Eventually the connection between two elements will disappear, or, as some have theorized, the connection strength may become negative or inhibitory. Elements that only occasionally co-occur will normally suffer the same fate (an exception will be discussed later). Figure 10.2 shows how a compensatory rule normally works. The high density of neural connections present in the brain of a young child reflects a world of possibilities where virtually anything can be connected to anything else (although other architectural factors discussed later in the chapter show that some kinds of connections are easier to make than others). Through associative learning, possibilities are culled and structures are formed that reflect experience in the world. A compensatory rule is crucial in this process.

A system with compensatory learning must be recency-based since new connections come at the expense of older ones. In a cognitive map this is a sensible heuristic. New information must be rapidly incorporated into the map. For example, let's say you go to a friend's house twenty times and the first ten times there is a tree in front of it. Before the eleventh trip the tree is removed. Common sense indicates that the probability of seeing the tree the twenty-first time should not be 50%. A more sophisticated process than a simple frequency-based strategy must be at work. Many computer learning algorithms try to achieve this effect by attaching a decay mechanism to connection weights. The problem with uniformly applying decay to a system's weights is that all old memories will eventually become undone. In a compensatory system the weight changes are locally limited so forgetting is not universal. This is one reason why certain unique environmental cues, such as particular smells, can so powerfully evoke old memories – in most cases forgetting occurs due to the similarity of events, the more unique an experience the more likely the memory is to be preserved.

Arousal

An often cited drawback of learning systems that rely so much on contiguity and repetition is lack of speed. People require repeated exposure to complex environments before they feel comfortable navigating in them. This can be frustrating from a practical standpoint when one's goal is to learn as quickly as possible. The system is, however, the product of a conservative behavioral strategy that implicitly places a premium on safety. Normally, strong connections are a product of repeated experience, and repeated experience implies familiarity and safety. Sometimes, though, it is necessary to short-circuit this basic paradigm when the lessons to be learned are too important to be forgotten.

Very few situations and environments are the same. Some are confusing or dangerous, while others can be dull and boring. In adaptive terms, some situations are more important than others. Consequently learning should be faster at those times. In a neural system with an associative learning rule this is accomplished by increasing the activation intensity of its elements. In most animals, including humans, accelerated learning is accomplished through the arousal system (for a discussion of some of the issues surrounding this see [5]). When someone is highly aroused their focus and intensity increases correspondingly with the effect of learning faster. From an adaptive perspective this is useful because it means that an encounter with a bear will stick out in one's mind much more than a quiet walk in the woods. In neural terms this increase in learning occurs because of the increase in the firing intensity of neurons. As neurons fire more rapidly the individual applications of Hebb's rule increase. In a sense arousal acts as a kind of gain mechanism for learning; when arousal is high learning is fast, when it is low learning is slow. This is a clear case of how human learning situationally short-circuits a purely probability-based approach.

In a sense the arousal system tracks importance. Factors that impact arousal range from the hard-wired (e.g. pleasure and pain), to the instinctive (e.g. snakes), to the predictive (e.g. anticipation of a big event). Heuristically the more important the event the better learned it should be.

Short-Term Memory

One of the great benefits of associative structure at all levels of cognition is that it can automatically fill-in missing information. At the level of an object this means that even if much of the object is obscured the cognitive system will treat it as though it is whole. This is true for cognitive maps as well. Seeing a familiar landmark can immediately and automatically call to mind an entire environment. This can be problematic, however, as the well-connected cell assemblies that comprise the cognitive map will tend to act as attractors and become active even when they should not. In the extreme case even a poor perceptual match might provide enough impetus to activate a well-learned assembly. Children sometimes display such behavior when learning new categories; a child that has just learned about zebras may be inclined to call anything with four legs a zebra. The problem of judging when something is really new or merely a poor example of something that is known has been termed the *plasticity/stability* dilemma by Grossberg [7]. The solution proposed by Kaplan et al. [11] was a mechanism called *short-term connection strength* (STCS). The basis for STCS is that once neurons reach a certain threshold of activity they become temporarily more sensitive to firing again. For a cell assembly this means a group of features that is not well-connected may, nevertheless, generate enough feedback to become a coherent active representation. The combination of direct perceptual stimulation and STCS is enough to provide a competitive edge against more well-learned assemblies.

STCS also provides the basis for short-term memory. Since active neurons are temporarily more sensitive, much less stimulation is required to reactivate a cell assembly once it has been active. Among the benefits of this is that the networks comprising cognitive maps can be more fluid than static network structures. Imagine walking into a familiar office where the housekeepers have rearranged everything in order to wax the floors. For you to effectively function, what is in your head needs to reflect the current state of the room rather than what you remember. It is unlikely your internal map of the room should change permanently, however, since the office is likely to be back in its old configuration the next time you are there. STCS provides a temporary structure that can override the long-term structure without the need to discard the long-term structure. This is one way that humans can cope with the issue of integrating knowledge into long-term structure, particularly when new knowledge seems to be at odds with what is known. This issue will be discussed in more detail in the discussion of the hippocampus in the next subsection.

The Functional Distance Principle

Some of the factors that impact activity in cognitive elements go under the heading of "wiring." One such factor, which will turn out to be extremely important to learning about large-scale space, is what Kinsbourne [12] called *the functional distance principle*. This principle has two parts; the first is that brain activities that are close together will tend to inhibit each other. The second part is that

Fig. 10.3. Functional Distance Principle. Nodes A through G are all very similar to each other, and therefore are processed near each other in the brain. Node H is slightly farther away because of its differences. Because it is away from the other nodes it will tend to become more active.

similar items will be processed in similar locations in the brain. This principle has a number of important implications for spatial learning.

First, the functional distance principle helps explain what makes something a landmark in the first place. Landmarks are central to virtually every cognitive mapping theory, yet very little is known about what makes something a landmark beyond "uniqueness." Imagine walking down a corridor full of red doors. Since all of these doors are perceptually similar they will all be processed in the same portion of the brain. The similarity of these representations means that they will tend to inhibit each other. This inhibition will, in turn, limit how active the representations become. Since activity is the basis of learning, these representations will be only weakly connected to the active cognitive map as shown in Fig. 10.3. Now imagine coming upon a distinctly different door. Perhaps it is blue. The perceptual difference will mean that the representation for the door will be shifted away, even if only slightly, from where red doors are processed in the brain. This will automatically reduce the amount of inhibition that the new representation faces, meaning it will naturally be more active than the representations of the red doors. Since learning comes as the result of activity, the representation of the blue door will be more strongly connected into the cognitive map.

In some cases the change in perception is even more dramatic. An example of such a change is when a view is occluded and subsequently opens up into a new space. Such locations serve as the basis for the gateway structures discussed in the introduction which in turn form the basis for survey maps in PLAN.

10.2.2 Survey Structures

By itself topology provides limited spatial information, specifically proximity and ordering. A pure topological map would not be useful for spatial operations that rely on the relationships of distal objects. In humans the representations that serve this purpose are survey maps. In PLAN the core representational element of a survey map is the local map. Gateways are usually the locations at which local maps are generated.

Functional Distance Revisited

Gateways occur where there is a significant change in what is perceived. With a great enough difference what is currently perceived will no longer be considered a part of what was previously perceived. The cognitive separation between what is perceived and what was percieved appears to allow humans to use gateways to parse large-scale space into smaller regions. As a person approaches an archetypal gateway their perception goes from being occluded, perhaps by wall, to opening up to a new area. Just as this opening usually marks a shift in one region of space to another, the locus of processing in the brain is marked by a shift from one area to another by means of the functional distance principle. The shift in the locus of processing automatically reduces the level of inhibition between processes thereby naturally increasing the amount of learning. In classical psychology this leads to learning effects known as *primacy* and *recency*: people tend to best remember the first and last things in new environments, lists, etc. In spatial terms the first and last things in an environment are entrances and exits, locations of extreme import for organisms in a dangerous world.

Entrances and exits have other implications. They serve to break up large-scale space into smaller chunks. The gateway construct takes its name from the design literature in honor of Christopher Alexander's construct of the same name [2]. Alexander et al. describe the importance of breaking a large space into smaller regions:

> Many parts of a town have boundaries drawn around them. These boundaries are usually in people's minds. They mark the end of one kind of activity, one kind of place, and the beginning of another. In many cases, the activities themselves are made more sharp, more vivid, more alive, if the boundary which exists in people's minds is also present physically in the world. (p. 277)

Alexander goes on to note that boundary crossings must be marked by gateways. Among the many uses of breaking a large space into smaller chunks is that it serves as the basis of hierarchy. The bound regions essentially become the nodes in a higher-level network. Long routes need not be extracted from a huge low-level network, but can be found in a two-stage process. First a path can be found from one region to the next. Then at the lower level the paths within regions can be extracted. This strategy mirrors what Parisean taxi drivers were found to do in a well-known study by Pailhouse [22].

Gateways also serve as the basis for an alternative to route maps. People are naturally inclined to stop and look around where gateways are considered to occur. Further, as entrances, they are places that people cannot help but visit when they go to that region of space. For these reasons, and because of the functional distance principle, people will have stronger visual memories at gateways than at other locations. These visual memories will not simply be landmarks, but are more like full-fledged views. Such views afford more spatial information than a purely topological map.

Interfacing Memories – The Hippocampus

One of the roles of STCS is to help differentiate what is perceived from what is merely remembered. This works well for landmarks partly because they tend to be very stable features of an environment. Views of an environment, on the other hand, are much less stable – individual objects move, people move about, etc. This raises a number of issues for view-based representations such as PLAN's local maps. One of the most crucial is that view recognition cannot work in exactly the same way as landmark recognition. Robots face a similar problem known as localization. The localization problem for a robot is characterized by the problem of reconciling what its sensors are providing with what it has stored in memory. In a nutshell: How does the robot know where it is based on what it sees? Even when a person recognizes where they are, their cognitive system still must reconcile the current state of that environment against what they have in their memory. Cognitive maps provide another way of looking at this: "maps" in a person's head are only approximations of real environments. While this is useful for planning in the abstract, as cognitive maps code the most likely environmental configurations, it is potentially confusing in practice. Cognitive maps may contain elements that no longer exist in the environment and may not contain newer elements that do exist. The compensatory learning rule will help the cognitive map better reflect this over time, but it does not overcome the need for an internal representation that reflects the current state of the world. Internal representations should have the ability to augment perception and occasionally even stand in for perception, but what is stored should never dominate what is perceived.

For humans localization appears to occur in the hippocampus. The hippocampus is connected to both the vision system (and specifically the "where" system) as well as to cortical memory structures and has been extensively studied for its role in spatial processing [20, 21] as well as its central role in memory [26, 17]. What is crucial to this discussion is that the hippocampus appears to resolve stored locations against perceived locations thereby solving the localization problem (for more on this see the Save et al. article in this volume). In many cases what is perceived will be similar enough to a stored memory that resolving them will be simple. Standing in the doorway of a familiar room, for example, should evoke a prototypical memory that is similar in configuration to what is being viewed. This is another reason why gateways are so important: gateways provide a canonical, repeatable, view of a space, so such resolution can normally be done without the need for transformations of the views.

Gateways make the localization problem simpler in two important ways. First, they reduce the number of views to be stored. Second, they occur at very specific and important locations. You cannot normally enter a space without passing through a gateway. Further, a view from a gateway can easily be matched against a previous view from the same gateway. For example, in the original robot experiments with gateways a robot using sonar was able to repeatably get within 3.5 degrees of orientation and 70 millimeters of position of its stored location [14]. Even though that robot used extremely simple "views" (it extracted vertical lines) it was able to effectively use that information for localization. For humans then, localization is a two-stage process. Individual regions are identified when gateways are passed through. Landmarks can then serve the localization function within a region. Note that landmarks are often not sufficient for global localization on their own. An oak tree might tell one where one is within a neighborhood otherwise bereft of oak trees, but simply seeing an oak tree with no other context is unlikely to provide enough information for localization.

Abstraction to Higher-Level Representations

One of the least well developed aspects of our understanding of how cognitive maps work is how the basic pieces of cognitive maps are abstracted into higher-level representations to become part of a larger hierarchy. This is particularly important with regard to survey maps because when local maps are abstracted and combined they can yield large-scale representations with a number of useful properties. One such property is sometimes called "where-to-look." In PLAN where-to-look information is naturally coded by the nature of the stored views. Each view in PLAN represents what a person would see from a particular point in space with a particular body and head orientation. In turn, relative positions of objects can be extracted from a given view. Because the local maps are associative in character activating the representation of an object will naturally activate the corresponding local map. This will automatically give a person the necessary body, head and eye positions for seeing the object from a particular view. This provides a basis for the ability to point to an object without looking at it. When scaled up from what can be directly perceived, a map of this character would give a person the ability to point at things that cannot be directly perceived because of distance, intervening obstacles, etc. This ability provides a basis for spatial reasoning at more than just a local level. For example, based on directional information a person might guess that there is a better route to their goal than the one they are familiar with. In robotics having information of this nature would help solve the *closed loop problem* discussed in several other chapters in this volume; which is yet another variation of localization and entails recognizing when one has re-entered a previously visited environment from a different route.

PLAN provides a rudimentary theory for how such maps are learned but has little concrete evidence to support it. The idea is grounded in associative learning theory and takes advantage of the predictive nature of associative representations. Chown et al. speculate that as people mentally run through journeys, as they might when they make a plan, the sequences can become very fast when the journey is a familiar one. Eventually the intermediate locations can be considered right along with where one is. This allows them to be incorporated with the current local map into R-Maps. The fact that people report having a kind of bird's eye view of familiar locations is seen as an artifact of having an egocentric representation from a particular location where distal objects can be "seen." Since the transformation from local map to regional map involves combining what is perceived and what is stored, it is reasonable to suppose that the transformation is yet another function of the hippocampus, but there is no concrete evidence to support the theory yet.

10.3 Related Work

The learning theories presented in this chapter are broader in scope than what can typically be found in the spatial representation literature. The closest analogs would come from the large community that works on the hippocampus and in particular the role the hippocampus plays in spatial processing. Many of the ideas about the role of the hippocampus in this chapter derive from a long line of research going back to O'Keefe and Nadel's seminal work [21]. In robotics learning is typically framed in terms of Kalman filters, Monte Carlo localization and the like. Such methods work exceptionally well for what they are designed to do. However, as robots become more general purpose and face more difficult tasks, such as learning on the fly in large, dynamic, environments, it will undoubtedly be necessary to incorporate many of the same learning heuristics as the human cognitive architecture. Outside of robotics there has been some effort to adapt symbolic systems, such as the Soar architecture, along these lines in order that they might become more flexible and realistic [9]. In terms of the cognitive mapping community there are also few analogs to this work. The focus in that community is generally on what is learned rather than the mechanics of learning. In that sense PLAN benefits from being embedded in a cognitive architecture called SESAME that contains theories on representation and learning independent of the spatial components. The added perspective of an entire cognitive architecture helps make connections that could otherwise be lost (for example, the development of gateways was helped immensely by having an architecture that already included the functional distance principle).

In terms of representational structure, what made PLAN unique was its conception of the structure of survey maps and how it related survey and route maps through the two paths of the visual system. As already noted, the survey maps in the PLAN architecture drew partial inspiration from Yeap's work on *Absolute Space Representations* (ASRs) [31]. In turn PLAN's local maps helped to inspire some of the development of Kuiper's *Spatial Semantic Hierarchy* (SSH) architecture [15]. In many ways SSH is built in the spirit suggested in this chapter in that it is strongly inspired by human cognitive mapping, but also with an eye towards the practical issues required for implementation on a mobile robot. The local map idea also has a great deal in common with the *view graph* representation that was developed around the same time as PLAN [24] and is described in several other chapters in this volume. Since Kortenkamp's initial work on R-Plan, a number of other systems have incorporated gateways into their architectures [1, 33] and hybrid topological-metric systems have become increasingly common in robotics including work by some of the contributors to this book [28, 32, 29].

Despite all of the advances in robot hardware and theory, there has been no major effort to continue the development of the robot implementation of PLAN until recently. The system described in Sect. 10.4 is exploratory in the same sense that R-Plan was. The main questions being asked this time have to do with how systems can cope with highly dynamic environments. The principles that the system is built on are that humans do this is by building relatively sparse representations that fluidly adapt to changing worlds both by the nature of the learning process as well as through specialized mechanisms such as short-term memory and the hippocampus.

10.4 Adapting Architectural Principles for Robotics

Many of the theories about spatial representation and learning discussed in the previous sections can be used to drive research in autonomous mobile robotics. As stated in the introduction, behavioral strategies for navigation and vision are a natural place to look for inspiration when designing robotic navigation capabilities: animals appear to effortlessly solve many of the problems that continue to plague robotics as a field. Several examples of these unresolved problems include: localization, environmental representation, visual object recognition, and, in general, operating with uncertain knowledge in real-world situations. A common misconception in robotics is that neurobiological strategies are ill suited to the practical problems of robotics. This is hardly surprising given the elaborate, messy, and impractical models that are often proposed for neural behavior. A more plausible way to draw on the lessons of biology is to extract strategies in an abstract sense and apply them to robotics in the manner that best suits the underlying hardware and the challenge at hand. One need not develop a detailed cortical map, for example, in order to implement the functional distance principle on a robot. The abstract idea – a major shift in the character of perceptual inputs can be interpreted as a conceptual break – is not beholden to a specific type of underlying structure. For example, a sonar sensor might consistently report readings indicating that a wall is three feet away. A sudden shift in the readings indicating the wall at twenty feet away might signify a conceptual break. This is exactly the sort of logic used to develop the use of gateways in robotics [6, 13]. In human cognitive mapping the gateway is a visual construct, but generalizing the concept allows one to translate the idea to any sensory modality for use in many different types of environments [4]. In this section we will discuss an example of how this idea of generalizing cognitive principles can be adopted to a robotic system with an eye towards the design of robots that can effectively cope with real-world situations in dynamic environments.

C-Plan (for Corner-Plan), is a robotic navigation system developed at Bowdoin College. It is inspired by PLAN in structure but deviates from PLAN in implementation in several important respects. The goal of the project is to design a "cognitive" architecture for a robot that can create representations of an indoor environment capable of supporting navigation in realistic (e.g. dynamic) domains. Rather than trying to create extremely detailed maps of the environment, as most simultaneous localization and mapping (SLAM) methods do, the robot will rely on relatively sparse representations similar to those proposed in human cognitive mapping theories. Objects in indoor environments range from the ephemeral (people that move around a great deal), to the relatively stable (furniture moves less frequently), to the more or less permanent (walls rarely change location). A good map of highly dynamic environments, including indoor environments, should be built with an emphasis on permanence. One benefit of this approach is that such maps are less likely to be wrong when the environment changes and, therefore, will need to be updated less often. What follows are some highlights of the system that were inspired by some of the cognitive principles outlined in the preceding sections.

The system was run on an ActivMedia Pioneer 2-DX mobile robot equipped with a SICK LMS-200 laser rangefinder and wheel encoders that serve as a primitive dead-reckoning system. The laser-based perceptual system differs significantly from the optical visual systems found in most mammals. Laser rangefinders, however, represent the state-of-the-art in robotic sensing. They are a significant improvement over sonar being both more flexible and more accurate. While still prone to errors and providing limited information (distance to a surface along a straight line), laser rangefinders are the most popular and reliable sensor available. In practice the software navigation system that we outline shares a number of features with the system described by Tomatis et al. in this volume. Like that system the crucial features extracted from the environment are corners. In most indoor locations corners are a cheap way of providing a relatively accurate description of the environment. A second important feature of the system is the use of a topological map that encodes the relative positions of simultaneously viewed corners. This map serves as the primary representation of the robot's environment and is used as a reference during localization and navigation. In C-Plan, "views" of the environment are efficiently encoded within the associative structure of the network and gateways occur at narrow points within the graph (see below for details). Local collision avoidance was performed without regard to the structure of the "map" that the robot was developing, making the explicit representation of walls and obstacles unimportant. The representation (cognitive map) generated ensured that the paths between gateways did not cross physical boundaries like walls.

Since the representation in storage is highly abstract, a crucial feature of the system is the ability to extract essential information from detailed laser readings of the environment. In the following description we will discuss the influence of cognitive mapping strategies on the design of C-Plan. Technical details about localization, how dead-reckoning resolves stored structures against perceived structures, the geometry of stored maps, etc. are left to a future article.

10.4.1 Features – Corners

2-D lasers are not well suited for landmark recognition or for extracting unique perceptual information from a three dimensional environment, so we chose to focus on useful points in the environment that lasers could reliably identify (in this case the environment consisted of the Searle science building on the Bowdoin College campus). Our choice was corners, a frequent and stable environmental

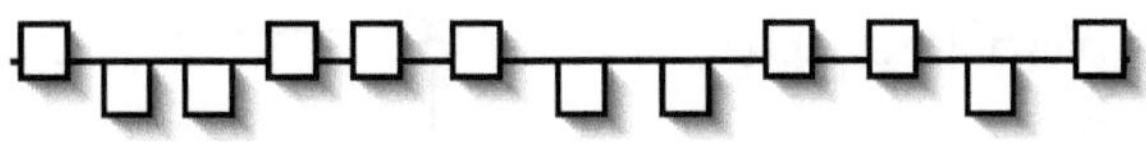

Fig. 10.4. Wall detection. Even when a surface is relatively smooth, sensor readings will rarely lineup perfectly. The straight line represents the wall being sensed. The boxes represent laser readings. Factors such as alignment, sensor error, small outcroppings, etc. all impact what is detected.

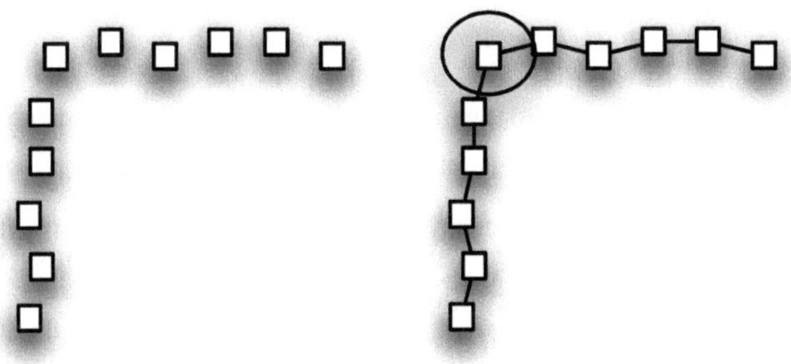

Fig. 10.5. Potential Corners. On the left the squares are the raw laser readings. The circled point on the right represents a potential corner that has been identified due to the angle between it and its nearest neighbors.

feature in most buildings. In order to determine when the robot was viewing a corner, a series of abstractions was performed on the raw laser data.

Corners exist where two flat surfaces come together, so one of the basic elements of our system is identifying walls. To call the walls in a typical building "flat" is actually misleading. Walls are rarely smooth and are often covered with outlets, pictures, etc. that stick out. Even more confounding is the fact that even perfectly smooth walls will not be percieved as such with laser readings (Fig. 10.4).

The robot attempts to recognize features in a hierarchy of abstractions with the aim of turning these jagged readings into walls. First the robot examines triples of laser readings. The readings are not adjacent due to the high degree of variation in adjacent readings, but are two readings away. This allowes a minimal amount of smoothing of the raw readings to take place before any sort of evaluation. If the angle formed around the middle of the three points is close to 90 degrees, that point is a good candidate to be a corner (Fig. 10.5).

If, however, the angle is close to 180 degrees the points are considered as candidates for a segment of a wall. Triples of this type are put into a line segment data structure. This data structure is a kind of sensory information store or "buffer" of what the robot has recently seen. If the line segment appears to be an extension of one already in the data structure the two are merged (Fig. 10.6). In a reflection of the attention and capacity limits of humans, the data structure is limited to a short period of time (line segments that continue to be updated stay in the buffer, ones that do not are removed).

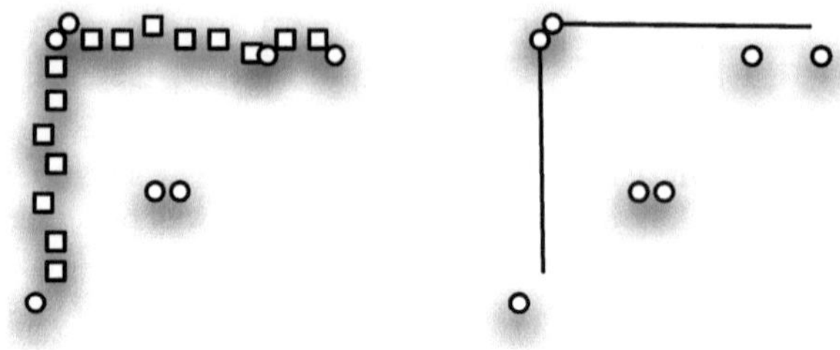

Fig. 10.6. Abstracting walls. Readings that are relatively well lined up are combined to form wall segments. In the diagram the laser readings are shown as squares. Points that have been identified as "interesting" are circles. These points are typically potential corners.

From this point it is easy to extract corners. Triples of points with the appropriate characteristics (i.e. an angle close to 90 degrees) are checked against the line segment data structure. If the triple exists at the junction of two line segments it is flagged as a corner.

It is important to note that this system relies on a fair number of assumptions about indoor environments – namely that they consist of straight walls and contain corners that are usually at 90 degree angles. This means that C-Plan will be ill-suited for some indoor environments. This, however, is not in conflict with human navigational abilities as people are not well-suited for certain types of environments and can be easily fooled by environments that play against expectations (e.g. funhouses).

In a series of tests run at Bowdoin College C-Plan was allowed to explore a building heavily used by students for classes. The robot's abstraction system was first calibrated when the halls were empty, purely to determine appropriate parameter settings. Then the robot was allowed to explore during a normal working day when classes were held in the building.

In one set of tests the robot was kept stationary for an hour while people walked past it in a hallway. Since people tended to move around they made it into the robot's sensory information store, but never into its long-term structures. In another set of tests the robot attempted to explore the building and create a map. The robot was able to correctly identify walls and corners with very little error while navigating around people and even while being herded by onlookers. Because people tended to clump together while watching the robot there were occasionally cases where groups would show up as walls, but only in one case did this lead to a mis-identified corner. In that case the corner did not make it to the long-term data structures (described later in this section) because it was not found in subsequent passes through the environment.

10.4.2 Memory Hierarchy

C-Plan implements a cognitive map for navigation and localization. This map consists of a network of nodes and edges. Recall that the corner is the only feature to make it out of short-term memory. A perceived corner is defined as two intersecting line segments at an angle close to 90 degrees. Each corner that is perceived directly corresponds to a node within the network. Corners that are viewed together, i.e. occur simultaneously within the sensory information store, are associated by means of a topological connection, helping to form the network. This topological connection encodes the geometrical relationship between corners. In this sense, the network can be thought of as efficiently encoding multiple "views" of the environment. The connection between corners is strengthened with repeated simultaneous viewings and weakened when only one of the corners is perceived at a time; an example of locally limited compensatory learning. This activity, taking place within the timeframe of the sensory information store, is the first and lowest level of what roughly corresponds to a memory hierarchy. The hierarchy consist of three states: the aforementioned *active* state, a *recently seen* state, and a *not seen in a while* state. These states resemble the sensory information store, short-term memory, and long-term memory respectively. When a node hasn't been directly perceived for a short period of time it moves into the recently seen state. This is a form of short-term memory that creates an expectation of perception thereby changing some of the thresholds involved in recognizing important points and lines in the sensory information store. It does not, however, lead to perception of corners in the complete absence of concrete sensory information. Finally, after a short time without being seen, corners move out of short-term memory and into long-term storage. Elements in long-term storage are used in planning and localization, but are not relied upon for execution.

10.4.3 Gateways

In C-Plan, the functional distance principle is at work within the robot's cognitive map, but gateways are not explicitly represented as they are in R-Plan. Despite this difference, conceptual breaks in the robot's spatial representation do occur and can be identified.

Corners located in open areas within the robot's environment tend to have a high degree of interconnectedness. This is because a large number of corners are usually visible in a relatively open unoccluded space, and the robot is recording many "views" within the network. Narrow openings in the environment including doorframes, hallways, and other structural constraints that partition space occlude the robot's perception and severely restrict the number of corners that can be simultaneuously viewed by the robot between open spaces. This cuts down on the number "views" that can be recorded. For example, a robot in a foyer may see many corners concurrently (including those created by a doorframe if the walls are thick enough), but not the corners in the room located through the door-frame. If the robot moves through the door-frame and into a room, it will see several corners in the room but not those outside of the room due to the occluding wall. Thus, the topological networks (cognitive maps) generated by the robot will have highly interconnected clusters of nodes linked to one another by a much sparser set of connections. These "narrow" portions of the network can be readily identified as gateways. Gateways in turn serve as important "landmarks": areas in the perceived environment that constrain space and limit the robot's navigational options.

10.4.4 Summary

C-Plan is still under development, but it shows a great deal of promise in being able to deal with highly dynamic indoor environments in a novel and interesting way thanks mainly to borrowing a number of ideas from theories of cognitive mapping and human learning.

10.5 Concluding Remarks

Advances in robotics and robot hardware afford tremendous opportunities to cognitive scientists interested in representations of large-scale space. There are numerous important questions about human spatial representations that robot implementations could help answer. For example, while it is widely agreed that humans have at least two modes of spatial functioning (based on route maps and survey maps) the developmental sequence involved is less understood as are issues surrounding how and why people switch from one mode to the other. Jefferies et al. argue in this volume, for example, that the metric information of survey maps can be acquired at least as early as route maps are. Others have argued that route maps probably proceed survey maps. Robot implementations allow exploration of such ideas. The majority of the learning theory in this paper comes out of psychology and neuroscience, but there are still gaps in our understanding of how people learn about space. Probably the most crucial missing area is concerned with how people build hierarchies, both in their route maps and, more importantly, in their survey maps. In the original PLAN paper Chown et al. speculated on how this might occur [6], but it has yet to be simulated and there is still no concrete theory to draw from neurophysiological data. The whole subject of learning is also a virtually untapped area where the understanding of cognitive maps could benefit from simulation on robots.

From the other side there is still a great deal for robotics to profit from cognitive theory and issues in learning and memory are one source of great potential gain. It is easy, for example, for a robot to extract information from an environment and keep it in storage as long as it wants. In highly dynamic environments, however, it is obvious that this is probably not a good strategy. Robots, even more than other artificial intelligence systems, face a tremendous knowledge integration challenge – the question of how representations should be updated in the face of new information, especially when that information might be contradictory, is far from trivial. This problem will only grow worse as robots are used in larger-scale environments. Purely mathematical approaches are unlikely to be effective in many situations because it is difficult to incorporate knowledge and context into them in a general purpose system. Humans have evolved a large number of factors that impact learning, each aimed at gleaning what is meaningful out of what is otherwise just information. Just the arousal system, for example, can be stimulated on a purely instinctual level or by predictions based upon experience. There are also special purpose systems, such as the hippocampus, that appear to be designed specifically to facilitate knowledge integration. Again, robotics offers a chance to explore how such systems work, but the field will also gain by such understanding as the answers will be beneficial for the increasingly complex tasks and environments that robots will be facing in the future.

Acknowledgements

This work was sponsered by the NSF under Grant No. 0092605. The authors gratefully aknowledges that support. Any opinions, findings and conclusions or recommendations expressed in this material are those of the authors and do not necessarily reflect the views of the National Science Foundation (NSF).

References

1. Adams, W., Perzanowski, D., Schultz, A.C.: Learning, storage, and use of spatial information in a robotics domain. In: ICML-2000 Workshop on Machine Learning of Spatial Knowledge, pp. 23–27 (2000)
2. Alexander, C., Ishikawa, S., Silverstein, M.: A pattern language. Oxford University Press, Oxford (1977)
3. Asada, M., Yasuhito, F., Tsuji, S.: Representing a global map for a mobile robot with relational local maps from sensory data. In: Proceedings of the 9th International Conference on Pattern Recognition, pp. 520–523 (1988)
4. Chown, E.: Making predictions in an uncertain world: Environmental structure and cognitive maps. Adaptive Behavior, 1–17 (1999)
5. Chown, E.: Reminiscence and arousal: a connectionist model. In: The proceedings of the 24th annual meeting of the Cognitive Science Society, pp. 234–239 (2000)
6. Chown, E., Kaplan, S., Kortenkamp, D.: Prototypes, location and associative networks: Towards a unified theory of cognitive mapping. Cognitive Science 19, 1–52 (1995)
7. Grossberg, S.: Competitive learning: From interactive activation to adaptive resonance. Cognitive Science 11, 23–63 (1987)
8. Hebb, D.O.: The Organization of Behavior. Wiley, Chichester (1949)
9. Henninger, A.E., Jones, R.M., Chown, E.: Behaviors that emerge from emotion and cognition: Implementation and evaluation of a symbolic-connectionist architecture. In: Proceedings of Autonomous Agents and Multi-Agent Systems 2003 (2003)
10. James, W.: Psychology the Briefer Course. Harper, New York (1892)
11. Kaplan, S., Sonntag, M., Chown, E.: Tracing recurrant activity in cognitive elements (trace): A model of temporal dynamics in a cell assembly. Connection Science 3, 179–206 (1991)
12. Kinsbourne, M.: Hemispheric specialization and the growth of human understanding. American Psychologist 37(4), 411–420 (1982)
13. Kortenkamp, D.: Cognitive maps for mobile robots: A representation for mapping and navigation. PhD thesis, The University of Michigan (1993)

14. Kortenkamp, D., Baker, L.D., Weymouth, T.: Using gateways to build a route map. In: IEEE/RSJ International Conference on Intelligence Robots and Systems (1992)
15. Kuipers, B.: The spatial semantic hierarchy. Artificial Intelligence 119, 191–233 (2000)
16. Murphy, R.R.: Introduction to AI Robotics. MIT Press, Cambridge (2000)
17. Nadel, L., Bohbot, V.: Consolidation of memory. Hippocampus 11, 56–60 (2001)
18. Nisbett, R., Ross, L.: Human Inference. Prentice-Hall, Englewood Cliffs (1980)
19. Norman, J.: Two visual systems and two theories of perception: An attempt to reconcile the constructivist and ecological approaches. Behavioral and Brain Sciences 25, 73–144 (2002)
20. O'Keefe, J.: Computations that the hippocampus might perform. In: Culicover, P., Nadel, L., Cooper, L.A., Harnish, R.M. (eds.) Neural connections, mental computation, MIT Press, Cambridge (1989)
21. O'Keefe, J., Nadel, L.: The Hippocampus as a cognitive map. Clarendon Press, Oxford (1978)
22. Pailhouse, J.: Representation de l'espace urbain at cheminements (representation of urban space and roadways). Le Travail Humain 3, 239–262 (1969)
23. Rochester, N., Holland, J.H., Haibt, L.H., Duda, W.L.: Tests on a cell assembly theory of the action of the brain, using a large digital computer. IRE Transations on Information Theory IT-2, 80–93 (1956)
24. Schölkpf, B., Mallot, H.A.: View-based cognitive mapping and path planning. Adaptive Behavior 3, 311–348 (1995)
25. Shelton, A.L., McNamara, T.P.: Multiple views of spatial memory. Psychonomic Bulletin & Review 4, 102–106 (1997)
26. Squire, L.R., Clark, R.E., Knowlton, B.J.: Retrograde amnesia. Hippocampus 11, 50–55 (2001)
27. Thrun, S., Beetz, M., Bennewitz, M., Burgard, W., Cremers, A.B., Dellaert, F., Fox, D., Hahnel, D., Roseberg, C., Roy, N., Schulte, J., Schulz, D.: Probabilistic algorithms and the interactive museum tour-guide robot minvera. International Journal of Robotics Research 19(11), 972–999 (2000)
28. Thrun, S., Gutmann, S., Fox, D., Burgard, W., Kuipers, B.J.: Integrating topological and metric maps for mobile robot navigation. In: Proceedings AAAI-98, pp. 989–995 (1998)
29. Tomatis, N., Nourbakhsh, I., Siegwart, R.: Hybrid simultaneous localization and map building: a natural integration of topological and metric. Robotics and Autonomous Systems 44, 3–14 (2003)
30. Ungerleider, I.G., Mishkin, M.: Two cortical visual systems. In: Ingle, D.J., Goodale, M.A., Mansfield, R.J.W. (eds.) Analysis of visual behavior, MIT Press, Cambridge (1982)
31. Yeap, W.K.: Towards a computational theory of cognitive maps. Artificial Intelligence 34, 297–360 (1988)
32. Yeap, W.K., Jefferies, M.E.: Computing a representation of the local environment. Artificial Intelligence 107, 265–301 (1999)
33. Youngblood, G.M., Holder, L.B., Cook, D.J.: A framework for autonomous mobile robot exploration and map learning through the use of place-centric occupancy grids. In: ICML-2000. Workshop on Machine Learning of Spatial Knowledge, pp. 13–20 (2000)

Part III

Cognitive Robot Mapping

Cognitive Robot Mapping: An Introduction

Benjamin Kuipers

It is now generally accepted that an adequate computational theory of the cognitive map must include multiple representations for knowledge of space. This third section of the book discusses implementations, with physical or simulated robots, of such theories of the cognitive map.

These papers deal with three levels of spatial representation — control, topology, and survey — each in its own way.

Chapter 11: Kuipers

In Chap. 11, Kuipers[1] describes the history of the development of his Spatial Semantic Hierarchy model of the cognitive map over more than three decades. The need for multiple representations of spatial knowledge was present from the beginning. However, as the TOUR model evolved into the Spatial Semantic Hierarchy (SSH) and then into the Hybrid SSH, different representations and capabilities were added to the model.

In addition, the original focus on the cognitive map — restricted to knowledge of *large-scale* space — was broadened to include knowledge of small-scale space within the agent's perceptual horizon. The research program also broadened to include an investigation of how foundational representations for space are learned in the first place. This research explores a thought experiment by assuming that an agent with uninterpreted sensors and effectors, exploring an unknown environment, can search for statistical regularities that motivate the creation of new descriptive terms. These new terms become a new ontology for describing space.

Chapter 12: Jefferies, Baker and Weng

Both Jefferies, Baker & Weng (Chap. 12) and Yeap, Wong & Schmidt (Chap. 13) work in the framework of the Absolute Space Representation (ASR) of Yeap and Jefferies [6, 5]. An ASR is a metrical representation of local space, with its own local frame of reference. Adjacent ASRs are linked together into a topological graph, connected by their *exits*.

Jefferies et al. (Chap. 12) show how a collection of linked ASRs can be merged into a non-local metrical map called the Memory for Immediate Surroundings (MFIS), with a single frame of reference. The MFIS has a limited extent, so it is not necessarily a global metrical map, but constructing the MFIS typically does involve closing large exploration loops by recognizing when the same ASR has

[1] In this introduction, I use the third person to discuss my own chapter.

M.E. Jefferies et al. (Eds.): Robot. & Cogn. Approach. to Spat. Map., STAR 38, pp. 239–242, 2008.
springerlink.com

been encountered through a different path. This paper discusses the use of both topological and metrical information to constrain the identification of matching ASRs and the construction of the MFIS.

It is interesting to compare the ASR/MFIS representation with the elements of the Hybrid Spatial Semantic Hierarchy [1, 4] described by Kuipers in Chap. 11. Like the ASR (which preceded it by several years), the Local Perceptual Map (LPM) in the Hybrid SSH is a local metrical map with its own local frame of reference. However, the LPM scrolls continuously with the agent as it moves through the environment. When a topological place is identified, a snapshot of the LPM is used to initialize a local metrical map of the place neighborhood. This place neighborhood map serves a role very similar to the ASR.

The ASR approach assumes that the agent's experience is tiled by a sequence of adjacent and overlapping ASRs. In contrast, the Hybrid SSH assumes that places and their neighborhoods are typically isolated, and are connected by path segments that are associated with trajectory-following control laws that take the agent from one place neighborhood to the next. In part, this difference seems to reflect the examples that motivated the researchers. Kuipers (Chap. 11) was initially inspired by descriptions of the urban cognitive map, such as Kevin Lynch's classic *Image of the City* [3]. Yeap and Jefferies [6, 5] are clearly inspired by their experience with robot exploration of indoor office environments. This difference is not fundamental, since each approach can express the other as a limiting case.

Kuipers' SSH approach emphasizes the use of topological information to make loop-closing decisions [2, 1], followed by global metrical map-building on the topological skeleton [4]. The ASR/MFIS approach of Jefferies et al. (Chap. 12) emphasizes the use of metrical information in the MFIS to constrain potential matches that could close loops.

Chapter 13: Yeap, Wong and Schmidt

Any approach to topological mapping must address the critical issue of robustness in abstracting the continuous environment to a set of discrete structures such as places or ASRs. That is, the set of places or ASRs that are identified on one occasion might be different from the set identified in the same environment on a different occasion.

Yeap, Wong & Schmidt (Chap. 13) take an interesting approach to this problem by demonstrating a method that makes robust estimates of a homing vector without requiring consistent identification of ASRs. In their experiments, the robot traverses a loop in a large environment, first in one direction, and then retracing its steps in the opposite direction. The ASRs identified in the two directions may or may not be the same. The first sequence of ASRs defines the topological/metrical map of the environment, and then the sum of the local distances along the ASRs identified on the return trip is used to estimate the current position in the map. That position in the semi-global MFIS is used to estimate the vector toward the starting point. In spite of arbitrary mismatches between the ASRs in the two map traversals, the errors in the estimated homing vectors are quite small.

Chapter 14: Franz, Stürzl, Hübner and Mallot

Franz, Stürzl, Hübner & Mallot (Chap. 14) summarize their extensively developed cognitive mapping approach based on *view graphs*. The authors discuss the capabilities of the cognitive map representation in terms of three major components: route navigation, topological navigation, and survey navigation. Central to this approach is the *view*: the sensory image or snapshot obtained at a particular pose during exploration. This sensory image could be a laser or sonar range-image or, as in these experiments, an omni-directional camera image.

The route navigation component describes local motion control, as the robot moves from the pose associated with one view to the pose associated with another. Local motion is along the gradient of image similarity between the current view and that of the goal. That is, the robot moves so as to increase the correspondence between its current sensory image and the goal view.

Topological navigation takes place in the *view graph*, a topological map where nodes correspond to views (or more properly, to the poses in the environment where the views are obtained), and the edges correspond to successful travel by route navigation between adjacent views. To simplify the identification of loops, Franz et al. (Chap. 14) assume that views are unique, so that observing a matching view after traveling along a new edge is sufficient evidence for closing a topological loop. It seems that this method could be strengthened by including loop-closing by hypothesis-testing using topological search (Kuipers, Chap. 11) or by including metrical structuring of the semi-global environment encompassing the current loop (Jefferies et al., Chap. 12).

This also raises the question of when new views are added to the view-graph. Franz et al. (Chap. 14) create a new view when the dissimilarity with the previous view (or with all other known views) exceeds a fixed threshold. This is necessary to allow the similarity-based route navigation strategy to work. The creation of new ASRs in Jefferies et al. (Chap. 12) follows a similar criterion. However, this contrasts with Kuipers' approach in the Spatial Semantic Hierarchy (Chap. 11), where places are defined according to local distinctiveness, and places can be linked by arbitrarily long trajectory-following control laws. Survey navigation is done by multi-dimensional scaling and relaxation of the layout of the topological map into a metrical map with a single frame of reference.

Chapter 15: Hafner

Hafner (Chap. 15) considers the same three cognitive map layers — route, topological, and survey — and discusses their relationship with recent results in cognitive neuroscience. She describes simulation results from a computational model of the cognitive map that predicts the activity of place cells as a function of the agent's pose in the environment.

The approach taken here is similar to the view-graph approach of Franz et al. (Chap. 14), with place cells representing sensory views. A self-organizing map (SOM) with adaptive connectivity is used to identify clusters of sensory images that can serve as views in the view-graph. One prediction from this model is that place cells will be denser near objects, and this appears to match experimental data from a variety of animals.

The learned place cells can be relaxed into a single frame of reference, using information about the connections between the cells as data about how close or far they should be from each other.

Hafner (Chap. 15) discusses evidence that distinguishes between place cells and "view cells" that respond to particular sensory images. This raises the question of whether there is a third category of "pose cells" that respond to position and orientation, but not primarily to the sensory view. Environments with perceptual aliasing (different poses with the same view) should be able to discriminate among the responses of these types of cells.

References

1. Kuipers, B., Modayil, J., Beeson, P., MacMahon, M., Savelli, F.: Local metrical and global topological maps in the hybrid spatial semantic hierarchy. In: ICRA 2004. IEEE Int. Conf. on Robotics & Automation, IEEE Computer Society Press, Los Alamitos (2004)
2. Kuipers, B.J., Byun, Y.-T.: A robot exploration and mapping strategy based on a semantic hierarchy of spatial representations. Journal of Robotics and Autonomous Systems 8, 47–63 (1991)
3. Lynch, K.: The Image of the City. MIT Press, Cambridge, MA (1960)
4. Modayil, J., Beeson, P., Kuipers, B.: Using the topological skeleton for scalable, global, metrical map-building. In: IEEE/RSJ Int. Conf. on Intelligent Robots and Systems (2004)
5. Yeap, W.K., Jefferies, M.E.: Computing a representation of the local environment. Artificial Intelligence 107(2), 265–301 (1999)
6. Yeap, W.K.: Towards a computational theory of cognitive maps. Artificial Intelligence 34, 297–360 (1988)

11

An Intellectual History of the Spatial Semantic Hierarchy

Benjamin Kuipers

Summary. The Spatial Semantic Hierarchy and its predecessor the TOUR model are theories of robot and human commonsense knowledge of large-scale space: the *cognitive map*. The focus of these theories is on how spatial knowledge is acquired from experience in the environment, and how it can be used effectively in spite of being incomplete and sometimes incorrect.

This essay is a personal reflection on the evolution of these ideas since their beginning early in 1973 while I was a graduate student at the MIT AI Lab. I attempt to describe how, and due to what influences, my understanding of commonsense knowledge of space has changed over the years since then.

11.1 Prehistory

I entered MIT intending to study pure mathematics. I was generally steeped in the ideology of pure mathematics, and I had every intention of staying completely away from practical applications in favor of abstract beauty and elegance. However, on a whim, in Spring of 1973 I took Minsky and Papert's graduate introduction to Artificial Intelligence. I was immediately hooked. I had always been fascinated by the idea of a science of the mind. But then in college I took a course in psychology, which was a crashing disappointment. The interesting parts weren't scientific, and the scientific parts weren't interesting. Now, in artificial intelligence, symbolic computation promised mathematical methods capable of rigorously modeling interesting aspects of the mind.

I spent that summer at the MIT AI Lab, reading papers and getting more and more excited. Marvin Minsky was circulating drafts of his "frames paper" [41], which advocated that research focus on representation and inference about complex symbolic descriptions of meaningful objects and situations, rather than on individual propositions and logical inference. Such a description was called a *frame*. It had a number of *slots*, which could contain *values*, and could be associated with symbol manipulation procedures for doing inference, including providing *default values* for empty slots. I recall telling Pat Winston once that I found the frames concept to be very compelling, but I wondered where the slots come from.

Minsky's classes introduced me to Piaget's theories of the development of children's knowledge of foundational domains, including space, time, causality, and so on. His and John McCarthy's writings [39, 42] also convinced me that the nature and representation of commonsense knowledge was a bottleneck issue for artificial intelligence. This was the problem I wanted to work on.

M.E. Jefferies et al. (Eds.): Robot. & Cogn. Approach. to Spat. Map., STAR 38, pp. 243–264, 2008.
springerlink.com

Following up on an idea of Minsky's for model-based object recognition, and using the edge-and-vertex representation from Blocks World vision, I wrote a paper showing how a vision system could discriminate among a small set of block models, tracing a hypothesis from vertex to vertex along edges, and using contradictory evidence to force a jump to an alternate hypothesis when necessary.[1] This paper earned me an invitation to spend Summer 1974 at Xerox PARC as a summer student working with Danny Bobrow and Terry Winograd. I implemented and demonstrated my recognition system in Smalltalk on the Alto, alternately marveling at the wonderful new technology and taking it totally for granted. The revised paper was named "A frame for frames" [23] in conscious homage to Fillmore's far more influential "The case for case" [11].

As the end of the summer approached, before returning to MIT, I met with Danny Bobrow to ask his advice on research topics. I explained that I had enjoyed working on model-based object recognition, but I really wanted to work on the problem of commonsense knowledge, and I didn't know where to begin. Danny suggested that I look at some work being done by Joe Becker and Bill Merriam at BBN on a simulated robot learning the structure of a simulated city [3, 4].

I knew immediately that this was the right problem: *How can a robot learn a cognitive map from its own experience of the environment?* It focuses on spatial knowledge, which is not only important, but is arguably the foundation for most other kinds of commonsense knowledge [34]. It also looked like it would factor well, in the sense that I could define interesting sub-problems that were small enough to solve, but which could be assembled into solutions to larger problems as I made progress. It would make a great PhD thesis topic, and I went back to MIT happy.

11.2 Cognitive Background

Quite a bit was already known about how human knowledge of space is structured, and how people use spatial knowledge to solve problems. I immersed myself in that highly diverse literature, reading papers from cognitive and developmental psychology, urban planning, geography, linguistics, and the visual arts. Two books that particularly influenced me were *The Image of the City*[2] by Kevin Lynch [37] and *Image and Environment*, a new collection of papers edited by Downs and Stea [9]. Also, among the more cognitively oriented denizens of the MIT AI Lab, Piaget's "genetic epistemology" approach to developmental psychology (e.g., [47]) permeated the atmosphere.

What quickly emerged from all this reading was a view of spatial knowledge consisting of several quite different types of knowledge. Some was procedural, "how-to" knowledge about getting from one place to another. Some consisted

[1] Only with the benefit of much hindsight do I recognize the similarity with the process of building topological maps.

[2] I later learned that both Lynch's *The Image of the City* [37] and Miller, Galanter, and Pribram's influential *Plans and the Structure of Behavior* [40] were inspired by Kenneth Boulding's seminal book, *The Image* [6].

of topological connections between places and travel paths. And some consisted of metrical layouts approximately analogous to the environment itself or to a printed map. But it was clear that accurate metrical layout descriptions came last, if at all, and depended on the earlier types of knowledge. Furthermore, spatial reasoning methods varied across individuals, with developmental stage, with experience in a particular environment, or simply with individual cognitive style. A year or so later, Siegel and White's masterful survey of the development of spatial knowledge [61] confirmed and deepened this view.

Since the differences between the representations for spatial knowledge are so central, I started collecting route directions and sketch maps from anyone available. These were informal probes, designed to elicit a wide range of behavior I could examine for qualitative features, not formal experiments designed to test or refute hypotheses. What I needed was to complement the literature review with an intimate sense of the phenomenon itself, as a basis for building a computational model.

One immediate conclusion was that there is a lot of individual variation in the amount, nature, and accuracy of spatial knowledge that different people have, and in how they express it. Another is that neither verbal directions nor sketch maps tend to be particularly accurate about absolute distances or directions. On the other hand, topological relations such as the order of places on a path, or the connections between paths at a place, tend to be represented accurately and when errors do creep in, they are usually detected.

A common style for drawing a map was to follow a mental route, drawing those places and paths needed for the route, and perhaps nearby structures. When the subject made an error in translating the route into the graphical map representation, the error was usually metrical, and could go unnoticed for quite some time as the map was elaborated in an incorrect direction. The error would be detected when it finally came time to close a loop, and two occurrences of the same place would be drawn far apart, sometimes separated by other structures. At this point, detecting the problem became easy, but identifying the specific error or correcting it could still be quite difficult.

Some subjects used a different style[3], sketching the overall structure of a region, such as the rectangular grid structure in Boston's Back Bay. Fortunately for my research, the geography of the Boston-Cambridge area abounds with interesting local structures that fail to generalize over larger regions, leading to easily detectable geographical fallacies and paradoxes in people's cognitive maps.

The overwhelming impression from both my own investigations and the published experimental studies is that human spatial knowledge consists of a number of distinct representations for different aspects of space. Some people have many of these cognitive modules, and they work together well, while others may have fewer of them, or they don't work together so well. As a working hypothesis, I took the position that there is a single "complete" structure for all of these modules, working well together, and that all the variants — with individual style,

[3] These two styles were also identified by Linde and Labov [36] in subjects' descriptions of their apartment layouts.

developmental stage, or amount of experience in a particular environment — are modified or restricted versions of the ideal. This is similar to both Piaget's "genetic epistemology" and to current notions of "ideal observer" models [12].

Since the target of my efforts was a structure of interacting modules, it was natural to do the research by identifying an interesting aspect of the phenomenon of the cognitive map, constructing and testing individual modules to explain that aspect, and then looking for further parts of the natural phenomenon not adequately explained by existing modules.

11.3 The TOUR Model

My doctoral thesis described the representation of knowledge of large-scale space — the *cognitive map* [24, 25]. Space is considered *large-scale* if its relevant structure is at a scale larger than the sensory horizon, so knowledge of the structure must be acquired from exploration within it. The focus on large-scale space allowed me to avoid the difficult problems of computer vision and scene understanding. I focused my attention on spatial representation and inference, and specifically, on the problem of how global spatial structure can be inferred from local sensory experience. The *TOUR model* is a computational model of this kind of knowledge, including in most cases how that knowledge is learned from experience.

The TOUR model describes an agent[4] that receives a sequence of experiences as it travels through the environment, and builds its own cognitive map of that environment. The cognitive map is a symbolic representation, consisting of a set of frames for describing different types of objects such as places, paths, and regions; each type with its own collection of attributes; each instance with values for some or all of those attributes.[5] A place includes an attribute for the set of paths it is on, and a path includes an attribute for the partially-ordered set of places on it. An agent on a path faces in one of two directions: up or down the place-ordering on that path.

As the agent receives experiences, it draws only those conclusions that can be inferred efficiently with information available at the time. This kind of "opportunistic" inference puts a premium on representations capable of expressing incomplete knowledge, so the results of small inference steps can be represented and stored, rather than being lost if attention moves elsewhere. Because of this strategy, inference is very efficient, but several travels along a particular route may be necessary for the TOUR model to infer all of the conclusions that follow logically from the experience.

[4] The TOUR model and the Spatial Semantic Hierarchy are intended to describe both human and robotic agents.

[5] The equivalence between frames and first-order predicate logic is now well understood [15]. James Crawford and I later formalized the intuitions behind this version of frames as "Access-Limited Logic" and its implementation, Algernon [7, 8].

The TOUR model divides spatial representation into three levels: procedural, topological, and metrical.[6] At the procedural level, experience is modeled as a sequence of GO-TO and TURN actions, with associated distance or angular magnitudes, respectively. The action description can be augmented with descriptions of the states before and after the action, each modeled as place, path, and direction along the path. When not provided explicitly, these may be inferred from context.

The inferential heart of the TOUR model is the "TOUR machine", a finite-state, rule-driven automaton. It has a set of registers called the "You-Are-Here pointer" describing the current place, path, direction, etc. Instead of an infinite tape, its memory is a potentially infinite set of frames reachable through the attributes of existing frames. Knowledge of the current state fills in the initial-state description in the current action. If the current place or path description can predict the final-state of the current action, it does; if not, new descriptions are created. In either case, the results update the You-Are-Here pointer, and they are stored as part of the action, place, and path descriptions, extending or confirming what was previously stored. Since the world itself is assumed to have a single consistent structure, and since the representation is supposed to be sufficiently expressive of incomplete knowledge for the results of opportunistic inference, contradictions between stored and newly-inferred information should be rare. The problem of more extensive reorganization and correction of the map when such an error is detected was beyond the scope of this research.

The sequence of GO-TO and TURN actions representing the agent's experience is provided by a simple natural language interface. The interface is based on Vaughan Pratt's elegant LINGOL parser [53], which allows context-free grammar rules to be annotated with semantic interpretation routines. The grammar makes it easy to describe the agent's experiences in natural-sounding route instructions, such as:

> Start on Broadway, at the intersection of Broadway and Prospect Street, facing Kendall Square.
> Turn right onto Prospect Street.
> Take Prospect Street to Central Square.
> Turn right onto Mass Ave.
> Take Mass Ave to Putnam Circle.

The topological level of the TOUR model is based on the connectivity of places and paths, the circular order of directed paths at each place, and the partial ordering of places on each path. It also includes boundary relations, whereby places can be described as "to the right" or "to the left" of a path. Boundary relations can be used to define regions in terms of bounding paths. All of these are learned by the TOUR model through opportunistic inference from experience in the form of GO-TO and TURN actions. Another form of topological knowledge is a region hierarchy, which allows the environment to be described, and

[6] This division into levels is updated to reflect the later perspective of the Spatial Semantic Hierarchy [33, 18].

route-planning problems to be solved, at many different levels of abstraction. For the region hierarchy, the TOUR model describes the representation and use of the knowledge, but provides no learning theory.

The metrical level of the TOUR model consists of attributes and relations with continuous values, like distance and direction. Analog spatial representations such as 2D occupancy grids [46] were still far in the future. Every GO-TO action includes a description of the magnitude of travel from one place to another along a given path. This provides a constraint on the relative location of the two places in the 1D frame of reference of that path. Enough observations of distances between pairs of places on the same path determines the layout of places within the path. Similarly, observations of TURN magnitudes at a given place provides a radial layout of the directed paths at that place. These radial layouts can be interpreted as defining the heading of an agent at that place, path, and direction, but only in a frame of reference local to the place, so headings cannot be compared from place to place. However, if the GO-TO action magnitude is extended to include a "net angular displacement" attribute $\Delta\theta$, then a single frame of reference can propagate along GO-TO actions to include multiple places. For places within a single frame of reference, GO-TO and TURN actions provide relative distance and direction measurements, from which a 2D layout of places can be inferred.

The TOUR model [24, 25] was the first computational model of the cognitive map that explicitly addressed the multiple types of spatial knowledge that must be represented. It specifically focused on the topological representations whose importance was well-understood by researchers deeply familiar with human cognitive mapping, but which was widely overlooked by many others in psychology, geography, and robotics. The major limitations of the TOUR model were the oversimplified interface to the agent's actual sensorimotor experience in the world, and the inadequate treatment of analog metrical representations.

11.4 Explicit Representation of Sensory Views

One problem with the original TOUR model is that the procedural level too thoroughly abstracts away the agent's sensory input from the environment. The route-direction-like input representation was unable to express either gaps in the sequence of experience or perceptual aliasing (different places that look the same). Part of solving this was to provide an explicit representation for sensory experience [26]. A *view* is an abstracted description of the sensory image experienced by the agent at a particular state (i.e., place, path, and direction). The TOUR model avoids the problem of interpreting input from any particular sensor (e.g., vision, sonar, laser) by treating views as atomic symbols that can only be used as retrieval keys or matched for identity. The specific representation or implementation of views is outside the scope of the theory (until later; see Sect. 11.7).

Given the concept of view we can define a more natural interface, representing the agent's experience as an alternating sequence of views and actions:

$$v_0, a_0, v_1, a_1, v_2, \cdots v_{n-1}, a_{n-1}, v_n.$$

An action a_i can have type Turn or Travel, with an associated magnitude.

We can now replace the procedural description of travel experience with a collection of causal schemas $\langle v, a, v' \rangle$, where the view v describes the context when action a is initiated, and v' describes the result after a has completed [26]. A schema $\langle v, a, v' \rangle$ has the declarative interpretation that in context v, after performing action a, one can expect resulting view v', and the imperative interpretation that if the agent experiences the context view v, it should do action a.

Knowledge of an experienced route is represented as a collection of schemas, indexed by their context views. This representation can express several very plausible states of incomplete knowledge. A gap in the route, perhaps due to inattention during exploration, corresponds to omitted schemas in the route description. If all the schemas $\langle v, a, v' \rangle$ in a route description are complete, they form a linked list, as the result v' of each schema allows retrieval based on the context v of the next schema along the route. However, incomplete schemas $\langle v, a, \_ \rangle$ can be constructed if working memory is disrupted during the possibly-extended time while a is taking place, before the result v' becomes available. Incomplete schemas still have their imperative meanings intact, and can still be used to traverse the route physically in the environment, since the environment will provide the result of each action. What is lost is the ability to review the route in the absence of the environment.

In these ways and others, the schema representation is very expressive of states of incomplete knowledge of a route. Variations may depend on developmental stage, amount of experience with this route, amount of computational resources available, and frequency of disruptions. We extended this concept to describe one aspect of individual variation in cognitive style, corresponding to the set of rules available for constructing partial schemas [27].[7]

As it happens, it took a while to recognize that a good formal structure for representing route experience is the familiar finite-state automaton, or more generally, the partially-observable Markov decision process (POMDP) [10, 1, 2]. We require a set of underlying states x, that are themselves unobservable, but which map to observable views v. The set of schemas $\langle x, a, x' \rangle$ represents the transition function for the automaton, and the relation $view(x, v)$ represents the mapping from unobservable state to observable view. In full generality, POMDP learning of automata with stochastic transition and observation functions is intractable. However, this direction of investigation takes us farther away from an understanding of human commonsense spatial knowledge.

[7] Starting around 1978-79, I decided to change research direction for a variety of reasons [19]. This led to a productive line of work on medical reasoning and qualitative simulation [32, 16, 17, 29]. Spatial knowledge became a secondary concern until the mid-1990s.

In the Spatial Semantic Hierarchy [18, 55], we assume that transitions $\langle x, a, x' \rangle$ among states are deterministic (reflecting the error-correcting capabilities of feedback control laws), and that the relation $view(x, v)$ is a function, though not necessarily one-to-one. With these assumptions, and when exploring physical space, learning a minimal underlying automaton from observational experience is generally feasible in practice.

11.5 Abstracting Continuous Experience to Discrete States

A second problem with the original TOUR model is that it presupposes that the continuous experience of the agent has already been abstracted to a discrete sequence of states and transitions. This was justified by Kevin Lynch's observation that humans tend to represent knowledge about decision points, with much less about the spaces between them [37]. Nonetheless, this unexplained abstraction remained a gaping hole in the theory, and it was a barrier to robot implementation.

My cognitive mapping research had been on hiatus for several years, with QSIM receiving all of my attention, when a new grad student named Yung-Tai Byun approached me in 1986, wanting to do research on robot exploration and mapping. In the course of our discussions, we ran directly into the problem of relating the robot's continuous behavior to the kind of discrete topological map that the TOUR model creates. When we contemplated the simplest non-trivial environment I could think of — two corridors joined to form a T (Fig. 11.1) — the concept of *distinctive place* became clear. If we overlay the obvious T-shaped topological map onto the continuous polygonal environment, the natural locations for the four topological places are at the dead-ends and the intersection, at locations equidistant from the nearest obstacles. The segments connecting places are clearly corridor midlines. These loci corresponding to topological places and topological paths naturally suggest the attractors of hill-climbing and trajectory-following control laws, respectively. This basic idea, of letting the attractors of continuous control laws define the topological features of large-scale space, led to several influential papers, including [30, 31]. Our focus was on the topological map, but we did show how the topological map (Fig. 11.2(a)) could provide a skeleton for a global metrical map (Fig. 11.2(b)), foreshadowing later work.

The selection of a control law couples the robot and its environment into a continuous dynamical system, which moves through its state space toward an attractor. The selection, execution, and termination of these control laws can be defined based entirely on sensory features available "from the inside" of the agent, without any appeal to the external semantics of the sensors or of the features. (It wasn't until later that we actually tried to *learn* the sensors, features, and control laws without appeal to external semantics [51]. See Sect. 11.8.) This method for defining symbolic entities referring to topological places and path segments in terms of the behaviors of control laws is a concrete example of a solution to the Symbol Grounding Problem [14].

By physically hill-climbing to the local optimum of a "distinctiveness measure" defined over the local neighborhood, the robot localizes itself within that neighborhood with minimal assumptions about the nature of its sensors (Fig. 11.3). Because the dynamical system defines motion over the robot's state space (location plus orientation), rather than over the work space (location alone), we came to realize that what is distinctive is the state, rather than the place, so we began to refer to *distinctive states* rather than *distinctive places*. For example, the single topological place at a T-intersection corresponds to four distinctive states, with the same location and different orientations. The Turn actions that link them correspond to trajectory-following control laws that change only orientation, followed by hill-climbing control laws to align with the walls of the corridors. (Later, in Sect. 11.7, we will see a new conception of places and place neighborhoods.)

Motion among distinctive states avoids the problem of cumulative error that typically plagues robot mapping. There is no attempt to maintain an accurate location in a single global frame of reference. Rather, the purpose of an action is to move reliably from one distinctive state to another one. Any error that accumulates during trajectory-following is eliminated by the hill-climbing step, as long as the error is not so large as to miss entirely the basin of attraction of the destination distinctive state.

11.6 The Spatial Semantic Hierarchy

We started with the idea that the cognitive map consists of different representations for knowledge of space. As we come to understand spatial knowledge more deeply, the actual representations have evolved. We can best organize these different representations by grouping them according to *ontology*: the types of objects that can be described and the relations that can hold among them.

The *Spatial Semantic Hierarchy* (SSH) describes the cognitive map as consisting of four different levels, each with its own ontology, and each level grounded in the ones below [33, 31, 18, 55].

- At the *control level*, the agent and its environment are described as parts of a continuous dynamical system. The agent acts by selecting trajectory-

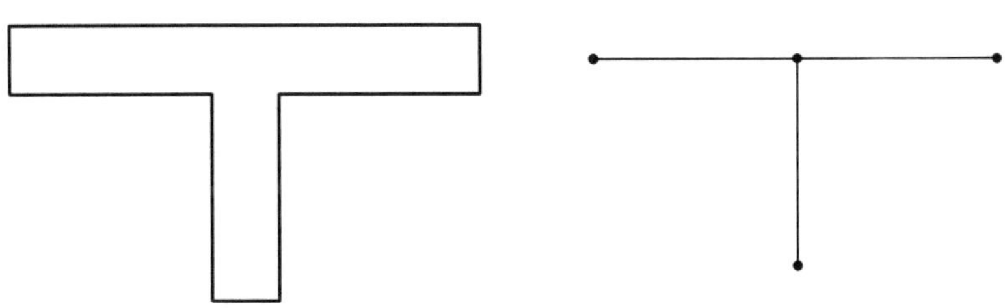

Fig. 11.1. A T-shaped space, and its topological model

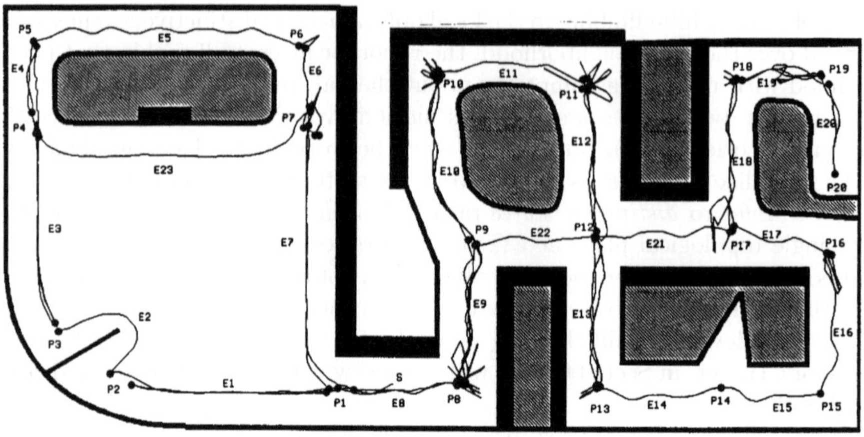

Fig. 11.2. A simulated robot applies the SSH exploration and mapping strategy [31]. (a) The robot's exploration trajectories (black spots and thin lines) reveal the locations where it identifies distinctive places and follows path segments. (b) Local metrical information about place neighborhoods and path segment length and curvature can be used to relax local frames of reference connected by a topological map into a single global frame of reference.

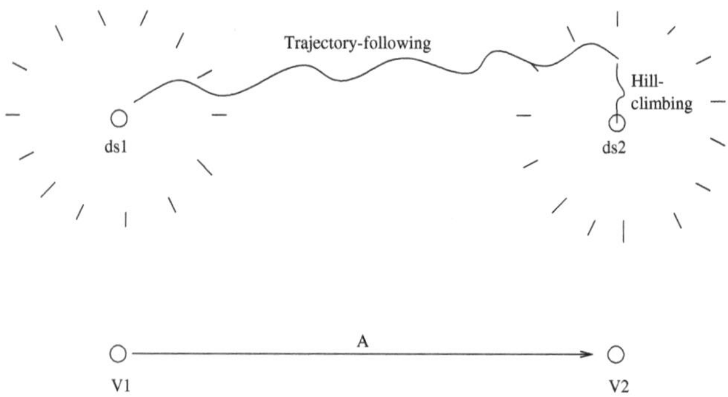

Fig. 11.3. Motion from one distinctive state to another via trajectory-following and hill-climbing control laws eliminates cumulative error. Reliable behavior can be abstracted to the causal schema $\langle V_1, A, V_2 \rangle$.

following and hill-climbing *control laws*, subject to their applicability and termination conditions, so the agent-environment system moves toward an attractor. The stable attractor of a hill-climbing control law is called a *distinctive state*.

- At the *causal level*, the agent and its environment are described as a partially known finite-state automaton, whose *states* correspond to the distinctive states identified at the control level, and whose *actions* correspond to sequences of control laws. *Views* are the observable properties of states. A discrete state transition at the causal level corresponds to the extended evolution of dynamical systems at the control level.

- At the *topological level*, the environment is described in terms of *places*, *paths*, and *regions*, with relations such as connectivity, order, and containment. A state of the agent, described at the causal level, corresponds to being at a place, on a path, and facing along the path in one of two directions. The topological map is created by a process of *abduction*, to explain the sequence of views and actions that represent the agent's experience at the interface between the control and causal levels [55].
- The *metrical level* has several different aspects. The causal and topological levels may include attributes with quantitative values, such as the magnitudes of actions, distances between places along paths, and angles between paths at places. A local place neighborhood can be described by a two-dimensional spatial analog such as an occupancy grid, with a single frame of reference. A spatial analog model of the large-scale environment can be created, based on the skeleton provided by the topological map.

There are logical dependencies (Fig. 11.4) among the levels, which constrain the combinations of representations that can occur. Different parts of the cognitive map may represent knowledge at different SSH levels, but each part of the map must respect the dependency structure. The agent's cognitive map may have a global metrical map of one portion of the environment, a topological map of another, simply causal knowledge of the sequence of actions to take in a third, and then use the control level to explore unknown territory. Or, when pressed for time or preoccupied with other concerns, the agent may access only causal knowledge to follow a familiar route even though topological and metrical knowledge may be available.

Emilio Remolina's doctoral work [55] provided a major step forward in the clarity of the SSH. He provided a formal axiomatization for the SSH causal and topological levels, plus the quantitative attribute portion of the metrical level. Since the topological map is the result of an abduction process, finding the best consistent explanation of the available observations, the formalization required a non-monotonic logic, in this case circumscription as embodied in Vladimir Lifschitz' nested abnormality theories [35]. The axioms express the consistency requirements for topological maps, and the nesting structure and the prioritized circumscription policy express the preference ordering on consistent maps. If a new observation should refute the current most preferred consistent map, then the preference ordering can be used to help select a preferred map from those still considered consistent.

This non-monotonic logical inference is implemented as an algorithm that creates a tree of all possible topological maps and imposes a preference order on the leaves.[8] At any point in time, the leaves of the tree represent the topological maps consistent with experience so far. After a travel action reaches and

[8] Strictly speaking, the abduction searches for the best set of equality and inequality axioms over the symbols representing distinctive states. The algorithm creates models of those sets of axioms, and tests them for consistency.

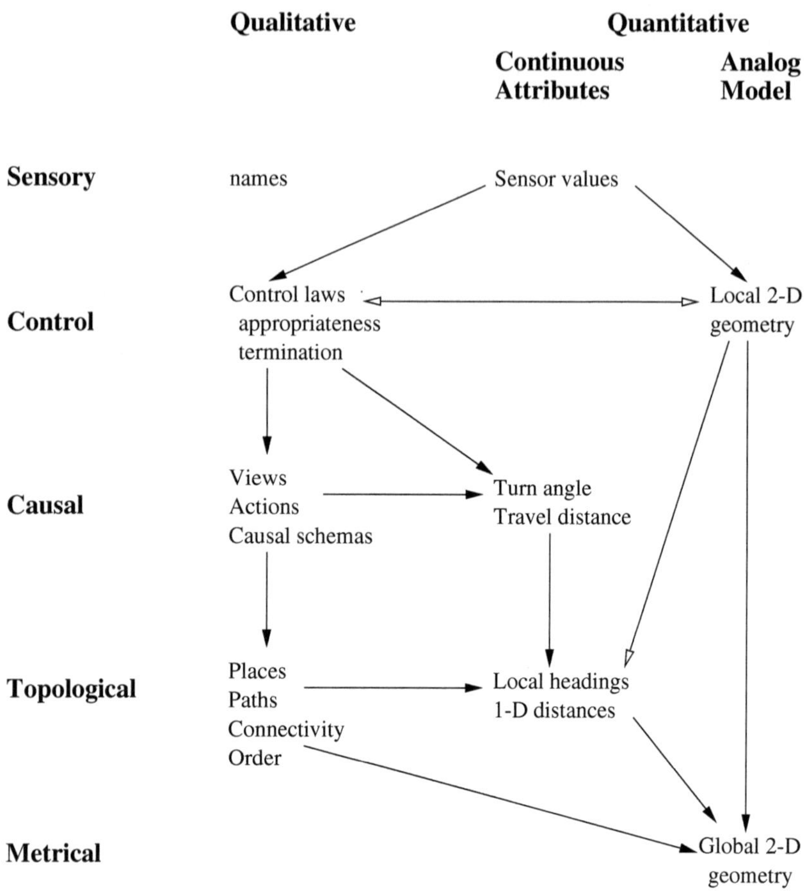

Fig. 11.4. The Spatial Semantic Hierarchy. Closed-headed arrows represent dependencies; open-headed arrows represent potential information flow without dependency.

describes a new place neighborhood, some maps at the leaves of the tree are refuted as inconsistent, some are confirmed as consistent, and others branch on all consistent extensions. Branches only take place when there is perceptual aliasing; that is, when different places can have the same appearance. Then if a travel action reaches a place that appears the same as a previously-known place, two hypotheses must be created: one that the new place really is the same as the old one, and a second that the new place is genuinely new, but has the same appearance as the old one.

By initially creating all possible consistent successors, and refuting only the inconsistent ones, we maintain the guarantee that the correct topological map is present in the tree [55, 22]. In subsequent work, Francesco Savelli augmented the existing topological axioms with a test for the planarity of the topological map, which could be applied either as a consistency requirement or as a preference criterion [60]. It will also be important to use probability as well as prioritized circumscription policies to order the consistent maps [13].

The SSH treats observations gathered during exploration as the fundamental source of experience for building a cognitive map of large-scale space. However, there are other ways to obtain information about the structure of the environment. Verbal route directions translate naturally into sequences of actions (and minimal descriptions of views) at the SSH causal level [38]. Informal sketch maps translate naturally into subgraphs at the SSH topological level. And precise graphical maps provide information at the SSH metrical level. These and other forms of spatial communication are a topic for active research in psychology, linguistics, and cognitive science. One role for the SSH is to provide a useful description of the target representation for such communication.

11.7 The Hybrid Spatial Semantic Hierarchy

The four levels of the basic SSH framework start to look pretty satisfactory. This lets us turn our attention to certain assumptions and issues whose resolution will help us broaden and improve the Spatial Semantic Hierarchy.

First, the basic SSH treats perception as a black-box process that returns "view" symbols, abstractions of the full sensory image, capable only of being matched for equality or used as retrieval keys. We are ready to break down the hard separation between large-scale space and small-scale perceptual space. A more realistic theory of perception of the local environment, with both laser range-finders and computer vision, needs to be integrated with the cognitive mapping process.

Second, the basic SSH assumes that distinctive states are identified through the agent's physical motion, hill-climbing to the location in the environment that maximizes the current distinctiveness measure. This physical motion seems awkward and unnecessary.

Third, there has been an explosion of successful work on the SLAM (simultaneous localization and mapping) problem, building metrical maps of increasing size directly from sensory input within a single global frame of reference [63]. This approach differs significantly from the human cognitive map and from the multi-representation approach of the SSH. Do the two approaches compete? Are they complementary? Is one suitable for modeling humans while the other is for building robots? We need to understand the relationship between these two approaches.

Fortunately, there is a synergy between these three concerns that leads to their resolution [22]. Having defined *large-scale space* as space whose structure is larger than the sensory horizon, it is natural to define *small-scale space* as space whose structure is within the sensory horizon. Small-scale space is described by a *local perceptual map* that is metrically accurate and is constructed directly from sensory input. Recently developed SLAM methods are well suited for creating such a local perceptual map. We avoid the problem of closing large loops by confining the map to the agent's local perceptual surround, where we can apply the strengths of existing SLAM methods. When reasoning about small-scale space, we are concerned only with the frame of reference of the local perceptual map, and not with its inevitable drift with respect to the world frame of reference. We call the resulting combined model of large-scale and small-scale space, the *hybrid SSH.*

Local SLAM methods continually maintain the agent's localization in the frame of reference of the local map. Accurate incremental localization supports accurate incorporation of observations into the local map, and accurate local motion planning. In the basic SSH, hill-climbing provides the same benefit of accurate localization under weaker assumptions about sensors and effectors, but at the cost of physical motion to the distinctive state. In the hybrid SSH, when the agent has enough knowledge about its sensors and effectors to maintain its localization within the local perceptual map, it no longer requires physical hill-climbing.

Where the basic SSH treats views as atomic symbols, matched only for equality, the hybrid SSH treats the local perceptual map as the observable manifestation of a topological place [22]. The local perceptual map of a place neighborhood is parsed to define a local topology that describes how directed path segments join at that place. Distinctive states in the basic SSH causal level correspond to *gateways* within the local perceptual map of the place. Two local perceptual maps are matched by first matching their local topology descriptions, and then matching their perceptual maps to give a probability that they correspond to the same state. The local perceptual map with its local topology description bind together the small-scale-space and large-scale-space descriptions of the same place neighborhood, and thus bind together the continuous sensorimotor ontology and the discrete topological ontology.

The agent's experience in the environment is an alternating sequence of views and actions. However, in the hybrid SSH, a view corresponds to a pose within the local perceptual map, a turn action corresponds to motion within the local perceptual map of the current place neighborhood, while a travel action moves from one place neighborhood with its local perceptual map, to another place neighborhood. In addition to fixed local perceptual maps of place neighborhoods, a scrolling local perceptual map is used by trajectory-following control laws as an "observer" process to model obstacles in the agent's immediate surround. A topological place is detected at a change in the qualitative properties of the local topology of the scrolling local perceptual map during execution of a trajectory-following control law [5]. The topological map is built by abduction to explain this sequence of experiences. Where it is possible to have *perceptual aliasing* (two different places look the same), we build a tree of topological maps consistent with the same sequence of experiences. After sufficient exploration, inconsistent maps are refuted, and a single simplest or most probable map can be identified.

At this point, we can combine the global topological map with local perceptual maps of place neighborhoods to build a global metrical map of the large-scale environment in a single frame of reference [43]. Each local perceptual map defines a local frame of reference for accurate metrical knowledge at a place neighborhood,

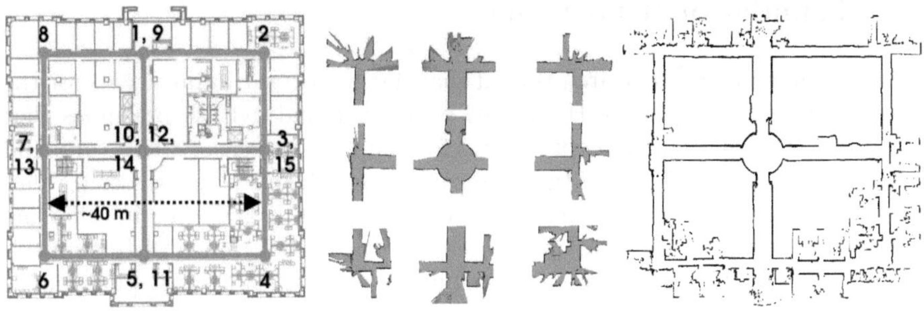

Fig. 11.5. The Hybrid SSH builds a global metrical map: (a) The robot explores an office environment with multiple nested large loops, identifying places in the sequence shown. (b) After inferring the correct topological map, the layout of local place maps in the global frame of reference. (c) The global map is created by localizing the trajectory poses in the global frame of reference, anchored by the poses in the local place maps, then creating the global map from the laser range-finder observations.

but the frame of reference will drift enough during travel to make it unusable globally. A consistent topological map hypothesis embodies a decision about which experiences of perceptually similar places were actually visits to the same place. Travel along each path segment between places can be used to estimate the displacement of each place in the local frame of reference of its predecessor. These local displacements between adjacent places can then be merged into a layout of the local place frames within a single global frame of reference, typically by applying a relaxation algorithm to the displacements. (The resulting probability of the global layout given the topological map and the displacements can be used as part of the preference ordering of topological maps in the tree of consistent maps.) The entire trajectory of robot poses can now be described in the global frame of reference, anchored by the poses at both ends of each path segment, which already have accurate localization within the local frames of reference. Finally, an accurate global metrical map can be constructed, given the accurately localized trajectory of poses. This factors the problem of global metrical mapping into three tractable steps.

Part of the original motivation for the TOUR model of the cognitive map was the observation that humans do *not* typically create an accurate global metrical map from observations during travel. However, with increasing experience in the environment, they can lean a cognitive map that is increasingly faithful to the correct Euclidean model of the world [47]. Furthermore, accurate global metrical maps are valuable engineering and scientific tools, so it is useful for a robot to be able to build them. We demonstrate the value of combining different representations of space by showing how to build a correct global metrical map on the skeleton provided by an accurate global topological map, using observations from experience in the local perceptual map.

11.8 Foundational Learning

We have jumped over a research thread that has important implications for the future. The Spatial Semantic Hierarchy, both basic and hybrid, presumes that the agent has a collection of control laws for coupling its sensors, effectors, and environment together. This, in turn, presumes that the agent possesses (or embodies) knowledge of which sensory features are useful, and how its effectors change those features. In an artificially constructed robot, much of this knowledge is built in by the designer. In a biological creature, some of this knowledge is innate. We ask, how can this knowledge be learned? Biologically, some of the learning is done by the species over evolutionary time, while the rest is done by the individual.

This question was inspired by a challenge problem proposed by Ron Rivest at MIT in 1984 [28]. Suppose an agent wakes up in an unknown world, with a sense vector and a motor vector, but with no knowledge of how they are related to its world. How can such an agent learn to predict the results of future actions? This challenge led Rivest, Sloan, and Schapire to a series of results about learning finite automata from observations [56, 59, 57, 58]. My own approach was to try to learn the sensorimotor foundation for the TOUR model from exploration experience [28].

Around 1988, David Pierce and I began to investigate this question for an agent with continuous experiences in a continuous world. After developing some preliminary pieces of the puzzle [49, 52, 50], we demonstrated a learning agent that started with an uninterpreted sensorimotor system in an unknown world, and learned: (a) to separate the sense vector into distinct sensory modalities; (b) to learn a low-dimensional spatial structure for the sense elements ("pixels") in a particular modality; (c) to identify primitive actions from the sensory flow fields induced on this spatial structure; (d) to identify a set of stable sensory features that can be extracted and tracked in the sensory image; (e) to learn which actions cause reliable changes to which perceptual features in which contexts; (f) to construct useful homing (i.e., hill-climbing) and trajectory-following control laws from those actions; and (g) to define distinctive states and actions linking them [48, 51]. Thus, by bootstrapping through a number of intermediate representations, the agent learned a sufficient foundation to reach the "bottom rung" of the SSH ladder. While there were a number of assumptions and limitations in this work, it genuinely demonstrated that a computational agent could learn its own sensorimotor grounding from its own interaction with the environment (Fig. 11.6).

This research thread returned to the back burner for several years, until Patrick Beeson and I started looking at the problem of place recognition [20]. A realistic robot receives a high-dimensional sensory image at any given moment. For the basic SSH causal level, that image must be abstracted to one of a discrete set of views. Our goal was to learn a view representation such that each view correctly determines a unique distinctive state. We build on the fact that perceptual aliasing of distinctive states can be overcome by continued

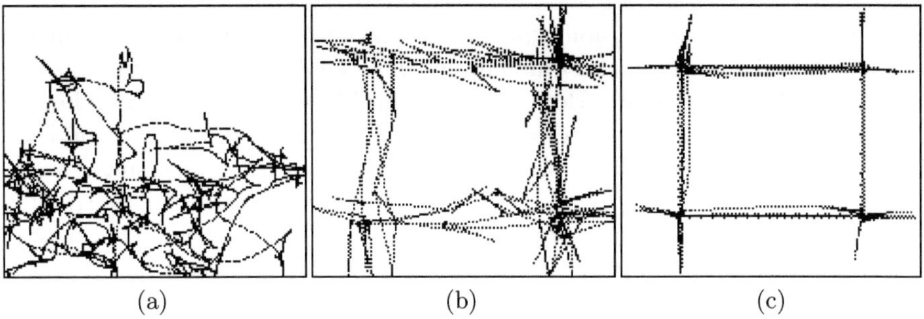

Fig. 11.6. Exploring a simple world at three levels of competence. (a) The robot wanders randomly while learning a model of its sensorimotor apparatus. (b) The robot explores by randomly choosing applicable homing and open-loop path-following behaviors based on the static action model while learning the dynamic action model. (c) The robot explores by randomly choosing applicable homing and closed-loop path-following behaviors based on the dynamic action model.

exploration, proposing candidate topological maps and refuting the incorrect ones when predictions are violated.

We gave the name *bootstrap learning* to the learning method we developed.[9] Start by creating an over-abstract but usable view representation: cluster sensory images aggressively enough that each distinctive state corresponds to only one view, even at the cost of multiple states having the same view (perceptual aliasing). Then the standard SSH exploration and mapping methods can converge to the correct topological map after enough exploration. The correct topological map provides a correct association between distinctive states and the high-dimensional sensory images, even if the views are aliased. So now we can use supervised learning (more powerful than unsupervised clustering), to learn correct associations between sensory images and distinctive states. In two experiments with rich sensors and real environments, the learning agents rapidly reached 100% accurate place recognition [20].

The generic structure of this bootstrap learning scenario is: (1) approximately abstract the problem using an unsupervised method; (2) use a much more expensive inference method to find the correct answer; (3) use supervised learning to find the correct level of abstraction. We believe that this pattern can be applied to other abstraction-learning problems [21].

Jefferson Provost and I have been investigating how an unsupervised agent can learn a high-level ontology of perceptual features, distinctive states, and extended actions from uninterpreted "pixel-level" experience. Provost's Self-Organizing Distinctive-state Abstraction (SODA) starts by training a self-organizing map (SOM) with adaptive topology to define a set of distinctive sensory images. The agent can then learn to climb the gradient of the activation

[9] We have since extended the term "bootstrap learning" to apply to this general approach to foundational learning.

level of the currently-leading SOM unit to move to, and define, distinctive states. Finally, it learns trajectory-following control laws to move from one distinctive state to the neighborhood of a different one, where hill-climbing brings it to the destination distinctive state. Both sets of control laws are learned by reinforcement learning [62] where the reward signal (e.g. the SOM unit activation level) is internally generated and has been autonomously learned. The result is an autonomous abstraction of a large continuous environment into a discrete set of distinctive states with actions ("options") for moving among them. This abstraction achieves large speedups when learning to solve high-diameter reinforcement learning problems [54].

Joseph Modayil and I have been investigating how a higher-level ontology of objects and actions can be learned from experience with a lower-level ontology of individual sense elements ("pixels") and motor signals [44]. This, too, requires a multi-stage learning process. It was developed and demonstrated using the range-sensor-based local perceptual map (implemented as an occupancy grid) used by our exploring robots. First, we identify those sensor returns in the current sensor image that are explained by static features of the environment, represented by cells in the occupancy grid that have high confidence of being occupied, and have never had high confidence of being free space. The remaining sensor returns are explained by cells whose occupancy has changed at some time in the past. Second, we cluster these "dynamic" sensor returns in the current sensory image frame; and third, we attempt to track these clusters from frame to frame over time. These trackable clusters are hypothesized to be explainable as images of objects. The fourth step is to collect a sequence of images of an object from different perspectives to describe its shape [45]; and the fifth is to create a classification hierarchy of object types based on this described shape. Ongoing work considers the abstraction of actions applied to these learned objects.

These foundational learning methods autonomously discover regularities in the agent's continuous sensorimotor experience that can be abstracted and described by symbolic expressions. That is, they provide tangible solutions to particular instances of the "Symbol Grounding" problem [14].

11.9 Conclusions

I began studying the cognitive map as a manageable subset of commonsense knowledge. I hoped that this problem would *not* be "AI Complete" — that is, it could be sufficiently separated from other major issues in AI and cognitive science that it would be possible to make useful progress without simultaneously solving every other major problem in AI. At the same time, knowledge of space is clearly a fundamental part of commonsense knowledge [47, 34], so progress in understanding the cognitive map contributes to the overall enterprise of understanding commonsense knowledge, and hence the nature of mind.

It seems to me that these hopes were well justified, and the research efforts have paid off. Boundaries separating one scientific problem from another are always artificial scaffolding, used to make a problem tractable for human minds.

Once enough progress has been made on one formulation of a problem, it becomes time to move the scaffolding so progress can be made on a larger formulation. The progress from the TOUR model to the Basic SSH and then to the Hybrid SSH seems to me to have exactly this character. Each problem definition served its purpose, led to an improved understand of the nature of spatial knowledge, and was replaced by a new, larger, problem definition. The focus of the TOUR model was primarily on the role of topological knowledge of space. The focus of the Basic SSH was on the role of control laws and dynamical systems. The focus of the Hybrid SSH is on the role of metrical knowledge and perception.

When I first learned about Minsky's frames for knowledge representation, I wondered where the slots come from. The multiple representations of the TOUR model and the Spatial Semantic Hierarchy are clearly distinct theories with distinct ontologies. The flexibility and robustness of commonsense knowledge depends on having multiple ontologies for the same domain of knowledge. The question of where the slots come from has been transformed into the question, *How can an agent learn, not just new knowledge within an existing ontology, but a new ontology it does not already possess?*

The foundational learning problem is not simply an enlarged version of the cognitive mapping problem. Rather, now that we have a reasonably solid theory of spatial knowledge in the cognitive map, we can ask questions about its foundation with a degree of specificity that was not possible before. We can also evaluate foundational learning methods according to their ability to support higher-level theories that we already understand. In my own case, the theory of the cognitive map serves this role. However, the learning methods we seek will serve as foundations for a much larger body of commonsense knowledge.

References

1. Basye, K., Dean, T., Kaelbling, L.P.: Learning dynamics: system identification for perceptually challenged agents. Artificial Intelligence 72, 139–171 (1995)
2. Basye, K., Dean, T., Vitter, J.S.: Coping with uncertainty in map learning. Machine Learning 29(1), 65–88 (1997)
3. Becker, J.D.: "Robot" computer problem solving system. Technical Report 2316, Bolt Beranek and Newman (September 1972)
4. Becker, J.D., Merriam, E.W.: "Robot" computer problem solving system. Technical Report 2792, Bolt Beranek and Newman (April 1974)
5. Beeson, P., Jong, N.K., Kuipers, B.: Towards autonomous topological place detection using the extended Voronoi graph. In: IEEE International Conference on Robotics and Automation, IEEE Computer Society Press, Los Alamitos (2005)
6. Boulding, K.: The Image. University of Michigan Press, Ann Arbor (1956)
7. Crawford, J.M., Kuipers, B.J.: Toward a theory of access-limited logic for knowledge representation. In: Proc. 1st Int. Conf. on Principles of Knowledge Representation and Reasoning, Morgan Kaufmann, San Mateo, CA (1989)
8. Crawford, J.M., Kuipers, B.J.: Negation and proof by contradiction in access-limited logic. In: AAAI 1991. Proc. 9th National Conf. on Artificial Intelligence, AAAI/MIT Press (1991)

9. Downs, R.M., Stea, D.: Image and Environment. Aldine Publishing Company, Chicago (1973)
10. Dudek, G., Jenkin, M., Milios, E., Wilkes, D.: Robotic exploration as graph construction. IEEE Trans. on Robotics and Automation 7(6), 859–865 (1991)
11. Fillmore, C.: The case for case. In: Bach, E., Harms, R.T. (eds.) Universals in Linguistic Theory, Holt, Rinehart and Winston, Chicago (1968)
12. Geisler, W.S.: Sequential ideal-observer analysis of visual discriminations. Psychological Review 96, 267–314 (1989)
13. Hähnel, D., Thrun, S., Wegbreit, B., Burgard, W.: Towards lazy data association in SLAM. In: ISRR 2003. Proc. Int. Symp. on Robotics Research (2003)
14. Harnad, S.: The symbol grounding problem. Physica D 42, 335–346 (1990)
15. Hayes, P.J.: In defence of logic. In: IJCAI 1977. Proc. 5th Int. Joint Conf. on Artificial Intelligence, pp. 559–565 (1977)
16. Kuipers, B.: Commonsense reasoning about causality: Deriving behavior from structure. Artificial Intelligence 24, 169–204 (1984)
17. Kuipers, B.: Qualitative simulation. Artificial Intelligence 29, 289–338 (1986)
18. Kuipers, B.: The Spatial Semantic Hierarchy. Artificial Intelligence 119, 191–233 (2000)
19. Kuipers, B.: Why don't I take military funding? (2003), http://www.cs.utexas.edu/users/kuipers/opinions/no-military-funding.html
20. Kuipers, B., Beeson, P.: Bootstrap learning for place recognition. In: AAAI 2002. Proc. 18th National Conf. on Artificial Intelligence, pp. 174–180. AAAI/MIT Press (2002)
21. Kuipers, B., Beeson, P., Modayil, J., Provost, J.: Bootstrap learning of foundational representations. Connection Science 18(2), 145–158 (2006)
22. Kuipers, B., Modayil, J., Beeson, P., MacMahon, M., Savelli, F.: Local metrical and global topological maps in the hybrid spatial semantic hierarchy. In: ICRA 2004. IEEE Int. Conf. on Robotics & Automation, IEEE Computer Society Press, Los Alamitos (2004)
23. Kuipers, B.J.: A frame for frames: representing knowledge for recognition. In: Bobrow, D.G., Collins, A. (eds.) Representation and Understanding, pp. 151–184. Academic Press, New York (1975)
24. Kuipers, B.J.: Representing Knowledge of Large-Scale Space. PhD thesis, Mathematics Department, Massachusetts Institute of Technology, Cambridge, MA (1977), http://www.cs.utexas.edu/users/qr/papers/Kuipers-PhD-77.html
25. Kuipers, B.J.: Modeling spatial knowledge. Cognitive Science 2, 129–153 (1978)
26. Kuipers, B.J.: Commonsense knowledge of space: learning from experience. In: IJCAI 1979. Proc. 6th Int. Joint Conf. on Artificial Intelligence, Tokyo, Japan, pp. 499–501 (August 1979)
27. Kuipers, B.J.: Modeling human knowledge of routes: Partial knowledge and individual variation. In: AAAI 1983. Proc. 3rd National Conf. on Artificial Intelligence, Los Altos, CA, Morgan Kaufmann, San Francisco (1983)
28. Kuipers, B.J.: The map-learning critter. Technical Report AI TR 85-17, University of Texas at Austin, Artificial Intelligence Laboratory, Austin, TX (1985)
29. Kuipers, B.J.: Qualitative Reasoning: Modeling and Simulation with Incomplete Knowledge. MIT Press, Cambridge, MA (1994)
30. Kuipers, B.J., Byun, Y.T.: A robust qualitative method for spatial learning in unknown environments. In: AAAI 1988. Proc. 7th National Conf. on Artificial Intelligence, Los Altos, CA, pp. 774–779. Morgan Kaufmann, San Francisco (1988)

31. Kuipers, B.J., Byun, Y.-T.: A robot exploration and mapping strategy based on a semantic hierarchy of spatial representations. Journal of Robotics and Autonomous Systems 8, 47–63 (1991)
32. Kuipers, B.J., Kassirer, J.P.: Causal reasoning in medicine: analysis of a protocol. Cognitive Science 8, 363–385 (1984)
33. Kuipers, B.J., Levitt, T.: Navigation and mapping in large scale space. AI Magazine 9(2), 25–43 (1988)
34. Lakoff, G., Johnson, M.: Metaphors We Live By. The University of Chicago Press, Chicago (1980)
35. Lifschitz, V.: Nested abnormality theories. Artificial Intelligence 74, 351–365 (1995)
36. Linde, C., Labov, W.: Spatial networks as a site for the study of language and thought. Language 51, 924–939 (1975)
37. Lynch, K.: The Image of the City. MIT Press, Cambridge, MA (1960)
38. MacMahon, M., Stankiewicz, B., Kuipers, B.: Walk the talk: connecting language, knowledge, and action in route instructions. In: AAAI 2006. National Conference on Artificial Intelligence, Boston (July 2006)
39. McCarthy, J.: Programs with common sense. In: Minsky, M.L. (ed.) Semantic Information Processing, pp. 403–418. MIT Press, Cambridge, MA (1968)
40. Miller, G.A., Galanter, E., Pribram, K.H.: Plans and the Structure of Behavior. Holt, Rinehart and Winston (1960)
41. Minsky, M.: A framework for representing knowledge. In: Winston, P.H. (ed.) The Psychology of Computer Vision, McGraw-Hill, NY (1975)
42. Minsky, M.L.: Matter, mind, and models. In: Minsky, M. (ed.) Semantic Information Processing, pp. 425–432. MIT Press, Cambridge (1968)
43. Modayil, J., Beeson, P., Kuipers, B.: Using the topological skeleton for scalable, global, metrical map-building. In: IEEE/RSJ Int. Conf. on Intelligent Robots and Systems (2004)
44. Modayil, J., Kuipers, B.: Bootstrap learning for object discovery. In: IEEE/RSJ Int. Conf. on Intelligent Robots and Systems (2004)
45. Modayil, J., Kuipers, B.: Autonomous shape model learning for object localization and recognition. In: IEEE International Conference on Robotics and Automation, IEEE Computer Society Press, Los Alamitos (2006)
46. Moravec, H., Elfes, A.: High resolution maps from wide angle sonar. In: IEEE International Conference on Robotics and Automation, pp. 116–121. IEEE Computer Society Press, Los Alamitos (1985)
47. Piaget, J., Inhelder, B.: The Child's Conception of Space. Norton, New York (1967). First published in French (1948)
48. Pierce, D., Kuipers, B.: Learning to explore and build maps. In: AAAI 1994. Proc. 12th National Conf. on Artificial Intelligence, AAAI/MIT Press (1994)
49. Pierce, D.M.: Learning turn and travel actions with an uninterpreted sensorimotor apparatus. In: IEEE International Conference on Robotics and Automation, Sacramento, CA, IEEE Computer Society Press, Los Alamitos (1991)
50. Pierce, D.M., Kuipers, B.J.: Learning hill-climbing functions as a strategy for generating behaviors in a mobile robot. In: Meyer, J.-A., Wilson, S.W. (eds.) Proceedings of the International Conference on Simulation of Adaptive Behavior: From Animals to Animats, MIT Press/Bradford Books, Cambridge, MA (1991)
51. Pierce, D.M., Kuipers, B.J.: Map learning with uninterpreted sensors and effectors. Artificial Intelligence 92, 169–227 (1997)
52. Pierce, D.M.: Learning a set of primitive actions with an uninterpreted sensorimotor apparatus. In: Birnbaum, L., Collins, G. (eds.) Machine Learning: Proceedings of the Eighth International Workshop, Morgan Kaufmann, San Mateo, CA (1991)

53. Pratt, V.: A linguistics oriented language. In: IJCAI 1973. Proc. 3rd Int. Joint Conf. on Artificial Intelligence (1973)
54. Provost, J., Kuipers, B.J., Miikkulainen, R.: Developing navigation behavior through self-organizing distinctive-state abstraction. Connection Science 18(2), 159–172 (2006)
55. Remolina, E., Kuipers, B.: Towards a general theory of topological maps. Artificial Intelligence 152, 47–104 (2004)
56. Rivest, R.L., Schapire, R.E.: A new approach to unsupervised learning in deterministic environments. In: Proceedings of the Fourth International Workshop on Machine Learning (1987)
57. Rivest, R.L., Schapire, R.E.: Inference of finite automata using homing sequences. In: Proc. 21st ACM Symposium on Theory of Computing, pp. 411–420. ACM Press, New York (1989)
58. Rivest, R.L., Schapire, R.E.: Diversity-based inference of finite automata. Journal of the ACM 41(3), 555–589 (1994)
59. Rivest, R.L., Sloan, R.: Learning complicated concepts reliably and usefully. In: AAAI 1988. Proc. 7th National Conf. on Artificial Intelligence, pp. 635–640. AAAI Press/The MIT Press (1988)
60. Savelli, F., Kuipers, B.: Loop-closing and planarity in topological map-building. In: IROS 2004. IEEE/RSJ Int. Conf. on Intelligent Robots and Systems (2004)
61. Siegel, A.W., White, S.H.: The development of spatial representations of large-scale environments. In: Reese, H.W. (ed.) Advances in Child Development and Behavior, vol. 10, pp. 9–55. Academic Press, New York (1975)
62. Sutton, R.S., Barto, A.G.: Reinforcement Learning. MIT Press, Cambridge MA (1998)
63. Thrun, S.: Robotic mapping: A survey. In: Lakemeyer, G., Nebel, B. (eds.) Exploring Artificial Intelligence in the New Millenium, Morgan Kaufmann, San Francisco (2002)

Robot Cognitive Mapping – A Role for a Global Metric Map in a Cognitive Mapping Process

Margaret E. Jefferies, Jesse Baker, and Wengrong Weng

Summary. In robotics it would be argued that we are closing the loop in a topological map using a global metric map. Drawing on our studies of human and animal cognitive mapping we proposed that a cognitive map comprises a topological map of metric local space representations [24]. Each local space defines a part of the environment that appears to enclose the animal/robot. Recently our Pioneer 2 robot has been computing such a map during its travels around our department. The advantage of such a map for a robot is that cumulative positional error is constrained to the local representation. Simpler localisation methods will often suffice for the local environment as global metric consistency is not required. The trade-off is that one cannot easily detect that one is re-entering a previously visited part of the environment via a new route (i.e. closing a loop) as is the case with a global metric map. The question we asked was: could we combine the local and global representations, exploiting the advantages of both - local representations for simpler localisation and no global metric consistency; global representation for easy loop detection. While a simple localisation method suffices for the local representation it would be inadequate for a global metric map. However the error could not grow unbounded if it were to be useful in the task of detecting loops. Our solution was to limit the size of the global map and have it move with the robot as it traversed its environment. We will describe the implementation of such a map and show that it can detect loops over a reasonable distance.

12.1 Introduction

For researchers grappling with the basic issues involved in representing an autonomous mobile individual's (robot or animal's) experience of their spatial environment, the nature of the underlying representation is at the core of the problem. Psychologists and geographers examine the behaviour of the animal or human and neuroscientists the behaviour of cells in certain parts of the animal's brain to determine the nature of the information that has been stored and how it is being used [7, 18, 22]. Artificial Intelligence and robotics researchers are concerned with computational issues, developing computational theories [11, 23, 24] and from a practical point of view how best to compute a representation for a robot with poor sensors and odometry [21]. Two themes which emerge from the studies of both groups are: (i) the notion of a representation for the local space i.e. the small area of the environment the individual occupies, versus (ii) a global representation in which conceivably their total experience could be represented using a single frame of reference. Related to these is the idea of a metric representation, where properties such as size, distance and location are explicitly

M.E. Jefferies et al. (Eds.): Robot. & Cogn. Approach. to Spat. Map., STAR 38, pp. 265–279, 2008.
springerlink.com

or implicitly represented, versus a topological representation where relationships such as connectivity between individual elements are represented. Global representations in the sense the term global is used in this paper are metric. The advantage of a global representation is that it is easy to detect when one is re-entering a part of the environment one has been to previously. The disadvantage is that complex error correction is required to eliminate accumulating sensing and odometric errors. The local space could be represented topologically, as in for example, the relationships between some key landmarks, or metrically where the structure of the space itself would be identified within some reference frame (see [25] for a discussion of object or landmark representations versus space-based representations of local space). One's total memory for the environment could be stored in a topological representation, as a collection of local space representations each with its own reference frame. The connections between pairs of local space representations would indicate that one could travel directly from one to the other. Representing the environment as a topological map of local spaces has the advantage that accumulating error is constrained to each local space. The trade-off is that one cannot detect, from one's location, when one is re-entering a previously visited part of the environment via a different route. It is this feature of global maps which makes them attractive (see Fig. 12.1). In robotics this is termed the closed loop problem. It is this problem which is the main focus of this paper.

This idea of a topological network of metric local space representations is central to the computational theory of cognitive maps developed by Yeap and Jefferies [24] and on which this work is based. It is argued in [24] that as one traverses an environment one must initially compute a representation which identifies the space one currently occupies. The algorithm we use emphasises the importance of detecting exits in view from the surfaces perceived and from these exits a boundary for the local space is computed. Each local space is computed using its own cartesian coordinate reference frame. The resulting representation is called an Absolute Space Representation (ASR), a term which emphasises the independent, local nature of each local space visited. See [24] for a discussion of the psychological support for the local space representation. If one remembers how one passed from one ASR into another they will be connected to form a topological representation of the traversed environment (see Fig. 12.1(b)), which we term a cognitive map. We have recently shown that the cognitive map is a suitable representation for an autonomous mobile robot mapping its environment. This implementation is described briefly in Sect. 12.2.

Our investigations into the closed loop problem have led us to propose the use of a size limited metric global map used in conjunction with the topological network [24]. We call this a Memory For the Immediate Surroundings (MFIS). We can thus exploit the advantages of both types of representation. The robot's location in the global map is used to detect that the robot is re-entering a part of the environment it has been to before; each time it detects this a loop in the topological map can be closed. Fig. 12.2 shows how the viewer can use the global MFIS to determine when a previously visited ASR is re-entered.

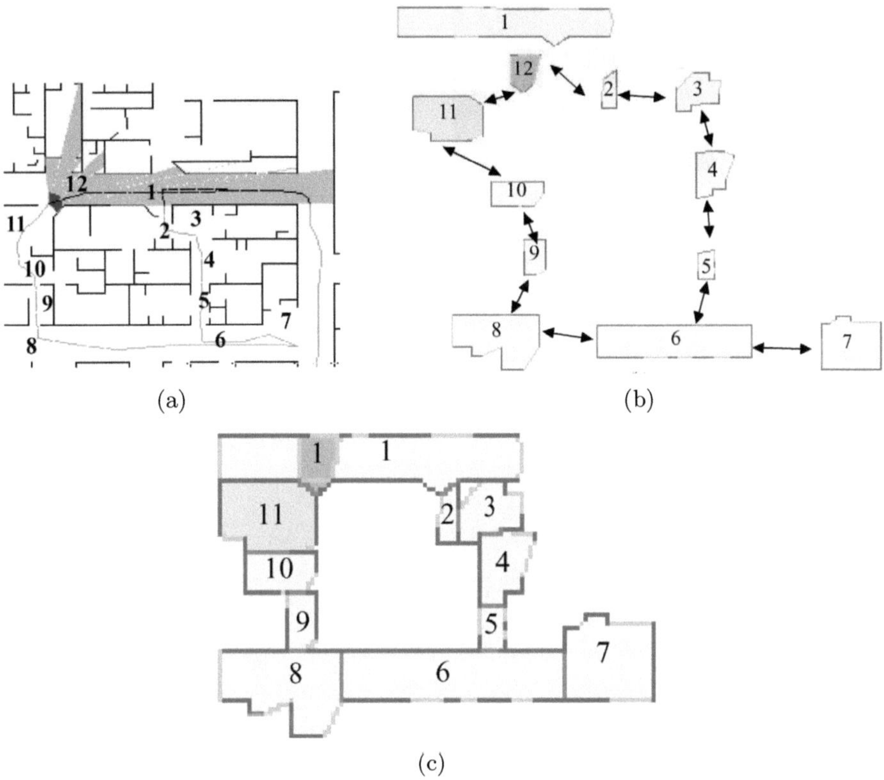

Fig. 12.1. The viewer has traversed a circuitous route arriving back in the same room from which it started. (b) Using a topological representation in which each ASR has its own coordinate system the viewer cannot use its location to determine that ASR1 is being revisited and it thus constructs a new ASR, ASR12, for the same physical space. (c) An MFIS. The ASRs are laid out side by side to indicate a single coordinate system. Using this global representation one can detect from the viewer's location that ASR1 is being re-entered. We use output from an earlier simulator which is error free to better illustrate the basic principles.

12.2 Constructing the Absolute Space Representation from 180° Laser Scans

The first step is to turn the raw laser data into lines representing surfaces in the robot's view of its environment. From this "view" the ASR algorithm firstly works out where the exits are. It does this by looking for surfaces which occlude other surfaces as it is here that the gaps in the boundary of the local space, i.e. the exits, occur. The occlusions are the lines labelled occ in Fig. 12.1(c). An occlusion map is constructed from the surfaces in view (see Fig. 12.2(c). The first occlusion map obtained for a local space is termed the master occlusion map as it is updated and used to recompute the ASR as the robot explores its

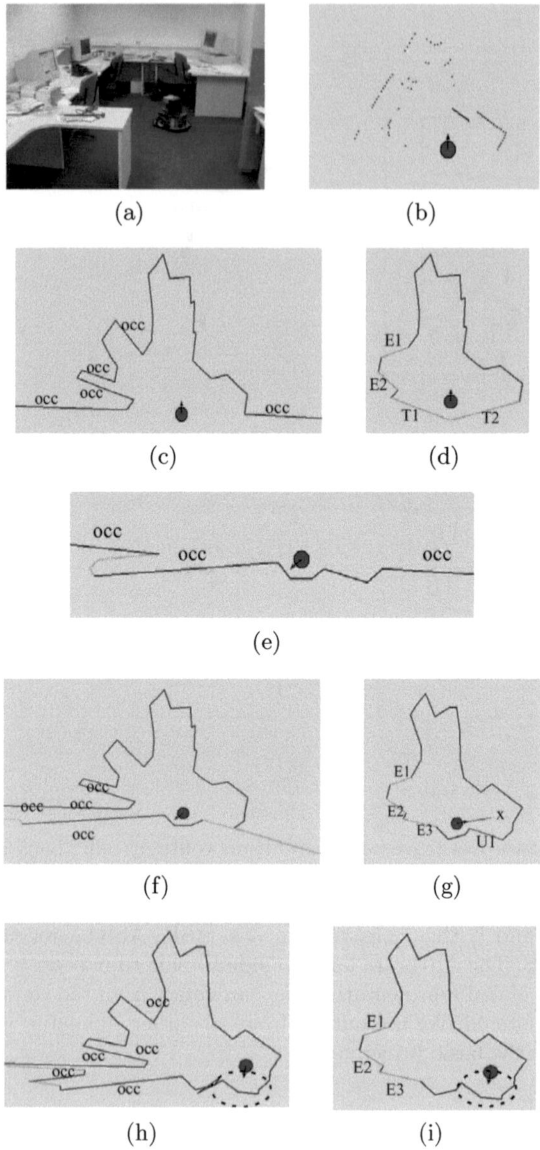

Fig. 12.2. Computing the first ASR. (a) The environment (b) The laser scan. (c) The first occlusion map constructed from the points in (b). The occlusions are marked occ. The black lines are surfaces formed by connecting points between the occluding points. (d) The ASR constructed from the master occlusion map in (c). E1 and E2 are known exits, T1 and T2 temporary exits. (e) the occlusion map obtained when the robot turns towards the temporary exits behind the robot. (f) The master occlusion map after (e) has been incorporated into (c). (g) The ASR generated from the master occlusion map in (f). U1 is an unknown exit. (h) the updated occlusion after the robot moves to position x to explore the unknown region U1. (i) The final ASR. Note different scales have been used to make the figures fit the space available.

local environment. The ASR depicted in Fig. 12.2 is the very first ASR the robot computed at startup. From its initial 180° view of its environment the robot has no notion of what is behind it. However, one can safely add a temporary point directly behind the robot to the occlusion map, so that the ASR algorithm will form a complete closure around the robot. As the robot enters subsequent ASRs the robot will have the exit just traversed directly behind it.

Exits are then created from the occlusions and surfaces in the master occlusion map. For each occlusion in the master occlusion map the algorithm determines which part of the gap associated with it is the actual exit. The exit computed is the shortest "virtual surface" which "covers" the occlusion. We refer the reader to [24] for an in depth description of this part of the algorithm. Surfaces outside the exit are eliminated. The point behind the robot ensures that two temporary exits are added to form a complete enclosure (see Fig. 12.2(d)).

Exits computed as above have a dual role, in the traditional sense to indicate where the robot can leave the current space and to indicate parts of the environment which are yet to be uncovered. These two roles are distinguished by labelling the latter as unknown (see U1 in Fig.12.2(g)) and the former as known (see E1–E3 in Fig. 12.2(g)). As the robot moves about the local space parts of it that were once unknown are no longer so, and the exits covering these areas are updated. See [9, 24] for a description of this updating process. Fig. 12.2(h) shows the updated master occlusion map and Fig. 12.2(i) the resulting ASR. Currently we limit the range of the laser scan to 8 metres. Gaps in the boundary which result from surfaces that are outside this range are marked as unknown exits as they perform a similar function as the unknown exits described above.

Lastly, a simple method is employed to correct for error in the robot's position. The disparity between corresponding occlusions in the overlap between successive views is used to determine the robot's true location. Note the ASR computed here is a robot-centred representation, i.e. ASRs do not look very "room like". Clutter such as desks and chairs are significant objects to a small robot and appear to enclose it. A typical cluttered laboratory could give rise to several ASRs.

12.3 A Global Memory for the Immediate Surroundings – The MFIS

There are two significant problems in maintaining a global memory of the immediate surroundings. The first concerns the definition of the immediate surroundings itself. When the robot moves a step, has its immediate surroundings changed? If it has, the representation could be computationally very expensive to maintain. The second concerns the amount of information that needs to be tracked. How could one then maintain a reasonably accurate representation given that one's perception of the world is inherently very noisy?

Technically, the first problem concerns what frame of reference is appropriate for the MFIS. One could use either an egocentric or an allocentric frame of reference. It is clear from an implementation viewpoint it is inefficient to use an egocentric reference frame. To use an allocentric reference frame, one has to specify where the reference frame should be centred. The choice of this external point need not be chosen arbitrarily if we use the current ASR as the frame of reference. An ASR has an extent and any part of it would be suitable. We chose the centre of the current ASR as the centre of the MFIS. When one moves out of the current local space, the origin of the MFIS is shifted to the centre of the next ASR.

The extent of the MFIS need not be defined exactly in metric terms. Varying its size is a trade-off between how much information is remembered and (i) how much error will accumulate and (ii) how much effort is required to compute it. Given an MFIS with a fixed size, part of an ASR will often be excluded as it lies outside the area covered by the MFIS. The advantage of having the MFIS is to help recognise that nearby local spaces were visited before. We, therefore, do not want to remove a part of an ASR (from the MFIS) because it falls outside the area covered by the MFIS. It is better to include the whole ASR if a part of it lies within the area covered by the MFIS.

The second problem, which concerns the vast amount of information that needs to be tracked, can now be solved by observing that the spatial arrangement of individual surfaces in each ASR is already maintained in the ASRs themselves. Tracking them becomes effortless as long as we treat the ASR as a whole when maintaining the MFIS. We thus propose a limited global memory (the MFIS) containing the last few local spaces visited. It contains the same basic representation for the local space as the topological map, but in addition it contains global location information that allows the viewer to determine when a recently encountered local space is being revisited.

Because the localisation method we use for ASRs is not perfect the MFIS will always contain a certain amount of error. This will result in some overlapping ASRs and possibly more than one ASR in the MFIS matching the newly entered ASR! To reduce the effect of this we cross match the centres of the ASRs. If the centre of one or more of these ASRs (in the MFIS) is contained inside this newly entered ASR we perform a check to ensure that the centre of the new entered ASR is also inside the matched ASR. The advantage of this method is that only the centre of each ASR needs to be recorded in the global coordinates of the MFIS.

Figures 12.3 to 12.6 show the MFIS being used to recognise previously visited ASRs. In Fig. 12.3 The robot has reached ASR7 and is about to go back into ASR3. In Fig. 12.3(b) it can be seen that there is already an overlap between ASR7 and ASR3 resulting from both ASRs claiming the underneath of a table as their own. This does not pose a problem as we do not look for a match until the robot leaves ASR7. In Fig. 12.4 it can be seen that the newly entered ASR covers ASR3. It also covers part of ASR4. In this case the centre cross check succeeds for ASR3 and fails for ASR4. However, it is feasible that the newly computed ASR could have extended across ASR4 to the extent that the centre cross check would have succeeded for ASR4 as well. Currently, the reason we can

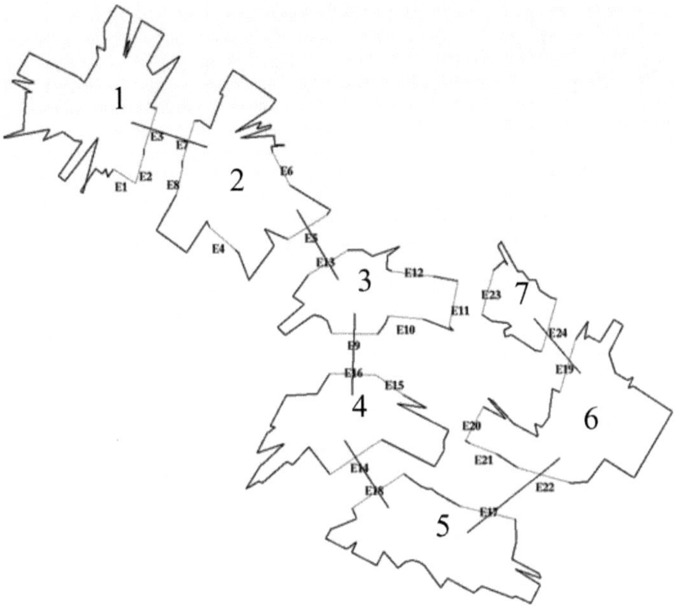

(a)

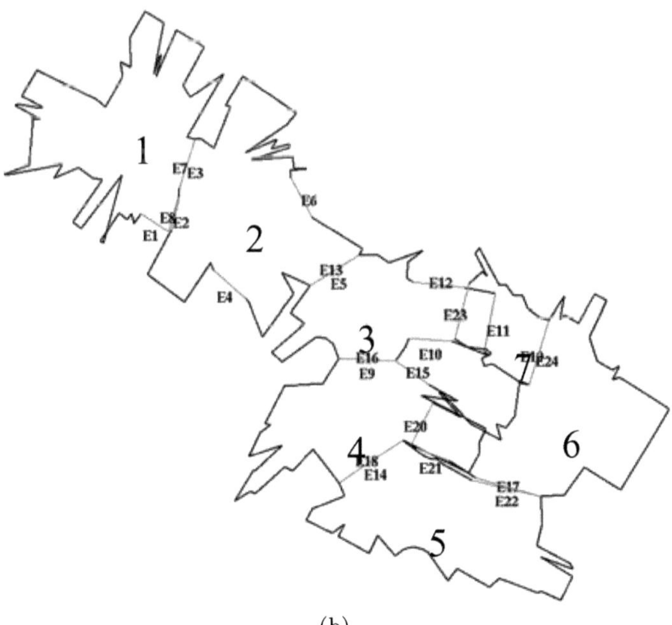

(b)

Fig. 12.3. (a) The topological map (b) The MFIS. The robot is about to re-enter ASR3 from ASR7.

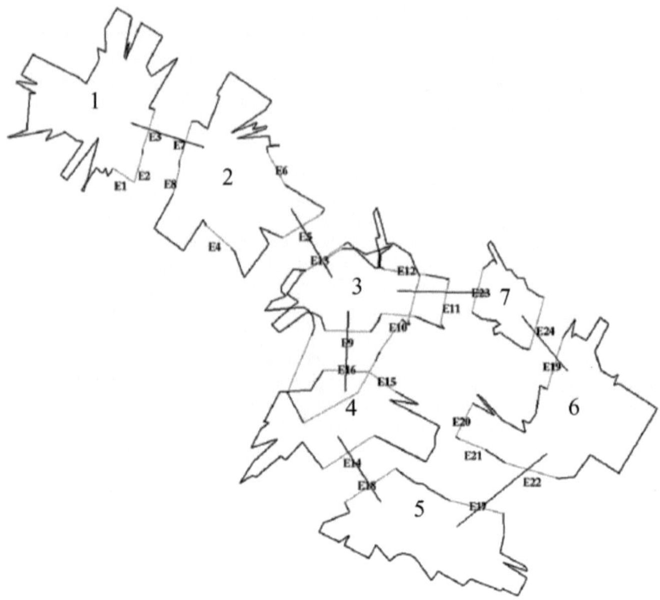

(a)

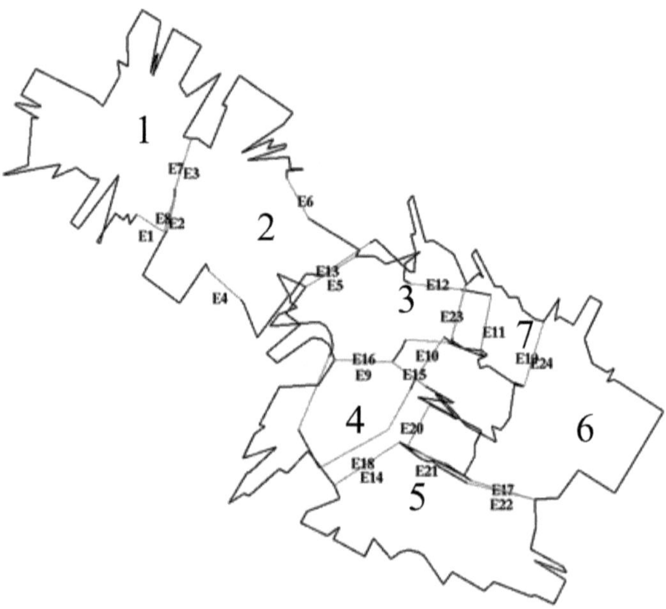

(b)

Fig. 12.4. (a) The topological map. (b) The MFIS. The robot has re-entered ASR3 from ASR7 and is detected in the MFIS. ASR3 and ASR7 are linked in the topological map.

compute such widely different ASRs for the same space is twofold. Firstly we are able to compute nonconvex ASRs where the robot can "see" around corners. Approached differently the corner will occlude and consequently a convex ASR will be computed instead. We don't see this as a problem. Spaces often look different when viewed from different sides. The trick is to get the relationships right to preserve the integrity of the topological map. Secondly, in our robot's environment the state of the environment changes rapidly. Chairs move; bags are dropped and picked up; students and colleagues invariably will try to confound the robot. The difficulty is that in the former case we want to preserve the different views, in the latter we don't -it is only the current view that counts. We are currently investigating this problem.

In Fig. 12.5 the robot has travelled from ASR1 to ASR12 and back again and then enters ASR13. At ASR13, ASR12 is about to re-entered. The drift in the representation can be seen in the discrepancy between ASR12 and ASR13. They should be aligned. In Fig. 12.6 the robot re-en-ters ASR12 and this is detected in the MFIS. However the problem is that in the earlier version of ASR12 the door linking ASR12 and ASR13 was closed. Currently we are maintaining the two representations simultaneously for the same space. When the door is open it is appropriate to use the later version, when it is closed the earlier version.

We are considering how to limit the size of the MFIS. Some issues to be considered are:

1. How much error is acceptable? The impact of error will be greater on small ASRs
2. How conservative should the approach be towards false positives. There is a trade-off between how much the MFIS is limited and the need to verify false positive matches.
3. Will need to consider both the number of rotations and distance travelled.

12.4 Related Work

12.4.1 Global Spatial Representations in Animals

In many animals including humans the mechanism by which the animal keeps track of its location in a global framework is called path integration [2, 4, 6, 12, 14]. It involves the animal maintaining a fix on its position by updating each change in its position in a geocentric coordinate system [7]. Using it animals are able to compute a direct path "home" after following a circuitous route on the way out. Gallistel and Cramer [7] describe how an animal constructs a global metric map by converting a landmarks' position vector in the animal's egocentric coordinate system to a position vector in the path integration's geocentric coordinate system.

However the usefulness of this global position in path integration coordinates is limited by its accuracy which gradually worsens as random and systematic errors accumulate. Gallistel and Cramer [7] suggest that to overcome the problem of accumulated error, the animal takes a positional fix computing the discrepancy

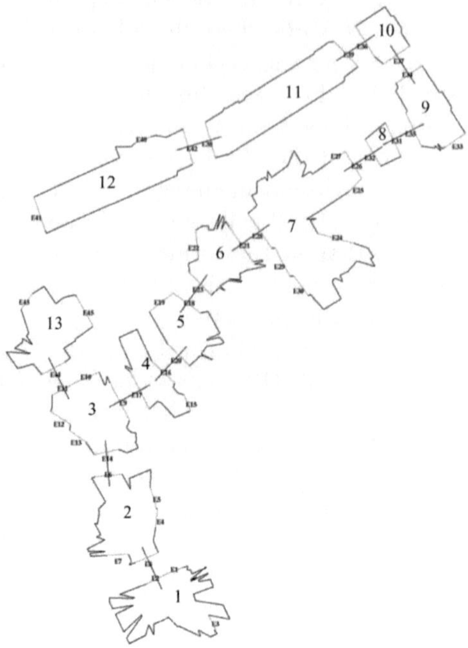

(a)

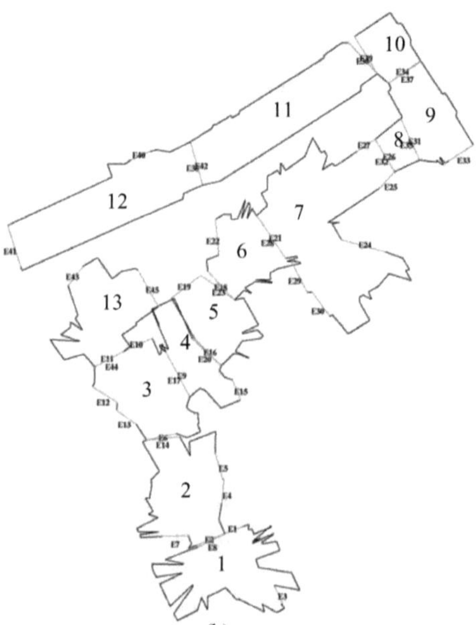

(b)

Fig. 12.5. (a)The topological map. (b) The MFIS. The robot travelled from ASR1 through to ASR12 and back to ASR3 from where it entered ASR3. It is about to re-enter ASR12 from ASR13.

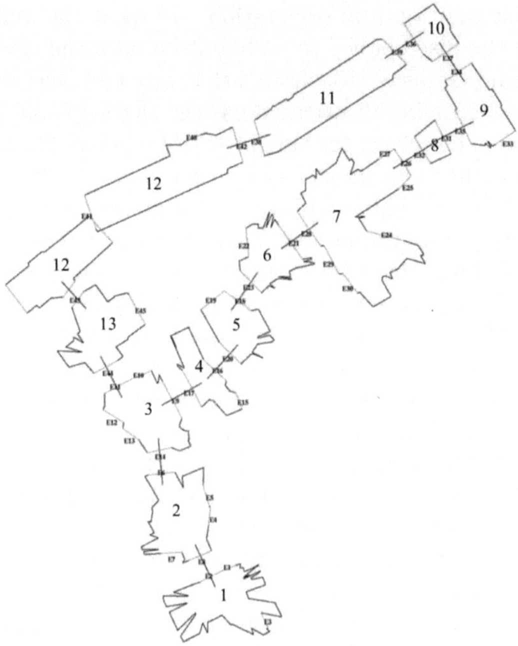

(a)

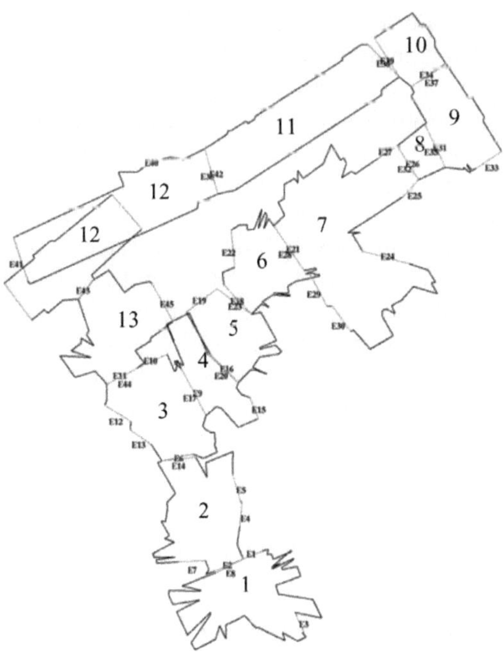

(b)

Fig. 12.6. (a) The topological map. (b) The MFIS. The robot has re-entered ASR12 from ASR13.

between its current position and orientation and what the animal thinks these values are. It uses the discrepancy to realign itself with the environment. Cheng [3] and Margules and Gallistel [13] showed that rats and Hermer [8] showed that young children, on becoming disoriented use the shape of the surrounding environment to reorient themselves. McNaughton [15] reports that research with rats suggests that when the environment is new the animal relies on path integration. As it becomes more familiar with the environment visual features are coded within the path integration framework. These features can be used to confirm or deny the results of the path integration system and can be used to realign the location obtained from path integration with the physical environment.

Several models of the path integration system have been proposed [1, 16, 17, 19]. In some way they all attempt to account for the mathematics of the integration process. Some even attempt to account for random [1] and systematic errors [17]. Redish and Touretzky [19] represent the different locations the rat visits by individual place codes. However, the fact that each one of these place codes is tied to the location represented in path integration coordinates, means that Redish and Touretzky have constructed a global representation of the environment with all the inherent difficulties of error accumulation. Significantly though, they have realised the importance of explicitly tying the path integration system to what is actually observed in the environment. While the definition of the path integration system is that it operates independently of vision, at some point the animal will be aware that where it thinks it is, is not where it really is. To know this, there must be a strong link between the location recorded, the structure of the surrounding space and its identifying features. Redish and Touretzky [19] and McNaughton et. al. [15] address this issue, proposing that the representation for the environment is built around the different locations the animal visits. It is difficult to imagine that an animal could remember its total experienced environment in this way within a single framework. The environments used by Redish and Touretzky in their simulations and by their robot, and those in the experiments McNaughton et. al. refer to, are extremely small compared to the part of the world the animal could be expected to be familiar with. McNaughton et. al., however, suggest that the animal would typically use several reference frames for different parts of its environment, and Redish and Touretzky discuss the significance of multiple reference frames for their model. This notion is closer to the idea we have for a global memory which is limited in size, and which follows the viewer traversing the environment constantly changing the frame of reference.

12.4.2 Robot Systems Combining Topological and Global Metric Maps

A few robotics approaches combine topological and global metric maps. Thrun's [21] global map is an occupancy grid with a single frame of reference. The topological map is derived from the occupancy grid using a partitioning algorithm based on voronoi diagrams. Using the global map the robot can always determine where it is in relation to other parts of its environment. The topological map is useful for planning. Kuipers [11] Spatial Semantic Hierarchy comprises several layers of interacting representations however it is the final two, the topological and metric layers which are relevant here. Kuipers stresses the importance of computationally less expensive and more robust representations, i.e. the topological layer, preceding the more expensive and error prone representations, i.e. the metric layer, in the computational process. To close loops in the topological map Kuipers employs topological matching [10]. They use local metric information to loosely match a robot's representation for the place it is currently at with those of places already visited. If it appears that a match is possible its validity is checked by getting the robot to follow known routes to adjacent places and back to the current place. If the routes can be followed as expected then the current place is identified as the one encountered previously. Franz et al. [5] employ a similar hierarchical scheme. It begins with routes which are then integrated into a "graph-like representation". A survey map, in other words a global metric map, follows when the elements of this graph can be expressed in a single frame of reference. Loop closing occurs with route integration. Like Kuipers, Franz et al. take a conservative approach so as to eliminate false positive matches. When the current view appears to match a previous view the robot tries to follow the route associated with this view to verify the match. Both Kuiper's and Franz et al.'s schemes follow a popular representation scheme propounded by Siegel and White [20] and which argues that landmarks are computed first, followed by routes of connected landmarks and lastly survey knowledge emerges. Such a progression is appealing, especially given the complexity of survey knowledge as opposed to pure landmark and route knowledge. However, we are argue that it is an oversimplification of how the different forms of spatial knowledge emerge. From a computational point of view the information required to compute metric knowledge (local and global) is as abundantly available as that for landmark and route knowledge (see [25] for a discussion on this).

12.5 Conclusion

We have described how a robot can utilise somewhat inaccurate global metric information to detect loops in its topological map. While it provides some useful recognition in its own right it could also be used in conjunction with other approaches to help verify matches, for example landmark matching.

References

1. Benhamou, S.: On systems of reference involved in spatial memory. Behavioural Processes 40, 149–163 (1997)
2. Biegler, R.: Possible uses of path integration in animal navigation. Animal Learning and Behaviour 28(3), 257–277 (2000)
3. Cheng, K.: A Purely Geometric Module in the Rat's Spatial Representation. Cognition 23, 149–178 (1986)

4. Etienne, A.S., Maurer, R., Georgakopoulos, J., Griffin, A.: Dead reckoning (path integration), landmarks, and representation in a comparative perspective. In: Golledge, R.G. (ed.) Wayfinding Behavior - Cognitive Mapping and other Spatial Processes, The Johns Hopkins University Press, Baltimore (1999)
5. Franz, M.O., Sturzl, W., Hubner, W., Mallot, H.: A robot system for biomimetic navigation - from snapshots to metric embeddings of view graphs. In: Proc Workshop on Cognitive and Robotics Approaches to Spatial Mapping (2003)
6. Gallistel, C.R.: The Organisation of Learning. Bradfors Books/MIT Press, Cambridge (1990)
7. Gallistel, C.R., Cramer, A.E.: Computations on metric maps in mammals: getting oriented and choosing a multi-destination route. The Journal of Experimental Biology 199, 211–217 (1996)
8. Hermer, L.: Internally coherent spatial memories in a mammal. Neuroreport 8, 1743–1747 (1997)
9. Jefferies, M.E., Yeap, W.K., Smith, L., Ferguson, D.: Building a map for robot navigation using a theory of cognitive maps. In: Proceedings IASTED International Conference on Artificial Intelligence Applications, Innsbruck, Austria (2001)
10. Kuipers, B., Byun, Y.T.: A robot exploration and mapping strategy based on a semantic hierarchy of spatial representations. Journal of robotics and autonomous systems 8, 47–63 (1991)
11. Kuipers, B.J.: The Spatial Semantic Hierarchy. Artificial Intelligence 19, 191–233 (2000)
12. Loomis, J.M., Klatzky, R.L., Golledge, R.G., Philbeck, J.W.: Human navigation by path integration. In: Golledge, R.G. (ed.) Wayfinding Behavior - Cognitive Mapping and other Spatial Processes, The Johns Hopkins University Press, Baltimore (1999)
13. Margules, J., G.C.R.: Heading in the rat: determination by environmental shape. Animal learning and behaviour 16(4), 404–410 (1988)
14. Maurer, R., Seguinot, V.: What is modelling for? A critical review of the models of path integration. Journal of Theoretical Biology 175, 457–475 (1995)
15. McNaughton, B.L., Knierim, J.J., Wilson, M.A.: Vector encoding and the vestibular foundations of spatial cognition: Neurophysiological and computational mechanisms. In: Gazzaniga, M.S. (ed.) The Cognitive Neurosciences, MA: Bradford, MIT Press, Cambridge (1995)
16. Mittelstaedt, H., Mittlestaedt, M.L.: Homing by path integration. In: Papi, F., Wallraff, H.G. (eds.) Avian Navigatio, Berlin (1982)
17. Muller, M., Wehner, R.: The hidden spiral: systematic search and path integration in ants, cataglyphis fortis. Journal of Comparative Psychology A 175, 525–530 (1994)
18. Burgess, N., Maguire, E.A., O'Keefe, J.: The human hippocampus and spatial and episodic memory. Neuron 35, 625–641 (2002)
19. Redish, A.D., Touretzky, D.S.: Navigating with landmarks: Computing goal locations from place codes. In: Ikeuchi, K., Veloso, M. (eds.) Symbolic Visual Learning, p. 27. Oxford University Press, Oxford (1997)
20. Siegel, A.W., White, S.H.: The Development of Spatial Representations of Large-Scale Environments. Advances in Child Development & Behavior 10 (1975)
21. Thrun, S.: Learning metric-topological maps for indoor mobile robot navigation. Artificial Intelligence 99(1), 21–71 (1998)

22. Wang, R.F., Spelke, E.S.: Updating egocentric representations in human navigation. Cognition 77, 215–250 (2000)
23. Yeap, W.K.: Towards a Computational Theory of Cognitive Maps. Artificial Intelligence 34(3), 297–360 (1988)
24. Yeap, W.K., Jefferies, M.E.: Computing a Representation of the Local Environment. Artificial Intelligence 107(2), 265–301 (1999)
25. Yeap, W.K., Jefferies, M.E.: On Early Cognitive Mapping. Spatial Cognition and Computation 2(2), 85–116 (2000)

13

Using a Mobile Robot to Test a Theory of Cognitive Mapping

Wai K. Yeap, Chee K. Wong, and Jochen Schmidt

Summary. This paper describes using a mobile robot, equipped with some sonar sensors and an odometer, to test navigation through the use of a cognitive map. The robot explores an office environment, computes a cognitive map, which is a network of ASRs [36, 35], and attempts to find its way home. Ten trials were conducted and the robot found its way home each time. From four random positions in two trials, the robot estimated the home position relative to its current position reasonably accurately. Our robot does not solve the simultaneous localization and mapping problem and the map computed is fuzzy and inaccurate with much of the details missing. In each homeward journey, it computes a new cognitive map of the same part of the environment, as seen from the perspective of the homeward journey. We show how the robot uses distance information from both maps to find its way home.

13.1 Introduction

A cognitive map is more than just a representation of the physical environment traversed. It should include, among other things, one's own experiences in the environment. For humans, this includes much of one's high-level interpretations of the environment itself. For this reason, roboticists seldom refer to the map they compute as a cognitive map. Yet, a mobile robot's ability to move about autonomously and sense its environment by various means implies we should use a robot for testing different theories about cognitive mapping. However, researchers interested in developing computational theories of cognitive mapping rarely test their theories on a mobile robot, at least not initially, because implementing such a theory on a mobile robot is not a straightforward task. A few attempts have been made recently [18, 21, 22].

When discussing cognitive maps, one is concerned with epistemic and semantic issues. This is true even at the level close to perception. To illustrate, consider the notion of an egocentric representation of space. Roboticists would think of an egocentric representation of space as the equivalent of any representation computed from the viewpoint of a robot. Furthermore, its transformation to an object-centred representation is nothing more than a straightforward mathematical transformation. The egocentric representation of space becomes a special case of an object-centred representation, where the object is the viewer's body. Such a mathematical interpretation of an egocentric view ignores the fact that a view relates how a cognitive agent perceives its immediate space.

Philosophers such as Evans [10] and Campbell [2] have argued that the importance of an egocentric view is not that the representation is centred on the

M.E. Jefferies et al. (Eds.): Robot. & Cogn. Approach. to Spat. Map., STAR 38, pp. 281–295, 2008.
springerlink.com

self but rather that the representation is action-guiding. Hence, although the representation is described mathematically as centring upon the self, the self is not identified in it. They argued that the actions afforded in each view are what matter most. Similarly, when humans perceive an object-centred representation of space, what is important is the realisation of the self as one of the objects in it and that space exists in an absolute sense. Space becomes something an individual moves into and space contains other objects. Interestingly, and as Pacherie [26] noted, since the self becomes one of the objects in such a representation, it could also then be used as its centre.

Out of a perceptual space, which is a relative view of space, animals (especially humans) are able to conceive space as absolute. Yeap [35] has argued that the cognitive mapping process should begin with the latter, which he referred to as an Absolute Space Representation (or ASR in short). However, investigations to date into the nature of ASR computations [16, 19, 20, 23] have focused primarily on its physical aspects, i. e., information about an ASR that can be derived at a perceptual level. Such information includes its shape, its boundary surfaces, surfaces inside or outside the ASR, exits, etc. Such ASRs are then shown to be interconnected as a network of traversable regions, thus forming a cognitive map of the environment.

From a roboticist's point of view, computing a (cognitive) map in this way is but another method of partitioning the environment into traversable regions. The role such a map plays as a cognitive map is little realized. This is particularly true if the (cognitive) map is then used like a cartographic map; successful use depends very much on the metric accuracy of the ASRs and less so on any kind of heuristic reasoning that animals (especially humans) aptly apply. So, where lies the cognitive sense of the map computed?

Just like the mathematical notion of egocentric representation, computing ASRs is indeed equivalent, mathematically, to the partitioning of space into separately identifiable regions. Since the introduction of the first mobile robots such as Shakey [25], roboticists have devised many such algorithms (see Part I of this book on robot mapping and for some classic examples see [3, 8, 28]). However, just like an egocentric view of space, what matters most in an ASR computation is not the partitioning of space per se into separately identifiable regions for the individual to traverse between them, but the formation of a foundation for developing a much higher-level description of space. If returning to a given place were important, animals could, and have been shown to, evolve more direct algorithms for arriving home (such as the various methods for path integrations, see [7, 9]). An ASR affords the development of one of the most basic elements in cognitive maps, namely the notion of a place.

The word "absolute" in the acronym, ASR, emphasizes the existence of the space itself into which one has entered. One computes a new ASR as soon as one enters a new local space, presumably to quickly identify the new local space into which one has just entered. Depending on the cognitive capabilities of the individual species, the new ASR would rapidly be transformed from a space-sense to a place-sense, i. e., from a purely perceptual sense to an increasingly conceptual one. For example, in humans, once the exits and boundaries are identified, the cognitive mapping process does not stop there. Boundaries could be interpreted as closed doors (which imply they are actually exits) and gaps that are perceived as exits might be interpreted as gaps between, say, two pillars inside the ASR (which imply they are not exits). Surfaces will be interpreted as objects and their functional significance realised. Objects could become landmarks. Events might be unfolding which then draw the attention of the individual further away from attending to the physical qualities of its environment. The notion of a place which begins with a single ASR could become a collection of ASRs, the map a network of places.

Ideally, for cognitive mapping, we have to show how ASRs are turned into places and the map used to solve various spatial tasks. Computing place representations provides a far more useful basis for reasoning about one's environment than computing the physical shape of ASRs alone and can only come with powerful reasoning capabilities, which, for now, the robot lacks. A blind rat can find its way in a maze. What we could investigate now are the different (cognitive mapping) ways in which the map could be used by the robot. Just like birds have a mapping process different from a rat, and rats from humans, robots can have their own cognitive mapping process [1]. By studying the cognitive mapping process of the different species (robots included), we gain better insights into the nature of the cognitive mapping process.

What would a cognitive mapping process of a robot be, given the sensors it has? Note that existing works (e. g., see Jefferies et al. in Chap. 12) have shown how a robot can compute a network of ASRs and use it to navigate in its environment. The ASRs computed are reasonably precise in metric terms and none is missing from the robot's memory. Such a robot is analogous to a cognitive agent navigating in a familiar part of the environment (where things are remembered fairly precisely).

In this paper, we show another example of how a robot computes and uses cognitive maps. In particular, we ask what would the process be like if, at the end of an initial exploration of a new environment, the robot does not have a well-formed network of ASRs? What if some of, or all the ASRs computed are incomplete and contain inaccurate information? Presumably, the robot could not localise itself in the environment using the map computed. What should it do to find its way home? Note that the ASRs computed when finding its way home could be very different from those computed during the initial exploration. How are the different maps, one in the memory and the other currently experienced, being used? How do we set up an experiment with our robot to investigate some of these questions?

The remainder of this paper describes an experiment with a mobile robot doing cognitive mapping. Our mobile robot is equipped with some sonar sensors and an odometer. Sonar sensors provide very inaccurate and unreliable

[1] Strictly speaking, a robot, not being a cognitive agent, cannot have a cognitive mapping process. However, rather than being verbose and say "simulating a cognitive mapping process", we will simply say the robot has a cognitive mapping process.

measurements. They were deliberately chosen so that the network of ASRs computed cannot function as an accurate map. Instead, these ASRs serve more like fuzzy memories of places visited. In the experiment, the fuzziness of the ASR is, of course, due to the poor sensors used. It is meant to simulate the typical kind of fuzzy memory of a place humans recall about the new environment they have just visited.

Section 13.2 describes how we set up our robot to explore its environment and compute fuzzy ASRs. Section 13.3 describes the problem faced by our robot. Section 13.4 describes the algorithms the robot used and the results of our experiment. Section 13.5 concludes with a discussion of cognitive and robotic mapping in the context of our experiments.

13.2 The Robot and Its Cognitive Map

The robot we used is a Pioneer 2 robot from ActivMedia and it came with a ring of 8 sonar sensors. The robot was positioned somewhere in the corridor in an office environment and was allowed to explore the environment until it was told to stop. No modifications to the environment were done. Everything already in the environment (such as rubbish bins, flower pots, cabinets, etc.) remained where they were and doors leading into offices were left as they were found, closed or open, at the time of the experiment.

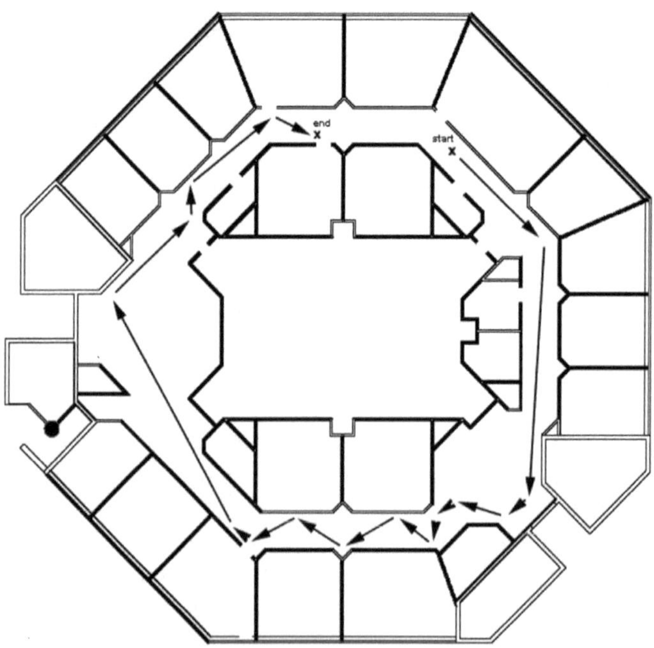

Fig. 13.1. The environment and the path traversed. The total distance traveled is about 70 m.

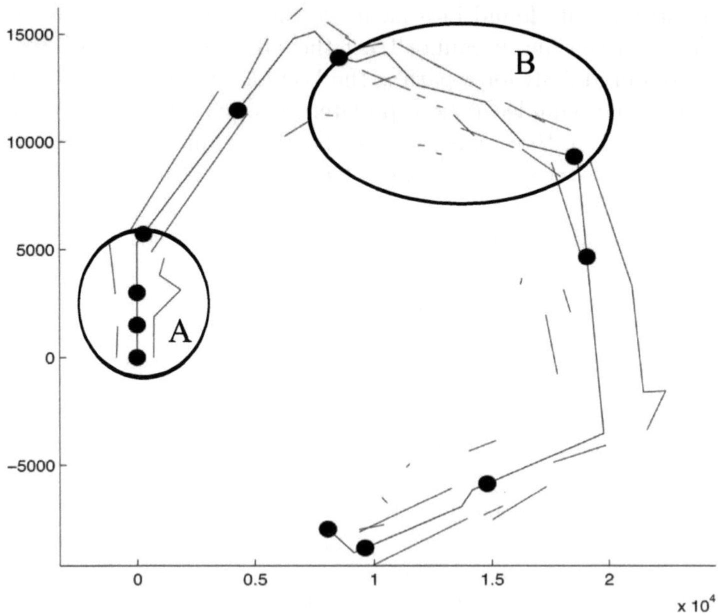

Fig. 13.2. ASRs computed for the journey as shown in Fig. 13.1. An ASR is between two adjacent dots and surfaces located to the left and to the right of the path inside the ASR are its boundary surfaces. The path is the solid line connecting the dots. (0, 0) indicates the starting position of the robot.

The environment used and one of the paths the robot took is as shown in Fig. 13.1. It does not use a wall-following procedure to navigate. It simply moves forward until it could not and then it "looks" for an empty space to move forward again. "Looking" is done using all eight sensors but information about the environment is sensed only via the two side sensors. That is, the robot uses its eight sensors to decide where to move next but only its two side sensors to gather information about the environment. The exploration algorithm used is as follows:

1. move in a "straight" line and collect sonar data from the sides;
2. stop when an obstacle is encountered; and
3. turn away from the obstacle but maintain a forward-going direction

The details of our new algorithm for computing ASRs for the experiments conducted here can be found in [29, 30, 32, 34]. Briefly, the key ideas underlying our new algorithm are:

1. ASRs are computed for each path traversed – a path is a single continuous movement of the robot through the environment (i. e., without any stopping or turning);

2. The important exits found in a path are the exits at both ends of it (i. e., given the poor sensing, it cannot trust the side exits detected). This means that the required ASR for a path is the bounded region for the path;
3. To compute the bounded region, preference is given to using the large surfaces as opposed to the smaller ones – the algorithm thus uses all the larger surfaces, say, greater than 700 mm in length, to compute a boundary. If the resulting boundary is greater than, say, 70% of the distance traveled, then that is an acceptable boundary for the current ASR; If not, more of the small surfaces are added until a reasonably sized boundary is obtained.
4. An ASR computed for a path represents an ASR computed from a single view of the robot. The next step is to merge or split ASRs obtained from individual paths into the final network of ASRs for the environment experienced.

Figure 13.2 shows the final ASRs computed for the journey shown in Fig. 13.1. The start and end point of an ASR are marked with a dark circle. The surfaces in between indicate the rough shape of the ASR computed. Note that in the area marked A the robot is shown moving through that part of the environment in a single path. That single ASR was split into three ASRs for that part of the environment. In the area marked B, it shows the robot moving through it using five paths. These five ASRs are later merged to form the final single ASR. Such merging and splitting is done in step 4 of the algorithm as discussed above.

13.3 The Experiment

You could imagine that our robot is a blind "rat" with a special sense. It could stretch out a pair of imaginary arms from its side to infinity, or until it "believes" its arm touches an object. Consequently, what it senses might not be correct. The object might or might not be there. In this manner, it gathers information about the shape of the environment as it moves down a path. ASRs are then computed from the information gathered. Without vision, ASRs are computed after the robot has left that part of the environment.

Like a typical rat experiment, our experiment is to let our robot wander freely between two points in an office environment, imagining them to be the home and food locations in a maze. The robot must then find its way home and the question is how. The robot has a cognitive map of its environment computed during its outward journey. During its homeward journey, it will compute another cognitive map.

Figure 13.3 shows an example of a network of ASRs generated during one of its homeward journeys. The network is not the same as that computed during the outward journey. In particular, nine ASRs are computed instead of ten and the sixth ASR computed in the outward journey is now perceived as three ASRs in the homeward journey.

Our robot cannot confidently re-compute incoming ASRs and match them with those in memory. Thus, it does not solve the widely accepted problem among robotics researchers, namely the simultaneous localization and mapping problem (famously known as the SLAM problem – see for example, [6]). Our

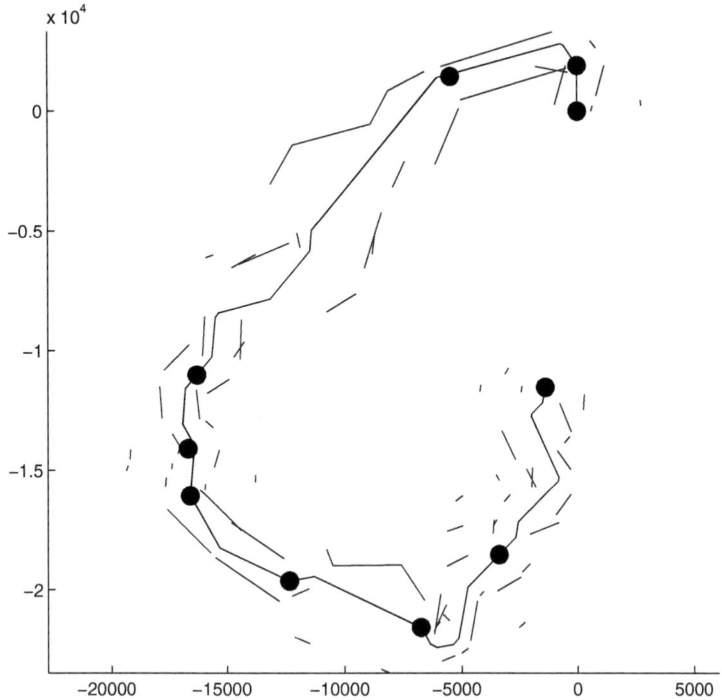

Fig. 13.3. ASRs computed in the homeward journey. (0, 0) indicates the starting position of the robot.

robot is at a very early stage of (cognitive) mapping of its environment. In many studies of cognitive mapping (for example, see [13]), much has been said about the use of distance and direction information in the process. We implemented an algorithm for our robot to find its way home that uses distance information implicit in each cognitive map. Given the cognitive maps computed, how good is the robot's sense of direction? We also implemented an algorithm for the robot to estimate its orientation in its current position to the home position.

13.4 Implementations and Results

Section 13.4.1 presents the implementation of our home-going algorithm that makes use of the ASR distance traveled as opposed to the actual distance traveled by the robot in its zigzag moves home. A total of ten experiments were conducted using this algorithm; the results for two of them are described in more detail. Section 13.4.2 presents an algorithm for the robot to estimate its orientation towards home and the result is compared with its actual orientation from where it is physically located in the environment.

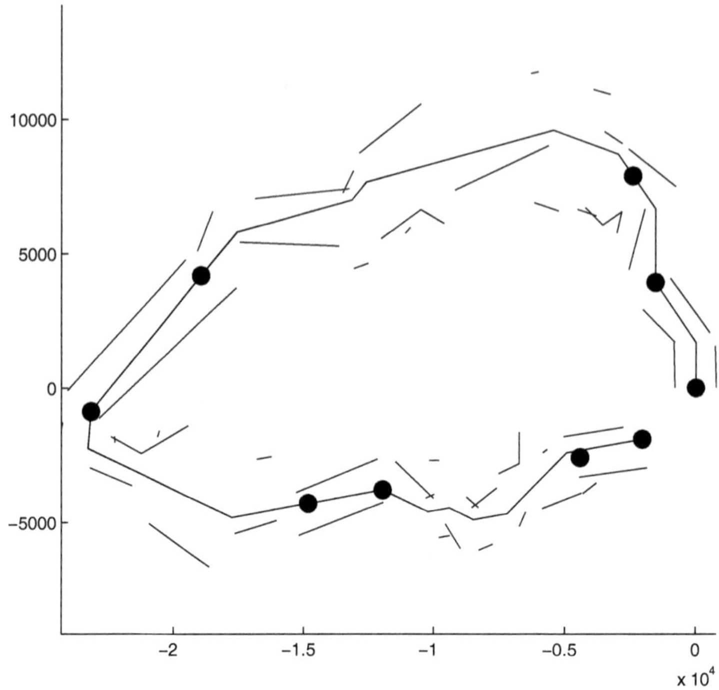

Fig. 13.4. ASRs computed in the homeward journey. (0, 0) indicates the starting position of the robot.

13.4.1 Going Home

The algorithm for returning home is:

1. Compute ASRs (up to the current position) in the homeward journey.
2. Measure the length of each ASR computed (as opposed to the actual distance the robot traveled).
3. Map the ASR-distance traveled onto the network of ASRs computed in the outward journey.

The last step provides an estimate for how far the robot has re-traced its steps towards home. The robot stops when it believes it has completely retraced its steps.

Figure 13.2 shows the cognitive map computed by the robot for its outward journey. Ten experiments were conducted using our "Going Home" algorithm for the robot to find its way home. Figures 13.3 and 13.4 show two cognitive maps computed in two different attempts to go home.

We measured the distance between the robot's final position and the real home position. For the two experiments presented in the figures, the robot was 1.5m short of the true home position in the experiment corresponding to Fig. 13.3, and 1m short for the experiment corresponding to Fig. 13.4. For the remaining

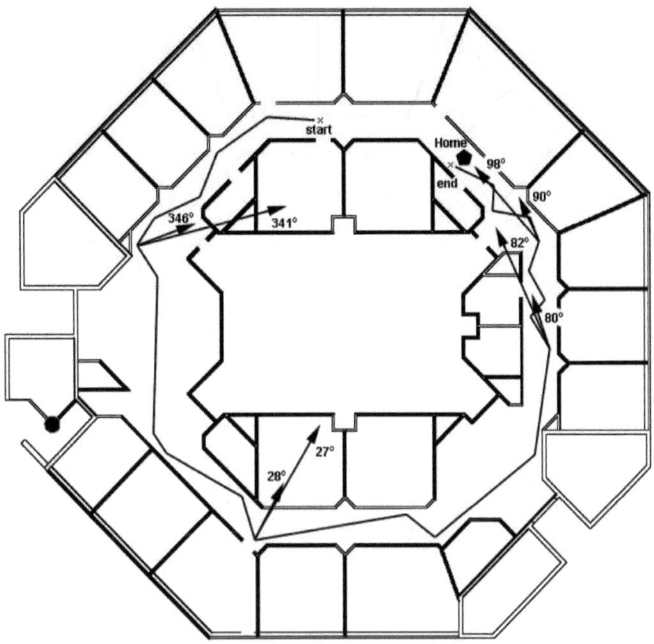

Fig. 13.5. Robot's estimations of home position (indicated by the shorter arrows) at four randomly selected positions during the first homeward journey. The longer arrow shows the correct orientation.

eight experiments, the robot ended within 3m of home, which is less than 5% of the total distance traveled. The robot's positions in the physical environment during the homeward journeys shown in Figs. 13.3 and 13.4, can be seen in Figs. 13.5 and 13.6 in the next section.

13.4.2 Orientation

During the homeward journey, the robot estimates where it is in the cognitive map it computed for the outward journey. It estimates its orientation to home from its current position using the information contained in the outward journey's map. This part of the experiment is to investigate how accurate is the robot's sense of home direction.

The robot can estimate the direction to home at any intermediate position. Four randomly-selected positions were chosen in each of the maps shown in Figs. 13.3 and 13.4, and the estimated home direction from each position was compared to the real world direction.

The results are visualized in Fig. 13.5 (corresponding to Fig. 13.3) and Fig. 13.6 (corresponding to Fig. 13.4), which show a map of the real environment containing the path the robot actually took to return home. The estimated direction to home is depicted as a short arrow, the correct one as a long arrow.

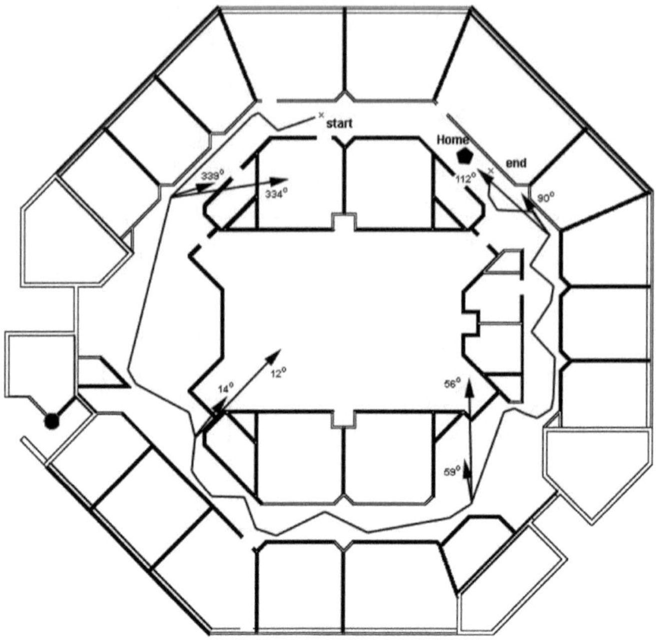

Fig. 13.6. Robot's estimations of home position during the second homeward journey

The estimated and correct angles with respect to the coordinate system of the map computed during the outward journey are given as well.

It can be observed that the direction estimated is fairly accurate; it has not been affected by errors due to odometer measurements and drift.

13.5 Discussion

Our robot has a very limited capability for sensing its environment. It uses cheap sonar sensors (as opposed to the more advanced sonar sensors used in [1, 5, 27] or those with powerful pre-processing software [31]. Nonetheless, even with such limited sensing, we have shown that it is possible to implement an algorithm for our robot to compute a network of ASRs. The notion of an ASR is versatile and is not restricted to having accurate or powerful sensors. Our work thus highlighted one significant difference between partitioning the environment into traversable regions (robot mapping) and into ASRs (cognitive mapping). Robot mapping is more concerned with dividing a large-scale space into smaller ones and their physical qualities whereas cognitive mapping is about the identity and character of each local space visited. For cognitive mapping, the more versatile notion of space is preferred.

We have implemented a very basic algorithm for the robot to find its way home, namely exploit ASR-distance traveled to re-trace its movements to return home. Much has been discussed with respect to the use of distance information in cognitive mapping. For example, numerous experiments with chickens and pigeons have shown that they are able to use both absolute and relative distance in their search for food (see [14]) for an example of recent work). Experiments with bees and ants have shown that they can perform internal calculations of the distance and direction traveled to perform path integration (see [7, 9] for a general discussion). Most of these experiments were concerned with the actual distance traveled and how the individual species deal with the errors in their measurements, as do most work on robot mapping to date. Using our robot, we have shown another way of using distance information, namely ASR-distance traveled as opposed to actual distance traveled. The method appeared to be useful.

ASR-distance is obtained from the shape of the ASR computed. In the past, there has been scant evidence that humans/animals do pay attention to the shape of each local environment (or, in our terminology, ASR) very early on in their initial exploration of a new environment [4, 11, 12, 15, 17, 24]. However, the debate has now intensified and this is especially true in the animal literature where the problem is commonly referred to as geometry in animal spatial behavior (see Cheng in Chapter 6). In a relocation task using a box-shaped environment, the principal axes of the environment appear to be most useful. However, Cheng questioned the general applicability of the principal axes and suggested other ways of utilizing the shape of "ASRs" computed. Our work here emphasized yet another possibility, namely using a straight line distance between exits of interests in an ASR.

Two remarks are worth making regarding the surprisingly good results obtained in our experiment. First, although our robot was allowed to wander on its own during all the trials, it managed not to enter any of the rooms. Consequently, the robot appears to be constantly moving forward along the corridor and this might have accounted for much of the success of the experiment. This was not planned. It would be interesting to see how the resulting ASRs would be if the robot enters the middle room and follows a path such as that shown in Fig. 13.7.

The ASR algorithm would have to be made more powerful so that it could reason about the overall shape of the ASRs computed. If no other kinds of sensors are used, this robot would not be able to learn much about its environment; it could not identify any objects in it. In the future we are planning to add other kinds of sensors (e. g., compass) to our robot to investigate how the extra information made available will enhance the robot's reasoning about its environment.

Second, it is interesting to note that in an earlier experiment [33], the following strategy is used:

1. Do not compute ASRs during the homeward journey.
2. Use the ASRs computed for the outward journey in reverse order.
3. Measure the length of the ASR that the robot thinks it is in and travel similar distances to reach the end of that ASR.

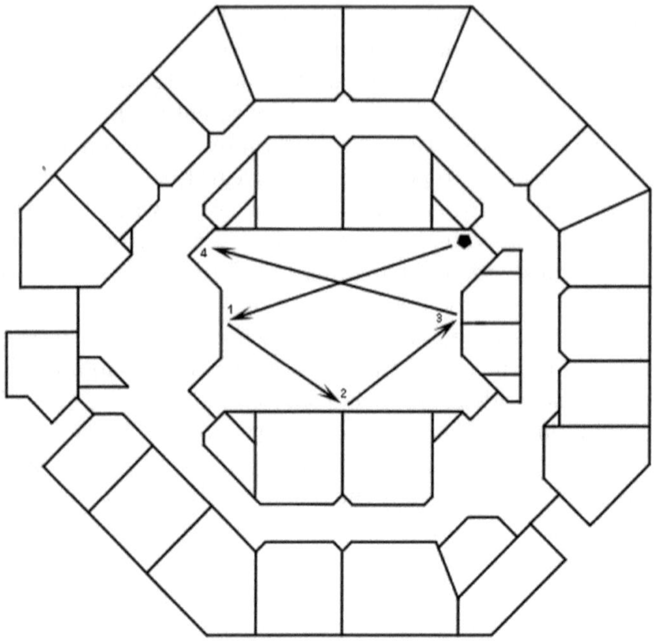

Fig. 13.7. Robot's possible navigation inside a room – Will it compute a single ASR for the room or multiple ASRs?

4. Once it believes it has reached the end of the ASR, search for the entrance to the next ASR. If the next ASR is on its left, turn left. Otherwise turn right.

The robot did not perform as well using the above strategy (4 out of 6 trials were successful). However, it does cause the robot to exhibit some interesting behavior at each decision point (step 4 in the above algorithm). In trying to find the exit, the robot makes small turns and movements, appearing to be cautious in its search for an exit. Figure 13.8 shows an example of a path taken by the robot using the above algorithm.

Given the current restricted paths through the environment and the small number of trials conducted, it is not interesting comparing the performance of the two algorithms. Rather, what is interesting is to observe that the two different algorithms represent two different approaches to using and updating a cognitive map. The first approach is to always compute an ASR from the input and then extract information from it for comparisons or updating with those held in one's cognitive map. The other is to directly use information from the input with those held in one's cognitive map.

We need to investigate many more different strategies before we can understand how the different strategies interact in a cognitive mapping process.

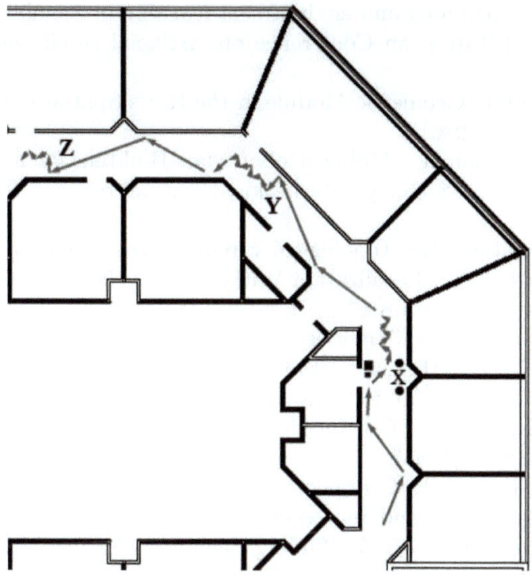

Fig. 13.8. The path the robot took on its way home. The points marked X, Y, and Z are critical decision points in the journey home.

Finally, we implemented an algorithm for the robot to tell us where it believes its goal is from its current position in its cognitive map. We compare that with the actual orientation of the goal from the robot's physical location in the building. The fact that the robot does not forget any of the ASRs along the way might help to explain the robot's ability to accurately orient itself. In the future, we would like to explore how the robot might use the orientation information to compute a short cut to home. It would also be useful to investigate means to orient itself if the network is not well-connected (i. e., with some ASRs missing, for example).

Acknowledgments

This research project is supported by a grant from the Auckland University of Technology. We are grateful to Ewing Caldwell for his comments on an earlier draft.

References

1. Barshan, B., Kuc, R.: Differentiating Sonar Reflections from Corners and Planes by Employing an Intelligent Sensor. IEEE Trans. on Pattern Anlysis and Machine Intelligence 12(6), 560–569 (1990)
2. Campbell, J.: The role of physical objects in spatial thinking. In: Eilan, N., McCarthy, R., Brewer, B. (eds.) Spatial Representation: Problems in Philosophy and Psychology, pp. 65–95. Blackwell Press, Oxford (1993)

3. Chatila, R.: Path planning and environment learning in a mobilerobot system. In: Proc. of the Fifth European Conference on Artificial Intelligence, Paris, France, pp. 211–215 (1982)
4. Cheng, K.: A Purely Geometric Module in the Rat's Spatial Representation. Cognition 23, 149–178 (1986)
5. Chong, K.S., Kleeman, L.: Mobile Robot Map Building for an Advanced Sonar Array and Accurate Odometry. International Journal robotics Research 18(1), 20–36 (1999)
6. Choset, H., Nagatani, K.: Topological Simultaneous Localization and Mapping (SLAM): Towards Exact Localization Without Explicit Localization. IEEE Trans. on Robotics and Automation 17(2), 125–137 (2001)
7. Cornell, E.H., Heth, C.D.: Memories of travel: Dead reckoning within the cognitive map. In: Allen, G. (ed.) Human spatial memory: Remembering where, pp. 191–215. Lawrence Erlbaum Associates, Mahwah, NJ (2004)
8. Crowley, J.L.: Navigation of an Intelligent Mobile Robot. IEEE Journal Robotics Research 1(1), 31–41 (1985)
9. Etienne, A.S., Maurer, R., Georgakopoulos, J., Griffin, A.: Dead reckoning (path integration), landmarks, and representation of space in a comparative perspective. In: Golledge, R.G. (ed.) Wayfinding Behavior: Cognitive Mapping and Other Spatial Processes, pp. 197–228. John Hopkins University Press, Baltimore (1999)
10. Evans, G.: The Varities of Reference. Oxford University Press, Oxford (1982)
11. Galliestel, C.R., Cramer, A.E.: Computations of metric maps in mammals: Getting oriented and choosing a multi-destination route. Journal of Experimental Biology 199, 211–217 (1996)
12. Gallistel, C.R.: Animal Cognition: The Representation of Space, Time and Number. Annual Review of Psychology 40, 155–189 (1989)
13. Golledge, R.G.: Human wayfinding and cognitive maps. In: Golledge, R.G. (ed.) Wayfinding Behavior: Cognitive Mapping and Other Spatial Processes, John Hopkins University Press, Baltimore (1999)
14. Gray, E.R., Spetch, M.L., Kelly, D.M., Nguyen, A.: Searching in the center: Pigeons encode relative distance from walls of an enclosure. Journal of Comparative Psychology 118(1), 113–117 (2004)
15. Hermer, L.: Internally coherent spatial memories in a mammal. Learning and Memory 8, 1743–1747 (1997)
16. Hill, R.W., Han, C., Van Len, M.: Applying perceptually driven cognitive mapping to virtual urban environments. In: Proc. National Conference on Artificial Intelligence, Edmonton, Canada, pp. 886–893. AAAI press, USA (2002)
17. Huttenlocher, J., Newcombe, N., Sandberg, E.H.: The Coding of Spatial Location in Young Children. Cognitive Psychology 27, 115–147 (1994)
18. Jefferies, M.E., Cree, M., Mayo, M., Baker, J.T.: Using 2D and 3D landmarks to solve the correspondence problem in cognitive robot mapping. In: Freksa, C., Knauff, M., Krieg-Brückner, B., Nebel, B., Barkowsky, T. (eds.) Spatial Cognition IV. LNCS (LNAI), vol. 3343, pp. 434–454. Springer, Heidelberg (2005)
19. Jefferies, M.E., Yeap, W.K.: Representing the local space qualitatively in a cognitive map. In: Proc. of the Cognitive Science Conference, Madison, USA, pp. 525–530 (1998)
20. Jefferies, M.E., Yeap, W.K.: The utility of global representations in a cognitive map. In: Proc. Conference on Spatial Information Theory, Morro Bay, US, pp. 233–246 (2001)
21. Kortenkamp, D.: Cognitive Maps for Mobile Robots: A Representation for Mapping and Navigation. PhD thesis, University of Michigan (1993)

22. Kuipers, B., Modayil, J., Beeson, P., MacMahon, M., Savelli, F.: Local Metrical and Global Topological Maps in the Hybrid Spatial Semantic Hierarchy. In: Int. Conf. on Robotics and Automation, New Orleans, USA, pp. 4845–4851 (2004)
23. Laviers, K.R., Peterson, G.L.: Cognitive robot mapping with polylines and an absolute space representation. In: IEEE International Conference on Robotics and Automation, vol. 4, pp. 3771–3776. IEEE Computer Society Press, Los Alamitos (2004)
24. Margules, J., Gallistel, C.R.: Heading in the Rat: Determination by Environmental Shape. Animal Learning Behaviour 16(4), 404–410 (1988)
25. Nilsson, N.J.: A Mobile Automaton: An Application of Artificial Intelligence Techniques. In: Proc. of the First International Joint Conference on Artificial Intelligence, Washington, USA, pp. 509–520 (1969)
26. Pacherie, E.: Perspectival and Objective Representations of Space and Time. In: Proc. of the 4th European Congress of Analytic Philosophy, Lund, Sweden (2002)
27. Peremans, H., Audenaert, K., Van, C.J.M.: A High-resolution Sensor Based on Tri-aural Perception. IEEE Trans. Robot Automation 9(1), 26–48 (1993)
28. Rueb, K.D., Wong, A.K.C.: Structuring Free Space as a Hypergraph for Roving Robot Path Planning and Navigation. IEEE Transactions of Pattern Analysis and Machine Intelligence 9(2), 263–273 (1987)
29. Schmidt, J., Wong, C.K., Yeap, W.K.: A split and merge approach to metric-topological map-building. In: ICPR. Int. Conf. on Pattern Recognition, Hong Kong, vol. 3, pp. 1069–1072 (2006)
30. Schmidt, J., Wong, C.K., Yeap, W.K.: Localisation and mapping with a mobile robot using sparse range data. In: Mukhopadhyay, S.C., Gupta, G.S. (eds.) Autonomous Robots and Agents. Studies in Computational Intelligence, Springer-Verlag, Berlin (to appear, 2007)
31. Tardos, J.D., Neira, J., Newman, P.M., Leonard, J.J.: Robust Mapping and Localisation in Indoor Environments using Sonar Data. International Journal robotics Research 21(4), 311–330 (2002)
32. Wong, C.K., Schmidt, J., Yeap, W.K.: Using a Mobile Robot for Cognitive Mapping. In: IJCAI. International Joint Conference on Artificial Intelligence, Hyderabad, India, pp. 2243–2248 (January 2007)
33. Wong, C.K., Yeap, W.K., Sapiyan, M.: A mobile robot that maps naively but plans intelligently. In: Proceedings of the Artificial Intelligence and Application Conference, pp. 562–567 (2005)
34. Wong, C.K., Yeap, W.K., Schmidt, J.: Computing a Network of ASRs Using a Mobile Robot Equipped with Sonar Sensors. In: RAM. International Conference on Robotics, Automation, and Mechatronics, Bangkok, Thailand, pp. 57–62 (June 2006)
35. Yeap, W.K.: Towards a Computational Theory of Cognitive Maps. Artificial Intelligence 34(3), 297–360 (1988)
36. Yeap, W.K., Jefferies, M.E.: Computing a Representation of the Local Environment. Artificial Intelligence 107(2), 265–301 (1999)

14

A Robot System for Biomimetic Navigation – From Snapshots to Metric Embeddings of View Graphs

Matthias O. Franz, Wolfgang Stürzl, Wolfgang Hübner, and Hanspeter A. Mallot

Summary. Complex navigation behaviour (way-finding) involves recognizing several places and encoding a spatial relationship between them. Way-finding skills can be classified into a hierarchy according to the complexity of the tasks that can be performed [8]. The most basic form of way-finding is *route navigation*, followed by *topological navigation* where several routes are integrated into a graph-like representation. The highest level, *survey navigation*, is reached when this graph can be embedded into a common reference frame.

In this chapter, we present the building blocks for a biomimetic robot navigation system that encompasses all levels of this hierarchy. As a local navigation method, we use scene-based homing. In this scheme, a goal location is characterized either by a panoramic snapshot of the light intensities as seen from the place, or by a record of the distances to the surrounding objects. The goal is found by moving in the direction that minimizes the discrepancy between the recorded intensities or distances and the current sensory input. For learning routes, the robot selects distinct views during exploration that are close enough to be reached by snapshot-based homing. When it encounters already visited places during route learning, it connects the routes and thus forms a topological representation of its environment termed a *view graph*. The final stage, survey navigation, is achieved by a graph embedding procedure which complements the topologic information of the view graph with odometric position estimates. Calculation of the graph embedding is done with a modified multidimensional scaling algorithm which makes use of distances and angles between nodes.

14.1 Way-Finding

The different types of navigation behaviour can be roughly divided into two groups: local navigation behaviours and way-finding behaviours. *Local navigation* behaviours such as aiming, guidance, path integration etc. are used to find a single goal by using only currently available sensory information, without the need of representing any objects or places outside the current sensory horizon [22]. Local navigation requires the recognition of only one location, namely the goal. In the Spatial Semantic Hierarchy (SSH) described by Kuipers in this volume, local navigation corresponds to the continuous control level. *Way-finding* involves the recognition of other places besides the goal, and the representation of relations between these places [19]. It relies on local navigation skills to move

M.E. Jefferies et al. (Eds.): Robot. & Cogn. Approach. to Spat. Map., STAR 38, pp. 297–314, 2008.
springerlink.com

from one place to another, but it allows the animal to find places that could not be found by local navigation alone. In the SSH, way-finding is associated with the causal and topological levels of the hierarchy.

Way-finding behaviours can be further categorized into three subsequent levels: recognition-triggered response, topological navigation and survey navigation.

Recognition-triggered responses. Connect two locations by a local navigation method, i.e., an association between a sensory pattern defining the start location and a motor action. In this context, a location is defined as a certain sensory situation in which the same local navigation method is selected. The recognition of the starting location triggers the activation of a local navigation method leading to the goal. There is no planning of a sequence of subsequent movements, only the selection of the very next action. Thus, the animal responds in an inflexible manner to the current situation. In spite of their apparent simplicity, recognition-triggered responses are already considered as way-finding behaviour since they need the recognition of two places (the start and the goal location), and the encoding of their spatial relation by a local navigation behaviour.

Several recognition-triggered responses can be concatenated to *routes*. Routes are sequences of recognition-triggered responses, in which the goal of one step is the start of the next. The local navigation method can be different in each step according to the local environment. Still there is no planning involved, as knowledge is limited to the next action to perform. If one route segment is blocked, e.g. by an obstacle, the animal has to resort to a search strategy until it reaches a known place again.

Topological navigation. An animal using recognition-triggered responses is confined to always using the same sequences of locations. Routes are generated independently of each other and each goal needs its own route. Navigation is more adaptive if the spatial representation is goal-independent, i.e. if the same representation can be used for navigating to multiple goals. To this end, the animal must have the basic competence of detecting whether two routes pass through the same place. Two possibly different sensory configurations associated with the different routes leading through the same place have to be merged by *route integration*. A collection of integrated routes thus becomes a topological representation of the environment. This can be expressed mathematically as a graph, where vertices represent places and edges represent a local navigation method connecting two vertices.

Any vertex can become the start or the goal of a route, so that, in the case of obstacles, alternative intersecting routes may be found. The fact that alternative routes may lead to one goal requires *planning* abilities which generate routes from the graph. Planning together with route integration are the capabilities required for *topological navigation*. The resulting routes are concatenations of sub-sequences from already visited routes. As a consequence, an animal relying on topological navigation cannot generate novel routes over unvisited terrain.

Survey navigation. Whereas for topological navigation different routes have to be integrated locally, *survey navigation* requires the *embedding* of all known places and of their spatial relations into a common frame of reference. In this process, the spatial representation must be manipulated and accessible as a whole, so that the spatial relation between any two of the represented places can be inferred. In contrast, topological navigation needs only the spatial relations between connected places. An animal using survey navigation is able to find novel paths over unknown terrain, since the embedding of the current location into the common frame of reference allows the animal to infer its spatial relation to the known places. Examples include finding of shortcuts in unknown terrain between unconnected routes, or detours over unknown terrain around obstacles.

Biomimetic navigation. Generally, each level of the navigation hierarchy requires new skills on top of the lower level skills. This could also indicate the direction taken during evolution, since new behavioural capabilities are usually built on pre-existing simpler mechanisms. A distinctive feature of a biomimetic robot way-finding system is, therefore, the use of a hierarchy of competences and their underlying mechanisms that should reflect an "evolutionary scaling" as discussed in [17]. Many navigation approaches in robotics (see, e.g., Thrun or Scheding et al. in this volume) are reminiscent to survey navigation since spatial knowledge is represented in a common global map. This contrasts with the above considerations in which survey navigation is the very last stage of the evolutionary development. Biomimetic approaches are therefore constructed in a bottom-up manner: Higher navigation abilities are used on top of simple, but reliable mechanisms. Sometimes these simpler mechanisms turn out to be sufficient for a given task, so that the higher levels need not to be implemented.

Several biomimetic navigation systems for recognition-triggered responses and topological navigation exist in the literature (see, e.g., Jefferies et al. in this volume, and the overview in [8]). The final step to survey navigation still awaits its robotic implementation. In the following, we present the building blocks for such a robotic survey navigation system that encompasses all three levels of way-finding. Route and topological navigation are already implemented on a mobile robot, survey navigation works so far only in simulations. All experiments were done using a Khepera miniature robot in a toy house arena of approximately $1\mathrm{m}^2$ size. We use a scene-based homing procedure as local navigation method (Sect. 14.2). The implementation and algorithms for the subsequent levels of recognition-triggered response, topological and survey navigation are described in Sects. 14.3, 14.4 and 14.5. We conclude in Sect. 14.6 by discussing the results obtained so far.

14.2 Scene-Based Homing

Bees or ants are able to use visual guidance (scene-based homing) as they find a location which is only defined by its spatial relationship an array of locally visible landmarks (for review, see [5]). The experimental evidence suggests that these insects store a relatively unprocessed snapshot of the surrounding panorama as seen from the goal. Cartwright & Collett [3] developed a computational model that allowed to find the goal by matching the snapshot with the current view. Computer simulations showed that the model could indeed account for the observed search behaviour of honeybees.

This simple form of visual guidance has inspired several robot implementations since no complex scene representations have to be handled to find an inconspicuous goal (overview in [8]). As robots usually move in the open space between obstacles, scene-based homing is especially suitable for robot navigation. Our own approach [10] used unprocessed panoramic images of the light intensities seen at the horizon. Under constant lighting conditions, our robot showed robust homing performance. However, when lighting conditions changed completely between taking the snapshot and homing (as, e.g., from sunlight to artificial illumination), the performance broke down [21]. This suggested that - instead of using unprocessed grey values - one could use a "snapshot" of the distances to the surrounding objects at the goal position since the distance distribution in a scene is invariant under illumination changes. There is also strong evidence that rodents, see e.g. [4], [6], and also humans, e.g. [14], use memorized geometric cues to return to already visited places.

The resulting homing algorithm used inverse distances (disparities) to the surrounding objects as snapshots for computational reasons (cf. Sect. 14.2.1). It showed robust performance with respect to changes in the lighting conditions. However, the area around the goal from which the goal can be found, i.e., the catchment area of the goal, was slightly smaller than in the original, grey-value based scheme [21]. Homing accuracy depends mainly on the noise properties of the imaging device, since a displacement can only be detected if it generates sufficient change in the image. In our experimental setup, this was usually the case at distances from the goal in the range of 1 to 3 cm, depending on the distances of the surrounding landmarks. The size of the catchment area for a single snapshot is mainly determined by the layout of the environment. In our toy house arena, maximum homing distances of 45 cm were achieved. The success rate was 95 % for homing distances smaller than 15 cm, and dropped to 50 % in the range of 20 to 25 cm. In the remainder of this section, we describe the disparity-based homing scheme in detail. Both homing schemes, view-based and disparity-based, are used as local navigation method in the way-finding system described in the subsequent sections.

14.2.1 Disparity Signatures of Places

In order to acquire geometric information of the robot's current place we have built a panoramic stereo sensor. Mounted on top of a Khepera miniature robot, a CCD-camera is directed vertically towards a bipartite conic mirror (see Fig. 14.1(a). It consists of two conical parts with slightly different slopes yielding an effective vertical stereo base line of $\approx 8\,\text{mm}$ (Fig. 14.1(b).

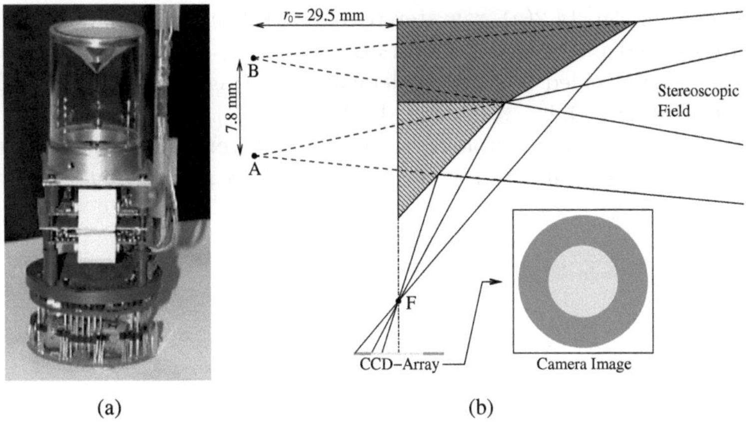

Fig. 14.1. (a) Khepera with panoramic stereo camera on top (diameter ≈ 5 cm, height ≈ 13 cm). (b) Schematic diagram of the bipartite mirror for an axial plane (not to scale). The imaging can be considered as "looking" through two vertically separated points (A, B) which are mirror images of the nodal point of the camera (F). The inset shows the resulting panoramic stereo image: The inner filled circle (light grey) depicts the part imaged through the lower cone; the outer part (dark grey) is imaged through the upper cone.

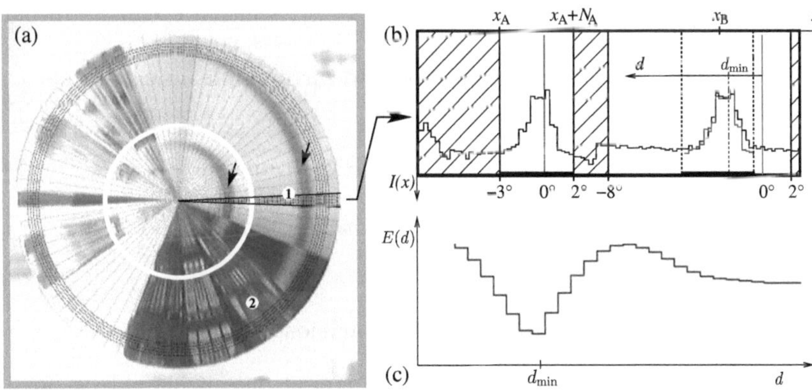

Fig. 14.2. Disparity estimation. (a) Raw stereo image. In the marked sector element (1), a horizontal line on a wall of the toy house arena is imaged twice (arrows). Images of toy houses can be seen in the lower right part (2). (b) Grey values corresponding to the sector element in a. Linear search for maximum correlation (error function plotted in (c) between the inner and outer part yields the disparity. The hatched parts are excluded because of low horizontal resolution in the image center (left) and because of imaging distortions at the transition area of the two different slopes of the mirror (middle).

As depicted in Fig. 14.2(a), raw stereo images, taken by the panoramic stereo sensor, are divided into $N = 72$ sectors (representing a 5° range horizontally). Each sector is subdivided into radial elements resulting in an array of 100 grey-scale pixels $I(x)$, $x = 0, 1, \ldots, 99$ (Fig. 14.2(b).

We have implemented a simple correlation based stereo algorithm to estimate the mean shift d (disparity) of the two image parts by minimizing the matching error (see Fig. 14.2(b), (c),

$$d_{\text{min}} := \arg\min_{d} E_{\text{m}}(d) \tag{14.1}$$

$$E_{\text{m}}(d) := \sum_{x=0}^{N_{\text{A}}-1} \left(I(x_{\text{A}} + x) - I(x_{\text{B}} - d + x)\right)^2 \ , \tag{14.2}$$

where $N_A = 20$ is the width of a window taken from the inner image, x_{B} is the outer image which has zero disparity with respect to x_{A} (start of inner image). Due to the setup of the imaging mirrors only a one-dimensional correspondence search is needed yielding a disparity range of $N_d = 30$ pixels, i.e. $d \in \{0, 1 \ldots, 29\}$.

For each estimated disparity $d_{\text{min},i}$, $i = 0, 1, \ldots, N-1$, we compute a value v ("variance") which depends on the uniqueness and reliability of the found match. Low values of v correspond to high reliability. After the stereo computation, the current place can be represented by $N = 72$ disparities and their corresponding v-values[1], $[\mathbf{d}, \mathbf{v}] = \{(d_i, v_i), i = 0, 1, \ldots, N-1\}$, which we call a "disparity signature" of the considered location.

Using elementary trigonometry, distances to surrounding objects can be computed according to

$$r(d) \approx \alpha/d - r_0 \ , \quad \alpha \approx 2010\,\text{mm} \ , \tag{14.3}$$

where $r_0 = 29.5$ mm is the distance between the effective view points (A, B) and the robot axis (see Fig. 14.1(b).

14.2.2 Homing Algorithm Using Disparities

By comparing the current signature with a stored one, it should be possible to return to the place where the signature has been memorized within a certain neighborhood. For this purpose we have extended the homing algorithm described in [10] for the use of disparities:

Using the current disparity signature $[\mathbf{d}, \mathbf{v}]$, we compute for several possible movements of the robot (rotations about an angle φ followed by a straight move of length l) predicted or expected signatures $[\mathbf{d}^{\text{e}}(\varphi_k, l_k), \mathbf{v}^{\text{e}}(\varphi_k, l_k)], k = 0, 1, \ldots N_{\text{e}} - 1\}$ using (14.3) and trigonometric calculus. In the current implementation the considered positions ($N_{\text{e}} = 132$) lie on a hexagonal grid within a radius of 30 cm.

[1] To simplify notation we omit the index 'min' in the following.

The similarities of the expected signatures to the stored signature at the home position, $[\mathbf{d}^{\mathrm{h}}, \mathbf{v}^{\mathrm{h}}]$, are estimated according to

$$E_{\mathrm{d}}(\varphi_k, l_k) = \min_{s=0,1,\ldots N-1} \frac{\sum_{i=0}^{N-1} w(i,s)\left(d_i^{\mathrm{h}} - d_{i[s]}^{\mathrm{e}}(\varphi_k, l_k)\right)^2}{\sum_{i=0}^{N-1} w(i,s)} , \tag{14.4}$$

$$w(i,s) := \left(v_i^{\mathrm{h}} + v_{i[s]}^{\mathrm{e}}(\varphi_k, l_k)\right)^{-1} ,$$

$$i[s] := (i+s) \bmod N .$$

Subsequently the robot moves to the position $(\varphi_{\mathrm{opt}}, l_{\mathrm{opt}})$, which minimizes (14.4). $(\varphi_{\mathrm{opt}}, l_{\mathrm{opt}})$ is called "homing vector".

To reduce the influence of single wrong decisions, the covered distance is limited to $l < 5\,\mathrm{cm}$. These steps are repeated until the position of highest similarity deviates only marginally from the current position, i.e. $l_{\mathrm{opt}} < l_{\mathrm{thresh}} = 5\,\mathrm{mm}$.

14.3 Route Learning

In our robot implementation, the recognition-triggered responses consist of pairs of panoramic views and scene-based homing steps [9]. The views can be one-dimensional 360° records of either the grey values at the horizon, or of the stereo disparities of the surrounding objects, depending on the used homing scheme. For simplicity, we use the terms snapshot or view for both types of place signatures in the remainder of the text.

The set of snapshots taken to represent a route should satisfy two criteria: First, a large distance should be covered with a small number of snapshots to keep processing requirements small. Second, the spatial distance of neighbouring views should be small enough to allow reliable navigation between them. If one intends to use the learned routes in a topological navigation system, a third criterion has to be added: the views should be distinguishable. In purely view-based routes, this is the only way to guarantee that route integration can be done properly. One way to fulfil this criterion is to incorporate only distinct views into the routes.

The selection of the snapshots is based on the current view and the stored snapshots. The criteria can be fulfilled by measuring the degree of similarity between views: Dissimilar views tend to be distant in space and are distinguishable by definition, and similar views often are spatially close.

Measuring similarity can be viewed as a pattern classification problem. We take a minimalistic approach by using the maximal pixel-wise cross–correlation as a measure of similarity. This is equivalent to the Euclidean distance of two view vectors (containing either grey values or disparities as entries), after first rotating one of them such as to maximize the overlap with the other one. Whenever a threshold of the view distance to all stored snapshots is exceeded by the current view, a new snapshot is taken. The threshold is chosen to ensure that the snapshots are both distinguishable and close enough to allow safe navigation between them. The number of snapshots that can be distinguished using this classifier usually falls in a range between 25 and 40, depending on the start position. Clearly, such a classifier can also be used to detect the proximity of already recorded snapshots and thus allows us to find already visited locations. We use this classifier for both tasks in our topological navigation system (see Sect. 14.4).

Using this simple classifier, the recording of routes is straightforward. If the view distance of the current view to the stored snapshots exceeds a threshold value, the robot takes a new snapshot and connects it to the last one. In this way, the classifier adapts the spacing between the snapshots to the rate of change in the optical input. Thus, areas which have to be covered by a denser net of snapshots, due to a rapid change of views, are also explored more thoroughly. After having taken a snapshot, the robot has to decide where to go next. The simplest conceivable rule is to choose a random direction and then to go straight until the next snapshot. The resulting Brownian motion pattern has the advantage that eventually every accessible point of the environment will be explored without the danger that the exploring agent is caught in an infinite loop. Good results can also be achieved if one uses a fixed turning angle. Using smaller angles distant areas are reached faster, whereas angles closer to π lead to a more thorough exploration of the local neighbourhood.

Distance sensors, together with low-level obstacle avoidance behaviours, are used to keep the robot away from obstacles. Typically, the visual input changes very rapidly near objects. Exploration of these areas thus requires a large number of snapshots which, in complex natural environments, would ultimately lead to a fractal graph structure near objects. To prevent the navigation system from becoming ineffective, the robot is not allowed to take new snapshots if nearby objects are detected by proximity sensors. The resulting routes tend to concentrate in the open space between obstacles.

```
REPEAT {
  compute view distance d of current
    view to all snapshots
  read out proximity detection
  IF no obstacle AND d < threshold THEN
    move into current exploration direction
  IF no obstacle AND d > threshold THEN {
    take new snapshot
    choose new exploration direction
  IF obstacle THEN
    modify exploration direction
  }
UNTIL dead end reached OR
    maxtime between snapshots exceeded
```

After a route has been recorded, using it for route navigation is again straightforward as the route consists of a chain of recognition-triggered responses: starting from the first snapshot in the route, the robot tries to find the next snapshot in the route by scene-based homing. As soon as the current view becomes sufficient similar to the goal snapshot, this event triggers a homing run to the next snapshot in the route as goal. This procedure is continued until the last snapshot of the route is reached.

14.4 View Graph

As pointed out in the introduction, a topological navigation system needs the capability of route integration to form a graph-like representation of the environment. In our case, detecting whether two routes run through the same place amounts to detecting identical views in two different routes. This, however, can only be done if all recorded views are unique. In our system, this is ensured by the view classifier which allows only sufficiently distinct snapshots to be recorded. The resulting graph-like representation is termed *view graph* [20] (see also Mallot et al. in this volume) with snapshots as vertices and connections traversable by scene-based homing as edges.

In principle, connecting two routes whenever two views are sufficiently similar would be enough for route integration. This, however, turned out to be sensitive to false positives since the views at the low resolution used by the robot tend to be similar in several places in the toy house arena. Therefore, we resorted to a more cautious strategy: Whenever the view distance between the current view and an unconnected snapshot drops below a threshold, the robot decides to home to this snapshot. If homing is successful, route integration is performed, i.e., a newly learnt edge is included into the graph. In cases where the robot gets lost or bumps into obstacles, we start a new exploration run, which will typically get connected to the old one in due course. Thus, the classifier has two tasks in our system: to decide when to take snapshots and to detect candidates for overlaps between routes.

Our navigation scheme is designed such that all vertices of the view graph remain in the catchment areas of their respective neighbours. This property can be used to choose the next exploration direction after a successful route integration: The system determines the directions of all neighbouring vertices and directs the next exploration step to the largest open angle. In addition, we use a several other routines that basically limit the connectivity of the vertices and prevents intersection of edges. This leads to an exploration behaviour that tends to concentrate on the least explored regions of the view graph, i.e., regions with a smaller number of snapshots and less connections between them. Further details can be found in [9].

The main loop of the route learning algorithm has to be expanded accordingly:

```
REPEAT {
  :
  compute view distance d2 to all
    unconnected snapshots
  IF no obstacle AND d2 < threshold THEN {
    home to snapshot
```

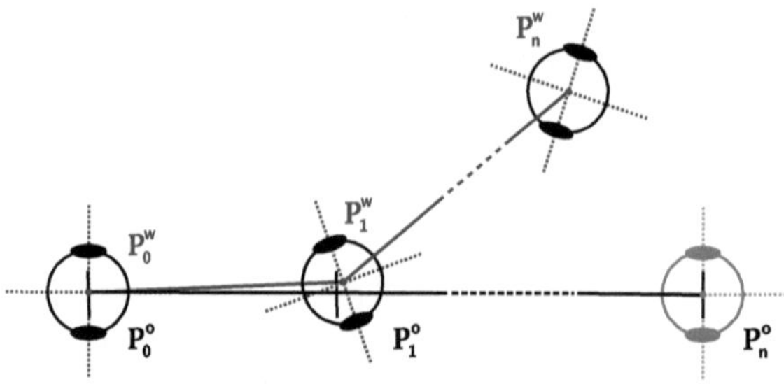

Fig. 14.3. Formation of random odometry errors. This figure illustrates the formation of random odometry errors during a translation. Due to non–systematic factors [1] the agent deviates from the straight path (black path). The instantaneous pose is derived from the integration of many small wheel revolutions, determined by the odometers resolution (indicated by the time index). It is assumed that each of these small steps is affected by a small translational and rotational error, causing the deviation (red path) from the intended path (black path). The deviation between the assumed position ($\mathbf{P}_n^o$) and the real world location ($\mathbf{P}_n^w$) could be corrected by scene based homing.

```
    IF vertex reached THEN {
      connect routes
      compute new exploration direction
    }
    ELSE start new graph
  :
  }
UNTIL ...
```

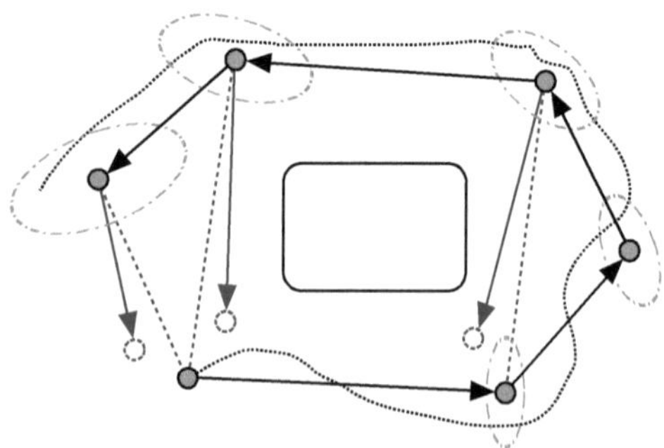

Fig. 14.4. Generalizing from path integration to pose relation networks. The agents odometry is used to measure recording poses of snapshots (gray dots) and their uncertainties (dashed ellipses) along a trajectory (dashed curve). Path integration is generalized to a pose relation network by adding additional links (dashed red lines). Estimating global poses, by integrating local pose relations along different routes may lead to inconsistent pose estimates for some nodes (red arrows).

For using the recorded view graph for topological navigation, one needs an additional planning module that can generate routes between a chosen starting view and a goal view. This can be achieved by standard graph search algorithms, e.g., as described in [20]. The generated routes can be navigated by using the route navigation module described in the last section.

The recorded view graphs typically contained 20 to 50 snapshots and 30 to 60 edges, covering about two thirds of the toy house arena. Since we required the snapshots to be distinguishable, a single graph could cover only areas with unambiguous view information. This general problem of topological navigation is known as *perceptual aliasing* [18]. One way to cope with this problem is to use context information, e.g., by embedding the view graph into a metric map with the help of additional metric information from path integration. This leads us to the final layer of the navigation hierarchy: survey navigation.

14.5 Metric Embedding of a View Graph

A possible way to distinguish between similar views seen at different locations is to label the snapshots with their respective recording poses. The consistent embedding of this pose information into a global metric map gives the agent the ability to perform survey navigation, i.e. the agent is able to find shortcuts apart from the learned routes. In the following, we assume that the robot collects pose information from its odometry in addition to the snapshots, such that each vertex of the view graph contains a snapshot and an odometric pose estimate.

To account for odometry errors, the state of the odometer is modeled as a three dimensional normal distribution $C_{\mathbf{P}_t}(\mathbf{P})$, with the assumed instantaneous pose ($\mathbf{P}_t = (x_t, y_t, \phi_t)^\top$) as the mean and the co–variance matrix $C_t \in I\!R^{3\times 3}$. Parameters for the distribution have to be updated after each movement of the robot. It is assumed that the robots movements are given as a sequence $M = \{\varphi_0, l_0, \ldots, \varphi_t, l_t\}$ of translations and rotations. Figure 14.4 illustrates the formation of random errors for one translation, using a simple model for a two wheel drive, as used in the Khepera–robot. These random errors have to be propagated along the sequence M in order to get an estimate of $\mathbf{P}_t$ and C_t. Using odometry and vision, the total state vector of the agent is extended to $\mathbf{S}_t = (\mathbf{I}_t, x_t, y_t, \phi_t)^\top$, where $\mathbf{I}_t$ is the instantaneously perceived snapshot.

If the robot returns to an already known place by scene-based homing, it closes a loop in the graph. Considering the cumulative error in the robot's odometry, it is clear that a simple vector addition will lead to erroneous position estimates along the path and to contradicting position estimates at the starting vertex (see Fig. 14.4). Instead of calculating path integration along single paths we use a graph–embedding procedure which takes all available routes into account and prevents the accumulation of errors [15].

14.5.1 Multidimensional Scaling

Multidimensional scaling (MDS) problems [2] are directly related to the problem of deriving globally consistent pose estimates from uncertain local pose relations.

A local pose relation is the change in the pose vector between two nodes (v_i, v_j), i.e. $\mathbf{\Delta P}_{ij} = \mathbf{P}_j - \mathbf{P}_i$. In this context, the MDS problem is defined as follows: Given a set of local pose relations $(D = \{\mathbf{\Delta P}^m_{ij} = (\Delta x^m_{ij}, \Delta y^m_{ij}, \Delta\omega^m_{ij})^\top\})$[2], what is the most probable global pose configuration $(X_0 = \{\mathbf{P}_i\})$ fitting into the local relations.

Mathematical this is described by a least square minimization:

$$Q(X,D) = \sum_{(i,j,m)} (\mathbf{P}_j - \mathbf{P}_i - \mathbf{\Delta P}^m_{ij})^\top C^{-1}_{ij} (\mathbf{P}_j - \mathbf{P}_i - \mathbf{\Delta P}^m_{ij}),$$
$$\text{and} \quad X_0 = \arg\min_X Q(X,D) \tag{14.5}$$

Since MDS solutions are only unique except for rigid body transformations [2] of the whole pose configuration, it is necessary to setup a fixed reference frame. Therefore, the first node is always used as the origin $(\mathbf{P}_0 = (0,0,0)^\top)$ and the second node is located on the x–axis $(\mathbf{P}_1 = (x_1, 0, \phi_1)^\top)$. The remaining vertices will be consistently integrated into the ego–centric reference frame, spanned by the first two vertices.

MDS solutions usually use local metric relations like distances or angles in order to define an error term for the least square minimization. In order to incorporate angles, it is necessary to derive pose relations over two edges sharing a common vertex. For three vertices v_i,v_j and v_k, with $(v_i, v_j) \in E$ and $(v_j, v_k) \in E$, two local pose relations, $\mathbf{\Delta P}_{ij}$ and $\mathbf{\Delta P}_{ik}$, are recorded.

The dot product is used as a dissimilarity measure, accounting for distances and the inner angle of the triangle patch spanned by the three vertices:

$$s(\mathbf{\Delta x}^m_{ij}, \mathbf{\Delta x}^m_{ik}) = \mathbf{\Delta x}^m_{ij} \circ \mathbf{\Delta x}^m_{ik} = \|\mathbf{\Delta x}^m_{ij}\| \|\mathbf{\Delta x}^m_{ik}\| \cos \alpha_{ijk} \tag{14.6}$$

The cross product is used in the same way, in order to make the inner angle unique

$$d(\mathbf{\Delta x}^m_{ij}, \mathbf{\Delta x}^m_{ik}) = \det(\mathbf{\Delta x}^m_{ij}, \mathbf{\Delta x}^m_{ik}) = \|\mathbf{\Delta x}^m_{ij}\| \|\mathbf{\Delta x}^m_{ik}\| \sin \alpha_{ijk} \tag{14.7}$$

Altogether, objective function (14.5) becomes:

$$\begin{aligned} Q_p(X,D) = & \sum_{(i,j,k,m)\in D} \frac{1}{\sigma^2_{s^m_{ijk}}} \left[s(\mathbf{x}_j - \mathbf{x}_i, \mathbf{x}_k - \mathbf{x}_i) - s(\mathbf{\Delta x}^m_{ij}, \mathbf{\Delta x}^m_{ik})\right]^2 + \\ & \frac{1}{\sigma^2_{d^m_{ijk}}} \left[d(\mathbf{x}_j - \mathbf{x}_i, \mathbf{x}_k - \mathbf{x}_i) - d(\mathbf{\Delta x}^m_{ij}, \mathbf{\Delta x}^m_{ik})\right]^2 = \\ & \sum_{(i,j,k,m)\in D} E(\mathbf{x}_i, \mathbf{x}_j, \mathbf{x}_k, \mathbf{\Delta x}^m_{ij}, \mathbf{\Delta x}^m_{ik})^2 \end{aligned} \tag{14.8}$$

Equations (14.6) and (14.7) could be interpreted as local dissimilarity measures, which are invariant under rigid body transformations. Therefore, this method

[2] The index m indicates the possibility to record different local measurements for one edge.

differs from other dissimilarity measures, e.g. used in [16, 7, 13, 11], which use coordinate transformations between local and global reference frames in order to define the mismatch between local measurements and the global map. Uncertainties $(\sigma^2_{s^m_{ijk}}, \sigma^2_{d^m_{ijk}})$ for local pose relations used in (14.8) depend on the trajectory followed by the agent while moving between the vertices. Therefore, recordings taken from straight paths are more certain than recordings taken during a homing trail.

Since (14.8) is independent of the global heading, it is possible to separate position estimates from estimating recording directions of the snapshots. Instead of assuming a compass, it is possible to estimate the global heading from allocentric landmark information, again by formulating the estimation process as an optimization problem. The question is, what is the best set of recording directions $\Phi_r = \{\phi_{r1}, \ldots, \phi_{rn}\}$ fitting into the set of local pose changes $\Omega = \{\Delta\omega^m_{ij}\}$. The solution is found by minimizing the following objective function:

$$Q_r(\Phi, \Omega) = \sum_{(i,j,m)\in\Omega} \frac{1}{\sigma^2_{\Delta\omega^m_{ij}}} \left[(\phi_{rj} - \phi_{ri} - \Delta\omega^m_{ij}) \bmod 2\pi\right]^2 \qquad (14.9)$$

In (14.9) it is assumed that the local heading changes $(\Delta\omega^m_{ij})$ result from a motion, which always turns the agent back into the original recording direction after a successful homing trial, i.e. $\Delta\omega^m_{ij} = \Delta\phi^m_{ij} + \mu^m$. The additional rotation (μ^m) could be derived from the home snapshot $(\mathbf{I}^{\mathrm{h}})$ and the instantaneous snapshot $(\mathbf{I}_t)$ according to:

$$\mu^m = (\phi^w_t - \phi^w_r) \bmod 2\pi = \arg \max_{\phi_s\in(0,2\pi)} (\mathbf{I}_t \circ \mathbf{I}^{\mathrm{h}}(\phi_s)) \qquad (14.10)$$

ϕ^w_r and ϕ^w_t are the true orientations in the world–coordinate system, which in fact are unknown to the agent. Equation (14.10) allows the difference between both values to be measured, which in the ideal case is only limited by the visual resolution of the snapshots.

14.5.2 Application to Large Graphs

The method described in the previous section is applicable only in the case where all local measurements are available. Since local measurements are collected during exploration and the complexity of minimizing (14.8) increases with the number of stored locations, a direct minimization of (14.8) is not applicable. It is crucial to have a good estimate of the map on hand any time during exploration, since the path integrator requires regular recalibration. This is often referred to as the problem of "Simultaneous Localization and Mapping" (see e.g. [12]).

In [7, 11] a relaxation method has been used to iteratively approximate the minimum of the objective function. Following this idea we rewrite the original objective function (14.8) in the following way:

$$Q(X_f, X_v, D') = \sum_{(i,j,k,m)\in D'} E(\mathbf{x}_i, \mathbf{x}_j, \mathbf{x}_k, \mathbf{\Delta x}^m_{ij}, \mathbf{\Delta x}^m_{ij})^2 \qquad (14.11)$$

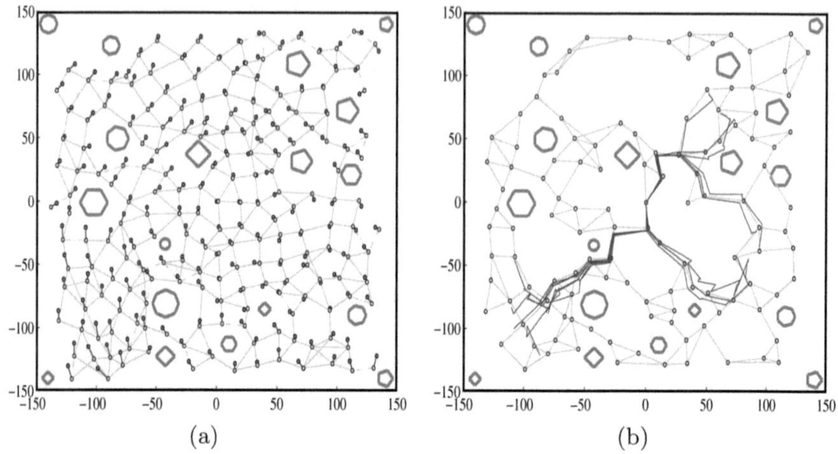

Fig. 14.5. Final map and clustering. (a) The final graph covers an environment of 9m^2 with a total of 191 nodes. Gray dots indicate the estimated node positions. Red dots indicate the true recording locations. (b) This figure shows trajectories from the route following behavior based on a graph, which has been extracted from the graph shown in Fig. (a). Edges in the reduced graph may correspond to shortcuts (see text for explanation). It can be seen that the agent corrects deviations from the planned course during route following.

$E(.)$ is the same error term as in (14.8). $X_v \subseteq X \backslash X_f$ is a set of vertices for which new position estimates are calculated according to: $\Delta X_v = \arg\min_{X_v} Q(X_f, X_v, D')$. $X_f \subseteq X \backslash X_v$ is a set of fixed vertices, which build a reference frame for locally integrating the vertices X_v. Depending on the available local measurements, the choice of X_f could be determined by X_v, so that X_f consists of all vertices which could make predictions of the locations of vertices in X_v. D' is a subset of D, selected according to the vertices $X_v \cup X_f$. As applied here, X_v consists, in one iteration step, only of the vertex for which a new local pose relation has been recorded. X_f includes all first and second neighbors of X_v.

After position estimates for X_v have been updated, the above procedure is repeated for all neighbors of X_v which have been significantly moved. This iteration cycle is aborted if the total movement of vertices is below a certain threshold, i.e. if the gradient of objective function (14.8) has entries close to zero for all elements of X_v.

14.5.3 Results: Exploration and Shortcuts

We use a simulation of a Khepera–robot in order to test the ability to explore environments which are much larger than the toyhouse–arena. Parameters for the simulated odometry (i.e. $\sigma^2_{s^m_{ijk}}$,$\sigma^2_{d^m_{ijk}}$ in (14.8) were adapted to the real robot. First, the odometry of the real robot has been calibrated in order to remove systematic errors [1]. Then, the remaining random errors were used to determine the parameters for the motion model (see Fig. 14.4).

Exploration is done in the same way as in topological navigation. In addition to the view distance, a metric distance to the recorded snapshots $d_M(\mathbf{x}_t, \mathbf{x}_i) = \min_{i \in V} ||\mathbf{x}_t - \mathbf{x}_i||$ is calculated at each time step. The classifier which decides when to take a new snapshot now also takes the metric distance into account. Even if the view distance stays below a certain threshold, a new vertex is added if d_M is greater than 18cm. After adding a new vertex the agent selects a new exploration direction by rotating about a fixed angle of 85°. As before, route integration is performed by homing when both view and metric distance fall below a certain threshold. After a successful homing run, a new edge is added. Finally, the position estimates are updated with the MDS–algorithm and the path integrator is reseted to the improved position estimation.

In the pseudocode examples given above, survey navigation leads to a further expansion of the main loop (besides complementing the view classificator with metric distances):

```
REPEAT {
  :
  IF no obstacle
     AND view distance < threshold1
     AND metric distance < threshold2 THEN {
    home to snapshot
    IF vertex reached THEN {
      connect routes
      compute new exploration direction
      update all positions using MDS-relaxation
      recalibrate path integrator
    }
  :
  }
UNTIL ...
```

Figure 14.5(a) shows the resulting map after an area of 9m^2 has been explored. Gray dots indicate the true recording locations, red dots illustrate locations estimated by the graph embedding procedure. Due to the combined homing scheme (see Fig. 14.4), i.e. the combination of path integration and scene based homing, position errors can be tolerated, as long as the estimated metric home vector guides the robot into the catchment area of the target snapshot.

Path planning, especially the calculation of shortcuts over graph meshes is problematic, since the embedded view graph does not contain information about the location of obstacles. Therefore, if the robot had to avoid obstacles near a vertex, this vertex has been labeled as an “obstacle vertex”. Figure 14.5(b) shows a subgraph, which has been extracted from Fig. 14.5(a). The node set

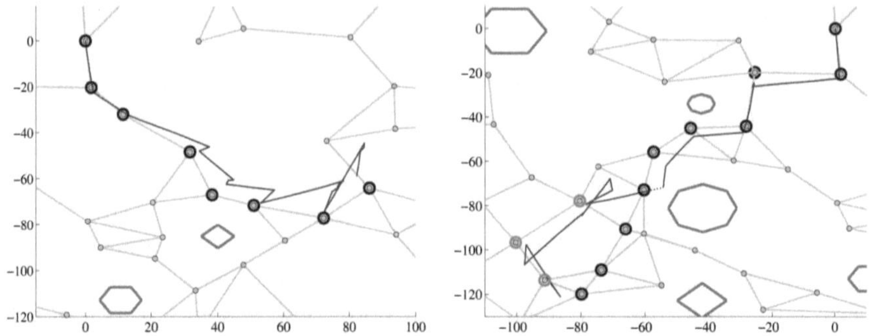

(a) (b)

Fig. 14.6. Example trajectories. These figures show two example trajectories taken from Fig. 14.5(b). Blue dots indicate the original planned route from the origin to the target node. (a) This example illustrates the ability of course corrections by scene based homing. (b) This example shows the ability to replan routes (green nodes) after avoiding an obstacle (dotted part of the trajectory).

of the reduced graph consists for the most part of "obstacle vertices"[3]. Each edge, which has been added from Fig. 14.5(a) to 14.5(b) represents a shortcut. Shortcuts are located more frequently in open space, since the reduced graph mostly consists of vertices near obstacles. This is an important property, since navigation in narrow passages requires more frequent recalibration than in open space.

In order to test the shortcut ability, the agent's task was to follow routes from the center node to a set of randomly selected target locations. The resulting trajectories are illustrated in Fig. 14.5(b). Figures 14.6(a) and 14.6(b) show two of these trajectories in more detail, illustrating course correction and replanning capabilities. The blue nodes in Fig. 14.6(a) show the planned path. Deviations from the route occur for three reasons. First, due to odometry errors the agent is not able to follow precisely a calculated path. Second, the global pose estimates are still erroneous (see Fig. 14.5(a)). Third, the scene-based homing algorithm has a limited spatial resolution. Therefore, the path integrator is not accurately rested at intermediate vertices and furthermore the trajectory does not end precisely at the desired goal location.

Figure 14.6(b) shows a second example where the agent tries to move through a small pathway and hits an obstacle. After avoiding the obstacle (dotted part of the trajectory) the agent relocalizes on the map and calculates an alternative route (green nodes) to the target location.

[3] The graph shown in Fig. 14.5(b) has been generated in three steps. First, the induced subgraph [23], with respect to the "obstacle nodes" has been calculated. Second, some additional vertices have been added, in order to allow localization in open space. Third, new edges have been added in order to make the graph connected. The last two steps have been done manually. Automatic generation of such reduced graphs is the issue of ongoing work.

14.6 Concluding Remarks

In the above section, we have presented all building blocks for a biomimetic survey navigation system, from scene-based homing as local navigation method to metric embedding of view graphs. All levels up to topological navigation have been implemented on mobile robots, only the final level, survey navigation, still runs only in a Virtual Reality environment.

The system is biomimetic in the sense that all behaviours of the navigation hierarchy are implemented in a bottom-up manner, such that each level of the hierarchy relies on the capabilities of the lower levels. Moreover, all the implemented behaviours can be observed in nature, although the biological algorithms and neural implementation will certainly be different from ours. In this sense, our work is not intended to model a specific animal, but to test whether the hierarchical layering of navigation behaviours can lead to a functional robot navigation system.

Although it is still a long way until such a biomimetic system can be used in practical applications, there are already some features that might be interesting from an engineering point of view: First, simple behaviours tend to be robust with respect to non-stationary environments and sensor errors. This inherent robustness is propagated in a certain sense to the higher layers since these are based on them and do not add low level behaviours on their own. Second, the lower layers provide a backup solution when higher levels fail. For instance, when the global metric map becomes incorrect, the robot still can use the graph structure to find its goal.

Acknowledgements

The authors wish to thank the workshop organizers for giving us the opportunity to participate. Financial support was provided by the Max-Planck-Gesellschaft and the Deutsche Forschungsgemeinschaft.

References

1. Borenstein, J., Feng, L.: Measurement and correction of systematic odometry errors in mobile robots. IEEE Transactions on Robotics and Automation 12(5) (1996)
2. Borg, I., Groenen, P.: Modern Multidimensional Scaling. Springer, New York (1997)
3. Cartwright, B.A., Collett, T.S.: Landmark learning in bees. J. Comp. Physiol. A 151, 521–543 (1983)
4. Cheng, K.: A purely geometric module in the rat's spatial representation. Cognition 23, 149–178 (1986)
5. Collett, T.S.: Landmark learning and guidance in insects. Phil. Trans. R. Soc. Lond. B 337, 295–303 (1992)
6. Collett, T.S., Cartwright, B.A., Smith, B.A.: Landmark learning and visuo-spatial memories in gerbils. J. Comp. Physiol. A 158, 835–851 (1986)

7. Duckett, T., Marsland, S., Shapiro, J.: Fast, on–line learning of globally consistent maps. Autonomous Robots (12), 287–300 (2002)
8. Franz, M.O., Mallot, H.A.: Biomimetic robot navigation. Robotics and Autonomous Systems 30, 133–153 (2000)
9. Franz, M.O., Schölkopf, B., Mallot, H.A., Bülthoff, H.H.: Learning view graphs for robot navigation. Autonomous Robots 5, 111–125 (1998)
10. Franz, M.O., Schölkopf, B., Mallot, H.A., Bülthoff, H.H.: Where did I take that snapshot? Scene-based homing by image matching. Biol. Cybern. 79, 191–202 (1998)
11. Frese, U., Duckett, T.: A multigrid approach for accelerating relaxation-based slam. In: RUR 2003. Proc. IJCAI Workshop on Reasoning with Uncertainty in Robotics, pp. 39–46 (2003)
12. Frese, U., Hirzinger, G.: Simultaneous localization and mapping – a discussion. In: Fox, D., Saffiotti, A. (eds.) RUR 2001. Proceedings of the IJCAI–2001 Workshop on Reasoning with Uncertainty in Robotics (2001)
13. Golfarelli, M., Maio, D., Rizzi, S.: Elastic correction of dead–reckoning errors in map building. In: IROS 1998. Proceedings of the IEEE/RSJ International Conference on Intelligent Robots and Systems, Victoria, Canada, pp. 905–911 (1998)
14. Hermer, L., Spelke, E.S.: A geometric process for spatial reorientation in young children. Nature 370, 57–59 (1994)
15. Hübner, W., Mallot, H.A.: Integration of metric place relations in a landmark graph. In: Dorronsoro, J.R. (ed.) ICANN 2002. LNCS, vol. 2415, pp. 825–830. Springer, Heidelberg (2002)
16. Lu, F., Milios, E.: Globally consistent range scan alignment for environment mapping. Autonomous Robots 4, 333–349 (1997)
17. Mallot, H.A.: Behavior-oriented approaches to cognition: Theoretical perspectives. Theory in Biosciences 116, 192–220 (1997)
18. Owen, C., Nehmzow, U.: Landmark-based navigation for a mobile robot. In: Pfeifer, R., Blumberg, B., Meyer, J.-A., Wilson, S.W. (eds.) From animals to animats, Proc. SAB 1998, vol. 5, pp. 145–240. MIT Press, Cambridge, London (1998)
19. Prescott, T.J.: Spatial representation for navigation in animats. Adaptive Behaviour 4, 85–123 (1996)
20. Schölkopf, B., Mallot, H.A.: View-based cognitive mapping and path planning. Adaptive Behavior 3, 311–348 (1995)
21. Stürzl, W., Mallot, H.A.: Vision-based homing with a panoramic stereo sensor. In: Bülthoff, H.H., Lee, S.-W., Poggio, T.A., Wallraven, C. (eds.) BMCV 2002. LNCS, vol. 2525, pp. 620–628. Springer, Heidelberg (2002)
22. Trullier, O., Wiener, S.I., Berthoz, A., Meyer, J.-A.: Biologically based artificial navigation systems: review and prospects. Progress in Neurobiology 51, 483–544 (1997)
23. West, D.B.: Introduction to graph theory. Prentice–Hall, Inc., Englewood Cliffs (1996)

15

Robots as Tools for Modelling Navigation Skills – A Neural Cognitive Map Approach

Verena V. Hafner

Summary. This chapter attempts to show how cognitive map models can be combined with robotic navigation strategies. A neural cognitive mapping strategy that is inspired by place cells but still abstract enough to be interpreted in a meaningful way is implemented in different experiments with both mobile robot and simulation experiments.

15.1 Introduction

The ability to navigate in a complex environment is one of the most challenging skills for every animal on this planet. It is crucial for survival, and is even seen as the evolutionary pressure to develop brains. The reason why plants do not have a brain is that they do not have to move [25].

There are different amounts of cognition involved in navigation when considering different species and different goals. This starts with simple aiming and obstacle avoidance already present in one-cell organisms, over route-following for example in insects, to high-level survey navigation which includes both topological and metric information. A useful categorisation of these different skill levels of navigation is presented by Franz et al. in this volume.

The approaches to navigation involving cognitive mapping that this book focuses on, are mainly found at the levels of topological and survey navigation. These approaches require some kind of internal representation of the space the animal is navigating in. However, complex navigation, such as finding back home from various locations on paths never travelled before, is also possible without a full representation of the environment, for example in insect visual homing. Here, the animal can reach the goal position from its current position by comparing visual snapshots in memory without knowing about the location of the goal position nor the current position [10].

There is also evidence from the neurosciences for an internal spatial map: In these experiments (see for example [16], cells have been found in the rat's brain that represent a certain place within the environment, and may therefore be part of a so-called cognitive map for the animal. These cells are called 'place cells' and are located mainly in the hippocampus, an area in the brain which is also responsible for memory [19]. Such cells have been found in rats, mice, and even primates including humans. However in primates, additional cells, called 'view cells', that are related both to a particular place and view have been found.

M.E. Jefferies et al. (Eds.): Robot. & Cogn. Approach. to Spat. Map., STAR 38, pp. 315–324, 2008.
springerlink.com

Biologists [23] and neuroscientists [16] have been studying navigation behaviour in animals for several decades, coming up with different hypotheses of how the navigation skills are acquired and implemented in the animal's brain and body. With the advent of behavioural [3] and biomimetic [22] robotics, a new field of research got interested in navigation behaviour. These subfields of robotics are inspired by biology, in particular using behavioural experiments. In these experiments, the interaction between the agent and the environment plays a major role. The morphology of the agent with the arrangement of different sensors along its body is also very important for the interaction [17]. Traditional or industrial robotics in contrast is only interested in fulfilling a task in a controlled and predictable way using methods such as planning. Along with biorobotics came several advantages which promised to strengthen or falsify the hypotheses of biologists by repeatable experiments using mobile robots. A good review on bio-inspired robot navigation can be found in [21]. Recently, the robotics community is focussing more and more on hybrid approaches [1, 20] to find a good balance between engineering and biological plausibility.

In this chapter, a neural cognitive map model inspired by place cells, implemented on a mobile robot is presented. In Sect. 15.2, the neuroscientific foundations of place cells are explained and some particular properties are discussed. Section 15.3 presents the neural cognitive map algorithm and structure, and different experiments on a mobile robot and in simulation are shown and their results discussed.

15.2 Neuroscientific Foundations: Place Cells

Place cells were first discovered in the hippocampus of rats [16]. They are cells whose firing activity depends on the spatial position of the animal in its environment. This implies some sort of internal representation of the outside environment in the brain. There is similar evidence of cognitive maps in humans. We present some of the experiments measuring cells in the human hippocampus in the next subsection. Spatial information is an important part of long-term memory, therefore it is also interesting that the place cell activity seems to be transferred into long term memory during sleep, which is presented in the subsection after that.

15.2.1 Place Cells in the Human Hippocampus

Place cells have also been found in the hippocampi of humans recently [6, 5]. Cells in the hippocampal and parahippocampal region of patients with epilepsy were recorded with implanted clinical depth electrodes. During the recording, the subjects were exploring and navigating a virtual town in a taxi driver computer game, searching for passengers and delivering them to fixed target locations. Out of 317 recorded neurons, 26% responded to a place, 12% responded to a view, and 21% to a goal. Eleven per cent of the cells were true place cells, which only responded to a place. From the 67 neurons measured in the hippocampus, 24%

were true place cells. These findings can be seen as evidence for a neural code of human spatial navigation, a cognitive map.

15.2.2 Place Cells During Sleep

Hippocampal place cells have also been recorded during rat's rapid eye movement (REM) sleep[1]. Wilson and McNaughton [24] discovered that cells that fired together when the rat occupied a particular location exhibited an increased tendency to fire together during subsequent sleep, in comparison to sleep episodes preceding the behavioural tasks. In these experiments, 50 to 100 single cells in area CA1 were recorded. They suggest that the neural states encoded within the hippocampus are "played back" as part of a consolidation process by which information is gradually transferred to the neocortex.

15.3 Robotic Experiments: Cognitive Maps

This section describes experiments performed with navigating artificial agents. The experiments are inspired by the findings of place cells in rats, and aim to produce a cognitive map of an environment during exploration. The question is, whether a neural representation can be found that explains the findings and properties of place cells for navigation in rats.

15.3.1 Experimental Setup

The experimental setup consists of an artificial agent (mobile robot or simulated agent) that performs random exploration tours within a newly encountered open environment with objects functioning as obstacles. The agent has omnidirectional sensory stimulation, either panoramic vision or distance sensors. A benefit of omnidirectional vision is that it approximates more closely the rat's field of view spanning 320°, making the sensory input more comparable between rat and robot. The agent gets proprioceptive feedback about its heading direction, but does not have access to any exact metric distance information, nor does it know its position within the environment. The learning of a cognitive map is purely based on neural plasticity (changing weights) within the agent's brain, using a variant of a self-organising map (SOM). The number of recruitable place cells, i.e existing cells that could function as place cells, is fixed at the beginning of the experiment. During exploration, the agent gets a new visual input every few time steps. The exploration of the environment is open-ended without being task-specific. There is no reward during learning nor is there a goal state. This setup is comparable to rats exploring a newly encountered environment before eating even if they are hungry [18]. The attributed motivation for this behaviour is curiosity.

The neural network representing the cognitive map is similar to Kohonen's [14] self-organising map (SOM), where the map layer neurons represent the place

[1] REM sleep: periods of mental activity during sleep.

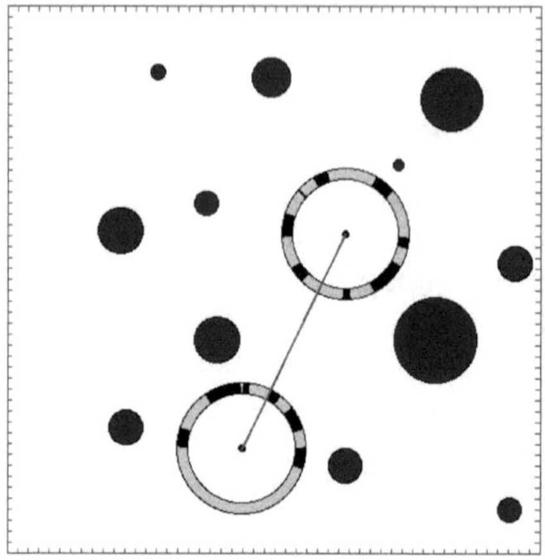

Fig. 15.1. Cylinder environment and visual information of the agent. In this figure, the omnidirectional binary view of an agent at two different positions is shown (connected by the line). Filled circles are obstacles of different sizes.

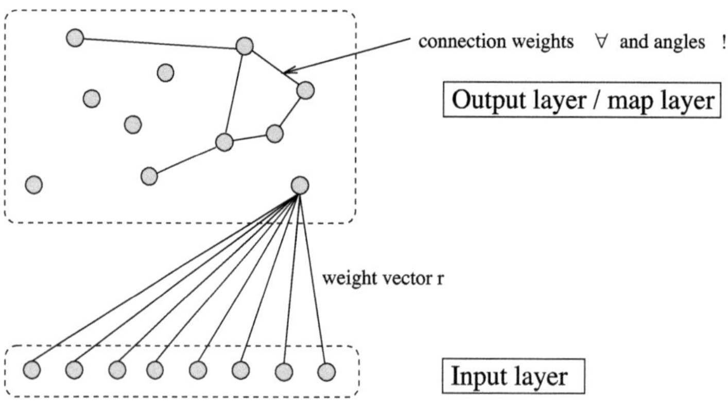

Fig. 15.2. Neural Network structure for learning of place cells and their connections. The cells in the output layer are place cells. The input layer consists of the sensory input from the robot.

cells. As in Kohonen's SOM, there is a winner neuron for each visual input with the strongest activation in the map layer whose connections to the input layer will get strengthened. The main difference, however, is that the neighbourhood relationship in this cognitive map model is not fixed. This follows from properties of neural place cells: they do not have a geometric connectivity as standard SOMs do have. During learning, the connections between the current winner cell w_t and the previous winner cell w_{t-1} are strengthened, resulting in a topological map of place cells with a variable number of connections per cell. One of the reasons for

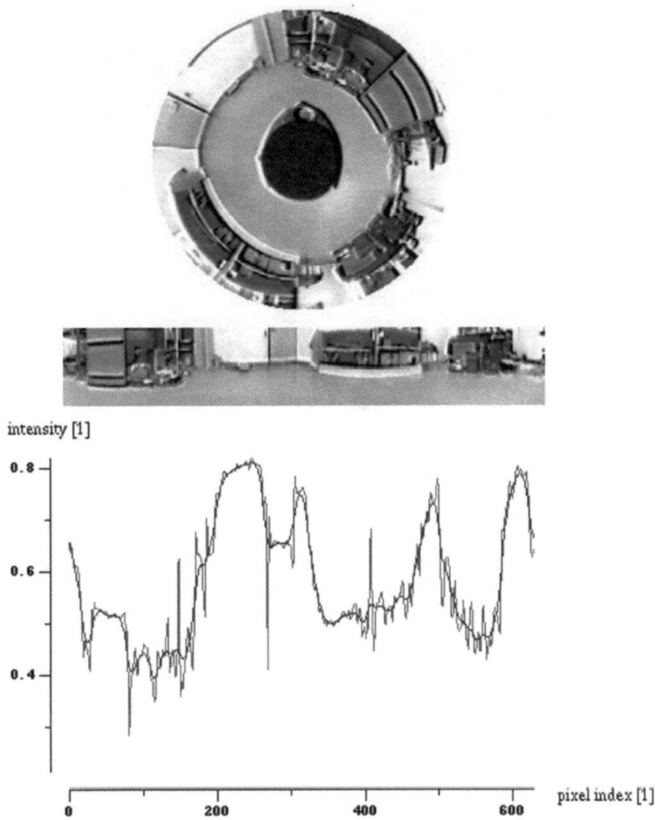

Fig. 15.3. Visual processing of the omnidirectional camera data from the mobile robot. The three steps are projection, horizontal averaging and low-pass filtering.

this choice is that there is no apparent relation between the spatial positions of place cells within the hippocampus and their corresponding place fields within the environment. Since the movement of the agent is continuous in space over time, place cells representing adjacent places get connected. The topological map represents a relationship within the sensory space of the agent, and does not explicitly map the two-dimensional Cartesian space (we only consider agents moving on a plane). The map learning parameters were selected empirically in one experiment [11], and evolved using evolutionary strategies in another one [12]. The parameters are the learning rates of both connection and input weights.

The available sensory information varied between the different experiments. In the case of the mobile robot that navigated in a standard office room using a compass and an omnidirectional camera, a 16-dimensional transformation of the omnidirectional camera image serves as sensory input to the neural network (see Fig. 15.3). These are equidistant visual input features based on an angular resolution of 22.5 degrees, horizontal averaging and low-pass filtering. In the

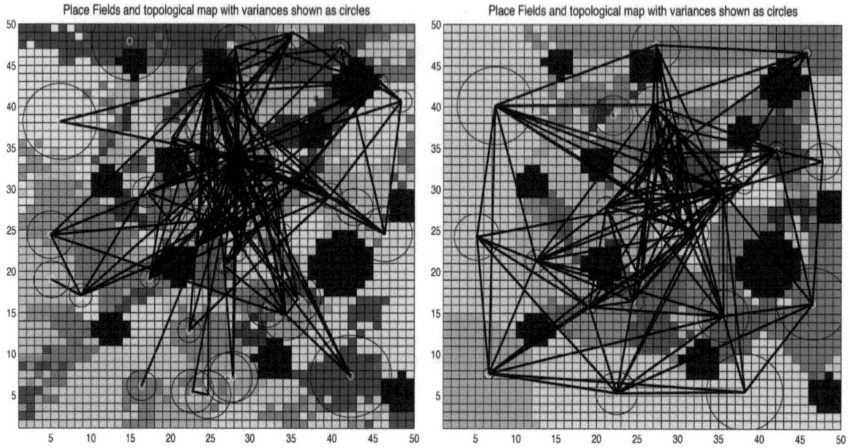

Fig. 15.4. Place fields after an exploration tour in a virtual environment with random learning parameters (left) and evolved learning parameters (right). The cylindrical objects (dark blue) are obstacles of different size. The coloured regions represent the different place fields. The black lines connect the centres of mass (the most sensible definition of a centre for an area with undefined shape) of the connected place cells within the environment.

simulation environment, 90-dimensional binary input from the cylinder world (see Fig. 15.1) has been used. Here we use only the information whether there is an obstacle in view at a certain angle or not. The problem of choosing the right sensory information is directly related to two complementary problems of reliable place recognition [15]: The first is perceptual aliasing, which means that different places may have similar or even identical sensory information. The second is image variability. The same position and orientation may have different sensory information at different points in time. Possible reasons are sensory noise, motor noise, or simply change in illumination.

In contrast to the sparse topological representation of this approach, Arleo and Gerstner [2] use a population of place cells with overlapping place fields. A similar approach has also been taken by Gaussier et al. [8].

15.3.2 Density of Place Fields

After an exploration tour in the simulated environment of Fig. 15.1, we can see that the same place cell is firing in certain restricted areas. We call these areas 'place fields'. An interesting aspect of the learned place fields after the exploration tour is that their number is significantly higher in the vicinity of objects (see Fig. 15.4 right, or other place field figures in [11]). This property has also been observed during electrophysiological recordings in rats, and tends to be explained by rats using a higher proportion of their place cells for 'interesting' places (for a review see [13]). Since neither the simulated nor the physical mobile robot have any concept of what counts as interesting, the explanation is simple:

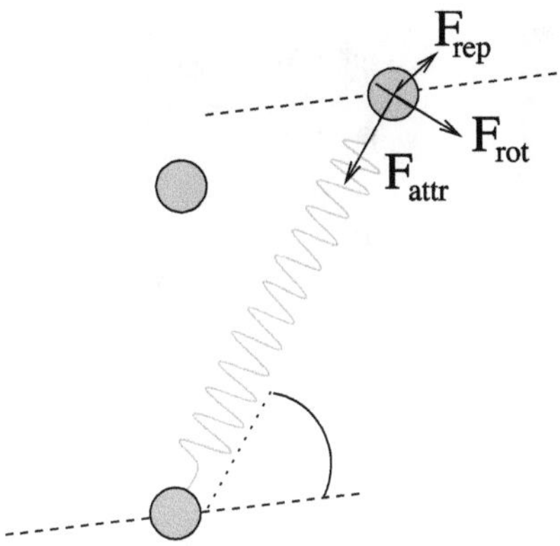

Fig. 15.5. Force model based on different forces applied on the place cells: an attractive spring force, a repulsive force and a third force caused by the preferred orientation of the connecting weight

The visual input (sensory information) is changing more rapidly when moving close to convex obstacles, and therefore more place cells will be recruited for this space. This effect has been explicitly encoded as a threshold for building new nodes in the graph algorithm by Franz et al. [7].

15.3.3 Extracting Metric Information

The map of the agent built from the place cell information and their connections is purely topological. However, it can be extended by using the additional information on heading directions between two place cells. Physiologically, this additional information might be accomplished by intersynapse connections between head direction cells and place cell connections. To enable the extraction of metric information from a topological map, a theoretical force model [9] is introduced, which is ideal for energy minimisation and assumes place cells as repulsive charges c_i and connections as springs s_{jk} connecting the cells (see Fig. 15.5). The spring constant for all springs is set to the same value based on the assumption that all directly connected places have the same distance. This is a generalisation since we have seen that there are place fields of different shape and size, however it is also clear that the distances of connected places are not too different, since two far-away places are linked by places between them. The initial position of the cells is random in $\mathbb{R}^2$. By repeatedly applying forces to the charges, their position converges to an energy minimum. The forces consist of an attractive spring force, a repulsive force and a third force caused by the preferred orientation of the connecting weight. Duckett et al. [4] proposed a slightly different algorithm called the 'relaxation algorithm' which is based on similar principles, but additionally assumes distance information between the nodes, and assigns a position likelihood to them. The algorithm is computationally cheap, and ensures that an optimal solution will be found.

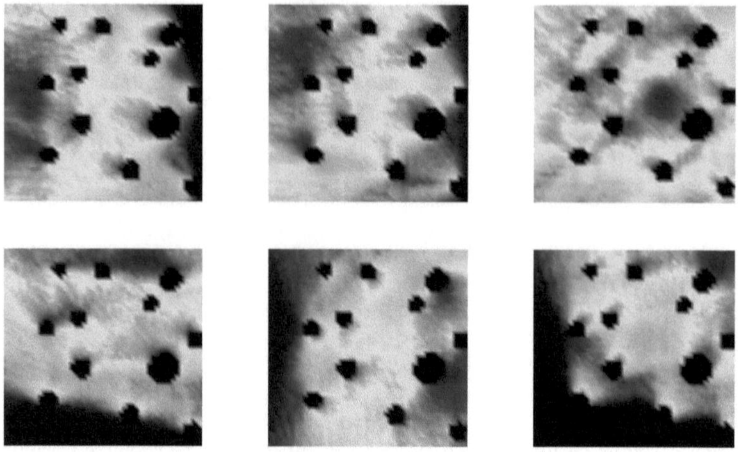

Fig. 15.6. Activation of six different place cells after an exploration tour in the virtual environment. The dark blue circles represent the obstacles.

15.3.4 Evaluation Methods

Evaluating the usefulness of the learned cognitive map for robot navigation is difficult. In principle, we can distinguish three main approaches to the problem of evaluating the cognitive maps. First, the evaluation can be based on the properties of the map itself, such as the shape of place fields or the properties of connections between place cells (see Fig. 15.4). Second, the learning strategy of the cognitive map can be evaluated by assessing the behaviour of a navigating agent after exploration. And third, both properties of the cognitive map and the resulting navigation behaviour can be directly compared with those of a navigating animal. Let us first consider the method of analysing properties of the cognitive map: Statistical properties such as density, shape, and number of place fields; activity and number of connections per place cell, or metric versus graph distance in the topological map can be collected and analysed easily in simulation. On a mobile robot, additional difficulties arise since these data can only be collected having an accurate tracking system and having access to the sensory input for every position of the robot in space, ideally requiring a large image database.

The activity shapes of single place cells after an exploration tour in the virtual environment using the optimised cognitive map learning strategy can be seen in Fig. 15.6. The place field shapes are very similar to place fields of cells in the rat hippocampus, also showing a general exponential decay of activity away from the centre of the place field. Place fields near walls also have a tendency to be more elongated than place fields in the centre of the area, which tend to be more circular.

Assessing the behaviour of an agent to evaluate the cognitive map gives a good fitness measure, since it is focused on the behaviour, but has additional difficulties, since the number of exploration runs and navigation runs between two arbitrarily chosen places within the environment has to be huge in order to be meaningful. The problem with comparing robot and animal behaviour is, that both the environment and the available sensory information should be comparable. An approach for evaluating a cognitive map learning strategy where the parameters have been evolved using evolutionary strategies can be found in [12].

15.4 Conclusions

This chapter has given an example of how neural cognitive maps can be implemented in robotic experiments. For simplicity, some of these experiments were performed in simulation. One of the important features of the resulting neural cognitive maps is that it includes both topological and metric information about places. The difficulty is to realise the integration of these informations with a restricted amount of memory (number of place cells and their connections), and without building a geometric world model.

On the one hand, research on cognitive maps for robots has the advantage of exploiting biological principles of navigation for building more reliable robots, and on the other hand, robots are ideal tools to test specific hypotheses on the underlying functions of navigation behaviour in animals. The first point is important in particular for robot navigation in dynamic environments, and in environments, where GPS is not applicable, such as in many indoor environments, underwater and extraterrestrial. To test biological hypotheses, the important argument is the repeatability of robotic experiments and the easy access to internal states.

References

1. Abreu, A., Correia, L.: Place energy and forget in maps for mobile robots. In: ROBIO 2004. Proceedings of the IEEE International Conference on Robotics and Biomimetics, IEEE Computer Society Press, Los Alamitos (2004)
2. Arleo, A., Gerstner, W.: Spatial Cognition and Neuro-Mimetic Navigation: A Model of Hippocampal Place Cell Activity. Biological Cybernetics 83, 287–299 (2000)
3. Brooks, R.A.: Intelligence Without Representation. Artificial Intelligence Journal 47, 139–159 (1991)
4. Duckett, T., Marsland, S., Shapiro, J.: Fast, On-Line Learning of Globally Consistent Maps. Autonomous Robots 12, 297–300 (2002)
5. Ekstrom, A.D., Caplan, J., Ho, E., Shattuck, K., Fried, I., Kahana, M.: Human hippocampal theta activity during virtual navigation. Hippocampus 15, 881–889 (2005)
6. Ekstrom, A.D., Kahana, M.J., Caplan, J.B., Fields, T.A., Isham, E.A., Newman, E.L., Fried, I.: Cellular networks underlying human spatial navigation. Nature 425, 184–187 (2003)
7. Franz, M.O., Schölkopf, B., Mallot, H.A., Bülthoff, H.H.: Learning View Graphs for Robot Navigation. Autonomous Robots 5, 111–125 (1998)

8. Gaussier, P., Revel, A., Banquet, J.P., Babeau, V.: From view cells and place cells to cognitive map learning: processing stages of the hippocampal system. Biological Cybernetics 86, 15–28 (2002)
9. Hafner, V.V.: Cognitive Maps for Navigation in Open Environments. In: IAS-6. Proceedings of the 6th International Conference on Intelligent Autonomous Systems (2000)
10. Hafner, V.V.: Adaptive Navigation Strategies in Biorobotics: Visual Homing and Cognitive Mapping in Animals and Machines. Shaker Verlag, NY (2004)
11. Hafner, V.V.: Learning Places in Newly Explored Environments. In: From Animals to Animats. SAB2000 Proceedings Supplement Book, pp. 111–120 (2000)
12. Hafner, V.V.: Evaluating cognitive maps for mobile robot navigation behaviour. In: Proceedings of the AISB 2003 Symposium on Scientific Methods for the Analysis of Agent-Environment Interaction (2003)
13. Hollup, S.A., Donnett, J.G., Moser, M.-B., Moser, E.I.: Accumulation of Hippocampal Place Fields at the Goal Location in an Annular Watermaze Task. Journal of Neuroscience 21(5), 1635–1644 (2001)
14. Kohonen, T.: Self-organizing formation of topologically correct feature maps. Biological Cybernetics 43(1), 59–69 (1982)
15. Kuipers, B., Beeson, P.: Bootstrap Learning for Place Recognition. In: AAAI 2002. Proceedings of the 18th National Conference on Artificial Intelligence, pp. 824–829. AAAI/MIT Press (2002)
16. O'Keefe, J.: Place units in the hippocampus of the freely moving rat. Experimental Neurology 51, 78–109 (1976)
17. Pfeifer, R., Bongard, J.C.: How the Body shapes the Way We Think. MIT Press, Cambridge, Mass., USA (2006)
18. Prescott, T.J.: Spatial representation for navigation in animats. Adaptive Behavior 4(2), 85–123 (1996)
19. Recce, M., Harris, K.D.: Memory for places: A navigational model in support of Marr's theory of hippocampal function. Hippocampus 6, 735–748 (1996)
20. Tapus, A., Siegwart, R.: Incremental robot mapping with fingerprints of places. In: Proceedings of the IEEE/RSJ International Conference on Intelligent Robots and Systems (2005)
21. Trullier, O., Wiener, S.I., Berthoz, A., Meyer, J.-A.: Biologically based artificial navigation systems: review and prospects. Progress in Neurobiology 51, 483–544 (1997)
22. Webb, B.: Can robots make good models of biological behaviour? Behavioral and Brain Sciences 24(6) (2001)
23. Wehner, R., Räber, F.: Visual spatial memory in desert ants: Cataglyphis bicolor Hymenoptera: Formicidae. Experientia 35, 1569–1571 (1979)
24. Wilson, M.A., McNaughton, B.L.: Reactivation of Hippocampal Ensemble Memories During Sleep. Science 265, 676–679 (1994)
25. Wolpert, L.: Love is a many-moleculed thing. The Observer (2002)

Index

Springer Tracts in Advanced Robotics

Edited by B. Siciliano, O. Khatib and F. Groen

Vol. 38: Jefferies, M.E.; Yeap, W.-K. (Eds.)
Robotics and Cognitive Approaches to Spatial Mapping
328 p. 2008 [978-3-540-75386-5]

Vol. 37: Ollero, A.; Maza, I. (Eds.)
Multiple Heterogeneous Unmanned Aerial Vehicles
233 p. 2007 [978-3-540-73957-9]

Vol. 36: Buehler, M.; Iagnemma, K.; Singh, S. (Eds.)
The 2005 DARPA Grand Challenge – The Great Robot Race
520 p. 2007 [978-3-540-73428-4]

Vol. 35: Laugier, C.; Chatila, R. (Eds.)
Autonomous Navigation in Dynamic Environments
169 p. 2007 [978-3-540-73421-5]

Vol. 34: Wisse, M.; van der Linde, R.Q.
Delft Pneumatic Bipeds
136 p. 2007 [978-3-540-72807-8]

Vol. 33: Kong, X.; Gosselin, C.
Type Synthesis of Parallel Mechanisms
272 p. 2007 [978-3-540-71989-2]

Vol. 32: Milutinović, D.; Lima, P.
Cells and Robots – Modeling and Control of Large-Size Agent Populations
130 p. 2007 [978-3-540-71981-6]

Vol. 31: Ferre, M.; Buss, M.; Aracil, R.; Melchiorri, C.; Balaguer C. (Eds.)
Advances in Telerobotics
500 p. 2007 [978-3-540-71363-0]

Vol. 30: Brugali, D. (Ed.)
Software Engineering for Experimental Robotics
490 p. 2007 [978-3-540-68949-2]

Vol. 29: Secchi, C.; Stramigioli, S.; Fantuzzi, C.
Control of Interactive Robotic Interfaces – A Port-Hamiltonian Approach
225 p. 2007 [978-3-540-49712-7]

Vol. 28: Thrun, S.; Brooks, R.; Durrant-Whyte, H. (Eds.)
Robotics Research – Results of the 12th International Symposium ISRR
602 p. 2007 [978-3-540-48110-2]

Vol. 27: Montemerlo, M.; Thrun, S.
FastSLAM – A Scalable Method for the Simultaneous Localization and Mapping Problem in Robotics
120 p. 2007 [978-3-540-46399-3]

Vol. 26: Taylor, G.; Kleeman, L.
Visual Perception and Robotic Manipulation – 3D Object Recognition, Tracking and Hand-Eye Coordination
218 p. 2007 [978-3-540-33454-5]

Vol. 25: Corke, P.; Sukkarieh, S. (Eds.)
Field and Service Robotics – Results of the 5th International Conference
580 p. 2006 [978-3-540-33452-1]

Vol. 24: Yuta, S.; Asama, H.; Thrun, S.; Prassler, E.; Tsubouchi, T. (Eds.)
Field and Service Robotics – Recent Advances in Research and Applications
550 p. 2006 [978-3-540-32801-8]

Vol. 23: Andrade-Cetto, J,; Sanfeliu, A.
Environment Learning for Indoor Mobile Robots – A Stochastic State Estimation Approach to Simultaneous Localization and Map Building
130 p. 2006 [978-3-540-32795-0]

Vol. 22: Christensen, H.I. (Ed.)
European Robotics Symposium 2006
209 p. 2006 [978-3-540-32688-5]

Vol. 21: Ang Jr., H.; Khatib, O. (Eds.)
Experimental Robotics IX – The 9th International Symposium on Experimental Robotics
618 p. 2006 [978-3-540-28816-9]

Vol. 20: Xu, Y.; Ou, Y.
Control of Single Wheel Robots
188 p. 2005 [978-3-540-28184-9]

Vol. 19: Lefebvre, T.; Bruyninckx, H.; De Schutter, J. Nonlinear Kalman Filtering for Force-Controlled Robot Tasks
280 p. 2005 [978-3-540-28023-1]

Vol. 18: Barbagli, F.; Prattichizzo, D.; Salisbury, K. (Eds.)
Multi-point Interaction with Real and Virtual Objects
281 p. 2005 [978-3-540-26036-3]

Vol. 17: Erdmann, M.; Hsu, D.; Overmars, M.; van der Stappen, F.A (Eds.)
Algorithmic Foundations of Robotics VI
472 p. 2005 [978-3-540-25728-8]

Vol. 16: Cuesta, F.; Ollero, A.
Intelligent Mobile Robot Navigation
224 p. 2005 [978-3-540-23956-7]

Vol. 15: Dario, P.; Chatila R. (Eds.)
Robotics Research – The Eleventh International Symposium
595 p. 2005 [978-3-540-23214-8]

Vol. 14: Prassler, E.; Lawitzky, G.; Stopp, A.; Grunwald, G.; Hägele, M.; Dillmann, R.; Iossifidis. I. (Eds.)
Advances in Human-Robot Interaction
414 p. 2005 [978-3-540-23211-7]

Vol. 13: Chung, W.
Nonholonomic Manipulators
115 p. 2004 [978-3-540-22108-1]

Vol. 12: Iagnemma K.; Dubowsky, S.
Mobile Robots in Rough Terrain – Estimation, Motion Planning, and Control with Application to Planetary Rovers
123 p. 2004 [978-3-540-21968-2]

Vol. 11: Kim, J.-H.; Kim, D.-H.; Kim, Y.-J.; Seow, K.-T.
Soccer Robotics
353 p. 2004 [978-3-540-21859-3]

Vol. 10: Siciliano, B.; De Luca, A.; Melchiorri, C.; Casalino, G. (Eds.)
Advances in Control of Articulated and Mobile Robots
259 p. 2004 [978-3-540-20783-2]

Vol. 9: Yamane, K.
Simulating and Generating Motions of Human Figures
176 p. 2004 [978-3-540-20317-9]

Vol. 8: Baeten, J.; De Schutter, J.
Integrated Visual Servoing and Force Control – The Task Frame Approach
198 p. 2004 [978-3-540-40475-0]

Vol. 7: Boissonnat, J.-D.; Burdick, J.; Goldberg, K.; Hutchinson, S. (Eds.)
Algorithmic Foundations of Robotics V
577 p. 2004 [978-3-540-40476-7]

Vol. 6: Jarvis, R.A.; Zelinsky, A. (Eds.)
Robotics Research – The Tenth International Symposium
580 p. 2003 [978-3-540-00550-6]

Vol. 5: Siciliano, B.; Dario, P. (Eds.)
Experimental Robotics VIII – Proceedings of the 8th International Symposium ISER02
685 p. 2003 [978-3-540-00305-2]

Vol. 4: Bicchi, A.; Christensen, H.I.; Prattichizzo, D. (Eds.)
Control Problems in Robotics
296 p. 2003 [978-3-540-00251-2]

Vol. 3: Natale, C.
Interaction Control of Robot Manipulators – Six-degrees-of-freedom Tasks
120 p. 2003 [978-3-540-00159-1]

Vol. 2: Antonelli, G.
Underwater Robots – Motion and Force Control of Vehicle-Manipulator Systems
268 p. 2006 [978-3-540-31752-4]

Vol. 1: Caccavale, F.; Villani, L. (Eds.)
Fault Diagnosis and Fault Tolerance for Mechatronic Systems – Recent Advances
191 p. 2003 [978-3-540-44159-5]

MIX
Papier aus verantwortungsvollen Quellen
Paper from responsible sources
FSC® C105338
FSC
www.fsc.org

If you have any concerns about our products,
you can contact us on
ProductSafety@springernature.com

In case Publisher is established outside the EU,
the EU authorized representative is:
Springer Nature Customer Service Center GmbH
Europaplatz 3, 69115 Heidelberg, Germany

Printed by Libri Plureos GmbH
in Hamburg, Germany